Biogeochemical Cycling and Sediment Ecology

NATO ASI Series

Advanced Science Institute Series

A Series presenting the results of activities sponsored by the NATO Science Committee, which aims at the dissemination of advanced scientific and technological knowledge, with a view to strengthening links between scientific communities.

The Series is published by an international board of publishers in conjunction with the NATO Scientific Affairs Division

A. Life Sciences **B. Physics**	Plenum Publishing Corporation London and New York
C. Mathematical and Physical Sciences **D. Behavioural and Social Sciences** **E. Applied Sciences**	Kluwer Academic Publishers Dordrecht, Boston and London
F. Computer and Systems Sciences **G. Ecological Sciences** **H. Cell Biology** **I. Global Environment Change**	Springer-Verlag Berlin, Heidelberg, New York, London, Paris and Tokyo

PARTNERSHIP SUB-SERIES

1. Disarmament Technologies	Kluwer Academic Publishers
2. Environment	Springer-Verlag / Kluwer Academic Publishers
3. High Technology	Kluwer Academic Publishers
4. Science and Technology Policy	Kluwer Academic Publishers
5. Computer Networking	Kluwer Academic Publishers

The Partnerschip Sub-Series incorporates activities undertaken in collaboration with NATO's Cooperation Partners, the countries of the CIS and Central and Eastern Europe, in Priority Areas of concern to those countries.

NATO-PCO-DATA BASE

The electronic index to the NATO ASI Series provides full bibliographical references (with keywords and/or abstracts) to about 50,000 contributions from international scientists published in all sections of the NATO ASI Series. Access to the NATO-PCO-DATA-BASE is possible via a CD-ROM "NATO Science and Technology Disk" with user-friendly retrieval software in English, French, and German (©WTV GmbH and DATAWARE Technologies, Inc. 1989). The CD-ROM contains the AGARD Aerospace Database.

The CD-ROM can be ordered through any member of the Board of Publishers or through NATO-PCO, Overijse, Belgium.

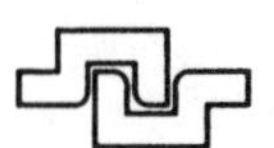

Series E: Applied Sciences – Vol. 59

Biogeochemical Cycling and Sediment Ecology

edited by

John S. Gray
Biologisk Institutt,
Universitetet i Oslo,
Blindern, Norway

William Ambrose Jr.
Department of Biology,
Bates College,
Lewiston, Maine, U.S.A.

and

Anna Szaniawska
Institute of Oceanography,
Gdansk University,
Gdynia, Poland

Kluwer Academic Publishers

Dordrecht / Boston / London

Published in cooperation with NATO Scientific Affairs Division

Proceedings of the NATO Advanced Research Workshop on
Biogeochemical Cycling in Marine Sediments
Hel, Poland
August 1997

A C.I.P. Catalogue record for this book is available from the Library of Congress.

ISBN 0-7923-5770-1

Published by Kluwer Academic Publishers,
P.O. Box 17, 3300 AA Dordrecht, The Netherlands.

Sold and distributed in North, Central and South America
by Kluwer Academic Publishers,
101 Philip Drive, Norwell, MA 02061, U.S.A.

In all other countries, sold and distributed
by Kluwer Academic Publishers,
P.O. Box 322, 3300 AH Dordrecht, The Netherlands.

Printed on acid-free paper

Printed in the Netherlands.

TABLE OF CONTENTS

PREFACE

Oceanographic discontinuities (e.g. frontal systems, upwelling areas, ice edges) are often areas of enhanced biological productivity. Considerable research on the physics and biology of the physical boundaries defining these discontinues has been accomplished (see [1]). The interface between water and sediment is the largest physical boundary in the ocean, but has not received a proportionate degree of attention. The purpose of the Nato Advanced Research Workshop (ARW) was to focus on soft-sediment systems by identifying deficiencies in our knowledge of these systems and defining key issues in the management of coastal sedimentary habitats.

Marine sediments play important roles in the marine ecosystem and the biosphere. They provide food and habitat for many marine organisms, some of which are commercially important. More importantly from a global perspective, marine sediments also provide "ecosystem goods and services" [2]. Organic matter from primary production in the water column and contaminants scavenged by particles accumulate in sediments where their fate is determined by sediment processes such as bioturbation and biogeochemical cycling. Nutrients are regenerated and contaminants degraded in sediments. Under some conditions, carbon accumulates in coastal and shelf sediments and may by removed from the carbon cycle for millions of years, having a potentially significant impact on global climate change. Sediments also protect coasts. The economic value of services provided by coastal areas has recently been estimated to be on the order of $12,568 10^9 y^{-1} [3], far in excess of the global GNP.

The ability of coastal sediments to continue to support important ecosystem and global services has been compromised by common anthropogenic disturbances to the coastal zone. Demersal fishing practices, alteration of watercourses and habitat, dredging, and organic loading resulting in anoxia and hypoxia all may effect benthic fauna and biogeochemical cycling. The ARW concluded that: *"Disturbance events affecting the benthic boundary layer probably have led to changes in species compositions, species loses and consequences for biogeochemical cycling and effects on large-scale carbon, nutrient, and contaminant fluxes"*. Participants also agreed that the frequency and magnitude of disturbance events in the oceans have increased in recent years.

In order to evaluate the influences of man-made disturbances on marine sediments and sedimentary processes, it is necessary to identify deficiencies in our knowledge of the structure and function of these systems and in our ability to manage coastal systems. Several contributors to the ARW review our knowledge of benthic marine ecosystems and conclude that our understanding of fundamental processes is inadequate to completely address central issues such as benthic-pelagic coupling, biodiversity, and sustainable harvest of benthic resources. Furthermore, it is noted that most of the research on coastal sediments has been concentrated in a few geographic areas, leaving us with little knowledge of basic processes for many coastal zones. Information on the Arctic Ocean, portions of the eastern Pacific and the Indian Ocean is particularly lacking in the western literature. Russian scientists have had a rather different tradition in oceanic research and have made comprehensive surveys oif large areas of the ocean. Much of this work still remains unknown to western scientists and a particularly important part of this ARW is to bring these major overviews into focus. Thes ereviews cover organic matter cycling (Romankevitch), the importance of marginal filters for fluxes of organic matter and elements in the coastal ocean (Lisitizin), the key processes in benthic-pelagic coupling in the Arctic (Gordeev) and sureveys of production processes in the Aral Sea (Orlova) and the Baltic Sea (Emelyanov). As editors we are aware that these reviews contain few references to relevant literature. Rather than requiring extensive rewrites we have accepted these papers after editing since they bring the attention of western readers to a vast source of relevant Russian literature.

Several contributors make it clear that to successfully manage coastal systems we need to appreciate their temporal and spatial variation and how reliably we can extrapolate from one system to others. Much of the variability in benthic dynamics and biogeochemical cycling among coastal systems may be a consequence of differences in the relative abundances of groups of organisms which affect the physical structure of the sediment-water interface and the rate of flux between the two systems. The use of functional groups as a means of categorizing often diverse benthic fauna has been a common practice among benthic ecologists for many years, and the ARW stresses the utility of this approach. The ARW emphasizes that some disturbances such as trawling may result in the virtual removal of some functional groups and consequently the loss of their contribution to bioturbation and nutrient and contaminant fluxes. Functional diversity may be more important than biodiversity in order for a system to provide critical services. In low diversity systems, such as the Baltic Sea, maintaining functional diversity may be particularly important if the system is to operate normally.

The benthic boundary layer is a key component of the sediment-water interface. Several contributors to the ARW emphasize that understanding processes at this boundary requires a multi-disciplinary approach on different scales. Models of the benthic boundary layer are most realistic when biological processes are combined with physical models of flux across the boundary. Small-scale laboratory experiments may be helpful in assessing the importance of including organisms in models. Particle and solute fluxes across the boundary can be spatially and temporally very variable, however. To be most useful, flux models need to combine fluxes on the scale of individual organisms and the larger scale of the community.

Most of the studies in marine sediments have been conducted at small scales, and scaling up the results of these studies to more meaningful scales typical of environmental disturbances is a continual challenge to benthic ecologists [4]. Our ability to assess and manage disturbances to coastal sediments may be increased by integrating processes over different spatial and temporal scales. Several contributors to the ARW promote the need for scale-related hypotheses and a landscape view of the benthos.

With this summary of the state of our knowledge of important sedimentary process as a beginning, the ARW then addressed what conferees considered to be the major anthropogenic disturbances to coastal sedimentary systems and approaches to addressing the impact of these disturbances on sedimentary processes and management of coastal sediments. These disturbances and the foci for key research questions are: 1) demersal fishing, 2) hypoxic and anoxic events and 3) dredging and disposal of dredged material.

The group recognized that the fishing industry imposes many different types of disturbance on marine sediments, but felt that trawling is perhaps the most severe and widespread. Three types of areas were identified for experimental studies: 1) non-fished areas, 2) set-aside areas where fishing could be manipulated and 3) gradients (or mosaics) of fishing intensity. Manipulative experiments and quantitative observational data are needed examining benthic spatial structure and benthic-pelagic coupling in relation to varying rates and scales of fishing disturbance. Rather than traditional species lists, size and age structure of benthic populations and benthic productivity should be evaluated as measures of restoration. Modeling is an integral part of addressing the impacts of fishing disturbance on benthic systems, and metapopulation analysis may be a promising approach. Experiments need to be done in areas varying in species richness so that comparisons of biogeochemical cycling processes and rates and restoration capabilities can be made among areas.

Hypoxia and anoxia may be permanent or intermittent and result from organic enrichment or other physico-chemical processes. The ARW acknowledged that the impacts of these disturbances on benthic systems are well documented. Less understood are the physical, chemical, and biological conditions necessary for restoration of the system and the temporal patterns of recovery. It was recognized that microorganisms dominate the initial recovery phase and that key biogeochemical processes driven by them need to be measured. Once macrofauna are established recovery often proceeds rapidly because macrofauna are known to enhance rates of organic matter degradation and biogeochemical cycling. Knowledge of how these processes vary over a range of spatial and temporal scales is necessary to understand patterns of recovery. A combined approach of field and mesocosm studies emphasizing the manipulation of key benthic groups was recommended to examine these questions. As with the impact of demersal fishing, studies from areas with different suites of bioturbating species are necessary to appreciate the range of recovery patterns and rates of biogeochemical cycling following an anoxic or hypoxic event. Modeling is also viewed as a valuable approach and population and diagenic modeling should be conducted in parallel with experiments.

Dredging and disposal of dredged material is a global problem. Many harbours are extremely contaminated, yet the spatial scale of contamination is limited allowing experimentation at manageable scales (from laboratory to field). Other disturbances (e.g. storms) may have large impacts on these sediments and it is therefore an urgent problem to determine how various disturbances interact and the environmental implications posed by the disturbance of contaminated sediments. The restoration process is likely to be very site specific and it is unlikely that generalizations will result from studies in one location. By relating important processes such as rate of carbon burial and benthic-pelagic coupling, and patterns such as chemoclines to temporal and spatial patterns of recovery, it may be possible to develop general models of the effects of dredging and disturbance of contaminated sediments on benthic systems.

The ARW felt that the best way to address the impact of these disturbances on sedimentary processes was to develop a research program on restoration of disturbed marine habitats. The ARW agreed that an ideal area to begin such a study was the Gulf of Gdansk. The gulf is a shallow sedimentary environment with documented problems of seasonal anoxia in the Bay of Puck and with contaminated sediments at the mouth of the Vistual River. The area was the site of a demersal fishery, but has not been trawled recently due to low catches. Consequently, it offers the opportunity for studies manipulating trawling. The biota has been studied for many years and the effects of anoxia and hypoxia and their temporal and spatial patterns are well documented. Furthermore, the benthos is comprised of few species making it a perfect place for initial manipulative experiments.

The ARW participants covered a wide range of expertise and provided interchange among groups of scientists that do not usually have scientific discussions. The interchange of ideas and approaches was one of the clear achievements of the workshop.

Acknowledgements

The organizers gratefully acknowledge financial support by NATO that enabled us to hold this Advanced Research Workshop and the help of graduate students and staff at the University of Gdansk. Eric Bonsdorff and Roman Zajac prepared a summary of the workshop was helpful in assembling the introduction to this volume. We also thank all those who reviewed manuscripts for this volume.

References

1. Mann, K.H. and J.R.N. Lazier 1991. Dynamics of marine ecosystems: Biological-physical interactions in the ocean. Blackwell Scientific Press.
2. Erlich, P. and H. Mooney 1983. Extinction, substitution, and ecosystem services. *BioScience* 33:248-254.

3. Costanza, R. et al. 1997. The value of the world's ecosystem services and natural capital. *Nature* 387:253-260.
4. Underwood, A.J. 1996. Detection, interpretation, prediction, and management of environmental disturbances: some roles for experimental marine ecology. *Journal of Experimental Marine Ecology and Biology* 200:1-27.

List of Contributors

WILLIAM AMBROSE JR.
Dept. of Biology, Bates College,
Lewiston, Maine 04240, USA

W.E. ARNTZ
Alfred-Wegener-Institut für Polar- und Meeresforschung, Columbusstrasse,
27568 Bremerhaven and Sylt, Germany

MAGDA J.N. BERGMAN
Netherlands Institute for Sea Research,
P.O. Box 59 1790 AB Den Burg, Texel,
The Netherlands

VONDA J. CUMMINGS
National Institute of Water and
Atmospheric Research, P.O. Box 11
115, Hamilton, New Zealand

J.M. GILI
Instituto de Ciencias del Mar (CSIC),
Passeig Joan de Borbo s/n 08039
Barcelona, Spain

THOMAS L. FORBES
Department of Marine Ecology and
Microbiology, National Environmental
Research Institute, Pob 358,
Frederiksborgvej 399, DK-4000,
Roskilde, Denmark

V. V.GORDEEV
P.P.Shirshov Institute of Oceanology
Russian Academy of Sciences
36, Nakchimovsky prospect
117851 Moscow, Russia

G. GRAF
University of Rostock, Department of
Marine Biology, Freiligrathstrasse 7/8,
18055 Rostock, Germany

JOHN S. GRAY
Biologisk Institutt, Universitetet i Oslo,
Pb 1064 Blindern,
0316 Oslo, Norway

JUDI E. HEWITT
National Institute of Water and
Atmospheric Research, P.O. Box 11
115, Hamilton, New Zealand

U. JANAS
Institute of Oceanography, Gdansk
University, Al. Marszalka J.
Pilsudskiego 46, 81-378 Gdynia, Poland

G.A.KORNEEVA
P.P. Shirshov Institute of Oceanology,
Russian Academy of Sciences
Nakhimovsky Prospect 36,
117851 Moscow, Russia

SARAH M. LAWRIE
National Institute of Water and
Atmospheric Research, P.O. Box 11
115, Hamilton, New Zealand

HAN J. LINDEBOOM
Netherlands Institute for Sea Research,
P.O. Box 59, 1790 AB Den Burg, Texel,
The Netherlands

A.P.LISITZIN
P.P.Shirshov Institute of Oceanology
Russian Academy of Scieces
36, Nakhimovsky prospect
117851 Moscow, Russia

A. NICOLAIDOU
Department of Zoology - Marine
Biology, University of Athens,
Panepistimiopolis, GR 15784, Athens,
Hellas

M. NORMANT
Institute of Oceanography, Gdansk
University, Al. Marszalka J.
Pilsudskiego 46, 81-378 Gdynia, Poland

M.I. ORLOVA
Zoological Institute, Laboratory of
Brackishwater Hydrobiology 199034.
Universitetskaya EMB. B. 1
St.Petersburg, Russia

J. A. NOTT
Plymouth Marine Laboratory, Citadel Hill, Plymouth, PL1 2PB, U.K.

K. REISE
Biologische Anstalt Helgoland
Wattenmeerstation
Hafenstraße 43
25992 List/Sylt, Germany

D. RAFFAELLI
Culterty Field Station, University of Aberdeen, Newburgh, Ellon, Aberdeenshire, Scotland, AB41 6AA, U.K.

A. SZANIAWSKA,
Institute of Oceanography, Gdansk University, Al. Marszalka J. Pilsudskiego 46, 81-378 Gdynia, Poland

E.A.ROMANKEVICH
P.P. Shirshov Institute of Oceanology, Russian Academy of Sciences
Nakhimovsky Prospect 36, 117851 Moscow, Russia

SIMON F. THRUSH
National Institute of Water and Atmospheric Research, P.O. Box 11 115, Hamilton, New Zealand

M. SZYMELFENIG
Institute of Oceanography, University of Gdansk, street Pilsudskiego 46, Gdynia 81-370, Poland

J.M.WESLAWSKI
Institute of Oceanology Polish Academy of Sciences, street Powstancow Warszawy 55, Sopot 81-712, Poland

A.A.VETROV
P.P. Shirshov Institute of Oceanology, Russian Academy of Sciences
Nakhimovsky Prospect 36, 117851 Moscow, Russia

ROMAN N. ZAJAC
Department of Biology and Environmental Science, University of New Haven 300 Orange Ave., West Haven, CT 06516, U.S.A.

GEOCHEMISTRY OF ORGANIC CARBON IN THE OCEAN

E.A.ROMANKEVICH, A.A.VETROV AND G.A.KORNEEVA
P.P. Shirshov Institute of Oceanology, Russian Academy of Sciences
Nakhimovsky Prospect 36, 117851 Moscow, Russia

Abstract

The results of analysis of data on the distribution and masses of organic carbon in the World Ocean obtained by the P.P.Shirsov Institute of Oceanology over forty years are presented. Dissolved and particulate forms are described in relation to climatic and structural-geomorphological zones, to the upper sedimentary layer (0-5 cm), biomasses and production of phytoplankton, input from land from river discharge, subterranean runoff, glacial runoff, eolian transfer, wave abrasion material. Maps resulting from models of TOC distribution in recent bottom sediments and fluxes of TOC to sediment-water interface were constructed. The total flux of organic carbon from land to the Ocean is caclulated as being approximately 670 x 10^{12} g. River discharge of dissolved organic matter is 210 x 10^{12} g C y^{-1} and particulate organic matter 250 x 10^{12} g C y^{-1} and eolian transfer 174 x 10^{12} g C y^{-1} which thus makes a major contribution to the overall flux. New data are based on bioproductivity of the Ocean (60 x 10^{15} g C y^{-1}) and the amount of carbon deposited in marginal and pelagic regions of the Ocean are respectively 150-240 x 10^{12} and 10 x 10^{12} g C y^{-1}. Fossilisation coefficients were estimated for bottom sediments of marginal regions at 0.8-1.3%, ocean floor 0.02% and the whole ocean 0.3-0.4. Results of OM enzyme destruction in various components of ecosystems are presented. Maximal values of biopolymer hydrolysis occurred in shelf sediments.

1. Introduction

The oceanic reservoir of carbon is one of four main reservoirs of the Biosphere (ocean, atmosphere, soils, earth's crust). Carbon, including organic constituents, is constantly exchanged between these reservoirs, which are usually measured as total organic carbon (TOC). This exchange influences fluxes, masses, cycles, processes of sedimentation and diagenesis of all kinds of chemical elements in the ocean mainly through biogeochemical cycling of living and non-living organic matter (OM) [1] [2] [3] [4] [5].

The goal of this paper is to consider some results on global carbon cycling obtained in the Laboratory of the Ocean Chemistry of the P.P.Shirshov Institute of Oceanology, Russian Academy of Sciences regarding to organic matter geochemistry

2. Organic carbon in recent bottom sediments of the ocean

Construction of geochemical maps of the distribution of chemical elements, as a rule, is complicated by the nonregularity and scarcity of sampling sites. Interpolation and extrapolation of data are unavoidable procedures in order to attain regional and global patterns. How successful depends on the set of factors controlling accumulation of these chemical elements.

Over many years, at the P.P.Shirshov Institute of Oceanology we have created a database comprising TOC measurements in surficial sediments (0-5 cm). It has allowed to construct new maps of TOC distributions in regions and in the whole ocean (Fig.1). These maps are more detailed than those previously published [3] [6] [7]. The maps take into account facial differences, climatic conditions, bioproductivity, lithologic and granular types of the sediments, depths, currents and terrigenous sources of OM. It was constructed using traditional mapping techniques.

J.S. Gray et al. (eds.), Biogeochemical Cycling and Sediment Ecology, 1–27.

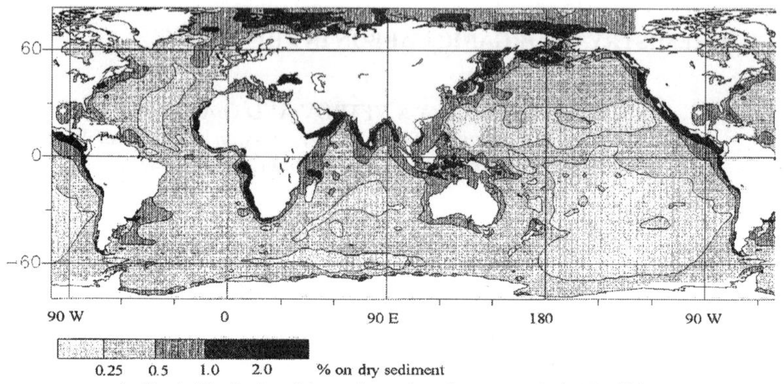

Fig. 1. Distribution of the total organic carbon content in the World Ocean.

The most common pattern of TOC distribution in sediments of the World Ocean is its circumcontinental zonality. Areas of high TOC (>0.5% to 16-20%) include marginal seas, oceanic shelf, continental slopes, continental rises, trenches. Here lithologic types of sediments (terrigenous, carbonate, siliceous, autigenic, volcanic) contain appreciably more OM than sediments more remote from land. Reduction of TOC moving away from continents to central halistatic areas is observed on meso- and microscales in relation to islands. In areas of high TOC content (>0.5-2-1, TOC Fig. 1), reveals a patchy, mosaic distribution. In the peripheral zones, two OM maxima can be recognised - a shallow-water one and relatively deep-water one, which are separated by a zone of low TOC content. The latter, as a rule, coincides with a zone of sediment transit and zone of condensed and more coarse grained relict sediments on the shelf. High TOC is observed in muds of the marginal and enclosed seas. It is usually higher in terrigenous than in carbonate muds, if one excludes iceberg sediments (low TOC) and carbonate oozes of bays, coral and mangrove environments (up to 20%).

Beside a circumcontinental zonality, the TOC distribution reveals latitudinal zonality, which is known as a feature of the settling particulate matter. Latitudinal zonality is expressed best in the Pacific Ocean by two high-latitudinal zones (Boreal and Antarctic) and one Equatorial zone of higher TOC contents, which are separated by the extensive tropical zones of low TOC content. The latitudinal zonality is less distinct in the Atlantic and Indian Oceans.

In this work an attempt was undertaken to construct maps of TOC distribution using computer programs taking into account sets of factors controlling accumulation of OM in sediments. Exponential decreases in TOC with depth and with distance from land were chosen as first approximations. They were defined separately for various facies ascribed to different geomorphologic structures (shelf, continental slope, continental rise, basin, ridge), granularity of sediments (pelite, silt, sand) and latitudinal belts (30° bands). Values of TOC were calculated for a regular grid at 1° intervals. The map (Fig. 2) shows circumcontinental and latitudinal character of TOC distribution. Addition of measured values of TOC to calculated ones supplies some extra details to the map, but does not alters its character. Thus two major factors have been determined infuencing TOC concentrations in sediments, depth and distance from land.

The map could be improved by using more precise approximations, of, for example, primary production. New models are needed to explain the patterns of TOC distribution in zones of high productivity, such as upwellings and fans of large rivers. This will be one issue for further study, as will construction of regional maps of TOC distribution.

Muds of gulfs, bays, enclosed and marginal seas reveal characteristically high TOC content, which is higher, in general, in terrigenous muds than in carbonate ones. The sediments of zones of hydrothermal activity have a higher content of OM with a highly specific composition confined to areas near the hydrothermal constructions. In mud zones and mud-volcanos, in general, increased TOC content occurs (e.g Mediterranean Sea, Black Sea etc.)

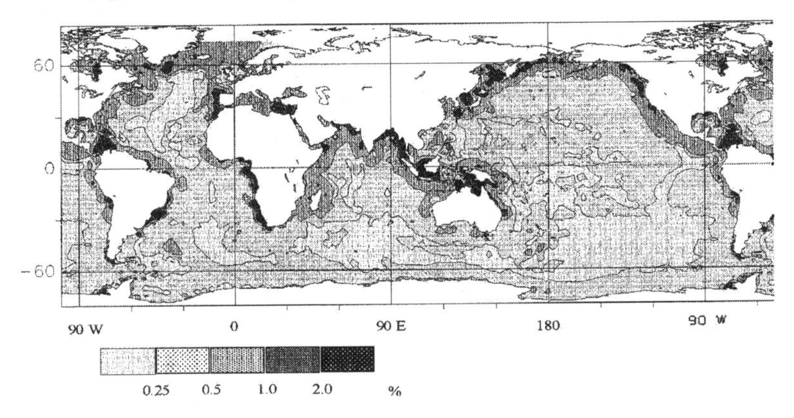

Fig. 2. Distribution of the total organic carbon in the World Ocean (computer mapping)

Direct correlation between TOC and sediment grain size is a feature of the sediments in near-continental areas of the ocean. It is not observed in deep-sea pelagic muds and indicates the facial and lithogenetic conditions of such environments. The very large sorption capacity of red clays and pelagic muds is rarely fully utilized, although DOC concentrations are comparatively high in near-bottom waters (about 1 mg $\cdot$ l^{-1}). The OM concentrations in near-continental and deep-sea pelagic regions of the Ocean are correlated with various regional TOC fluxes to the bottom, low sedimentation and TOC burial rates, and mineralization of most of the labile OM. The main physico-chemical, biochemical and biogeochemical processes (mechanisms) governing the accumulation of organic carbon in sediments are the sorption capacity of sediments, monolayer-equivalent loading or greater-than-monolayer loading under productive nearcontinental regions, rates of enzymatic OM destruction connected with these processes, microbiologic and benthic metabolism.

3. Organic carbon in sedimentary cover of the pacific and indian oceans

The accumulated organic carbon in surficial sediments of the ocean reflects the geological record of conditions on the Earth on a time scale from thousands to millions years. A global perspective of TOC distribution in surficial sediments will enable us to obtain a picture of the evolution of bioproductivity in the ocean. However, consideration of this issue needs better understanding of ket processes such as climate change, the balance between oxygen and carbon in the Biosphere, the paleoecology of benthos, accumulation of large masses of OM material used for oil and natural gas generation and their role in lithogenesis.

We have analysed the vast material on the TOC content in Cenozoic and Mesozoic deposits of the Pacific and Indian Oceans (DSDP and ODP, holes 1-869, [8] [9]). Paleomaps were constructed using average TOC content for each hole taking into account distribution of lithologic types of sediments [10] and paleobathymetric maps kindly provided by L.P. Volokitina. The maps of TOC distribution in deposits of Upper Cretaceous (88-65 million years), Paleocene (65-53), Eocene (53-33.7), Oligocene (33.7-23), Miocene (23-5.3) and Pliocene (5.3-1.75 mln. years) were constructed on the base of palinspastic reconstruction [11] respectively for 70, 60, 45, 30, 15, 5 x 10^6 yr as maps, (on mercator projection), showing location of the continents, islands and the holes 1-533. The positions of the holes 534-869 were calculated by one of authors as described in [12].

Estimated average TOC contents for the Mesozoic and Cenozoic epochs (Table 1) reveal three stages of deposition and burial of OM. These are:

The Early Cretaceous, deposits of which conspicuously differ from later deposits by high TOC content, peculiar conditions of OM accumulation (warm uniform ocean), heterogeneous composition of OM (terrigenous + planktonogenous in various ratios). Anoxic conditions in the Pacific and Indian Oceans were less developed than in the Atlantic Ocean.

The Late Cretaceous - Paleogene is transitional epoch from a warm ocean to a cold one. It is characterized by minimum TOC content in sediments. Higher TOC content in sediments of the Upper Cretaceous in the Indian Ocean is explained by increased input of OM from allochtonous and autochtonous sources into young spreading basin.

The Neogen - Holocene is a period of steady increase of TOC in deposits.

There are similar trends in alterations of TOC in surficial sediments of the Pacific and Indian Oceans in the distribution of dispersed OM and oil deposits of the continents and the oceans [13] [14] [15].

In the Late Cretaceous (Fig. 3), the equatorial band of increased TOC values corresponds to a broad equatorial current, that flowed unrestricted from the Atlantic Ocean to the Pacific Ocean and then to the Indian Ocean.

In the Indian Ocean, terrigenous muds containing rather high TOC content (0.25-0.5%) continued to be deposited in small spreading basins, soon after Hindustan had moved away from the Australia-Antarctica continent (in the Early Cretaceous). Increased TOC content, as a rule, attached to submarine continental margins of the very young Indian Ocean.

In the Paleocene in the Pacific Ocean (Fig. 4), two zones of higher TOC content arenoteworthy. These are a broad equatorial-tropical zone and an area of the developing Tasman Sea between Australia and New Zealand. They demonstrate the latitudinal and circumcontinental character of the distribution and accumulation of OM. The increased depth in the central part of the Pacific Ocean and the rise of the calcite-compensation depth led to enlargement of the area occupied by the red abyssal clays characterized by minimal TOC content.

TABLE 1. Organic carbon in the sedimentary cover of the Pacific and Indian Oceans, % on dry sediment.

·och	Number of sites	Number of samples	Max	Average of averages for sites	SD*	Number of sites	Number of samples	Max	Average of averages for sites	SD*
	The Pacific Ocean					The Indian Ocean				
Holocene	1801	1801	21.2	0.70	1.09	608	608	8.07	0.64	0.72
Pleistocene	222	3904	11.5	0.71	0.96	61	782	7.36	0.60	0.60
Pliocene	165	2571	10.8	0.58	1.21	67	565	8.30	0.66	0.93
Miocene	171	3606	9.90	0.36	0.63	63	926	4.99	0.34	0.44
Oligocene	78	1070	3.40	0.14	0.30	32	177	3.00	0.14	0.19
Eocene	59	612	5.70	0.23	0.51	30	215	1.11	0.10	0.11
Paleocene	16	87	1.40	0.17	0.20	17	74	1.80	0.14	0.16
Cretaceous K_2	24	387	2.80	0.16	0.50	20	221	2.00	0.16	0.11
Cretaceous K_1	13	258	8.60	0.97	1.62	15	482	10.5	0.87	1.15

*Standard deviation ** Surface sediments (0-5) cm sampled by box corers and tubes. *** Sediments from drill sites mainly pleistocene age (pleistocene and holocene are not divided)

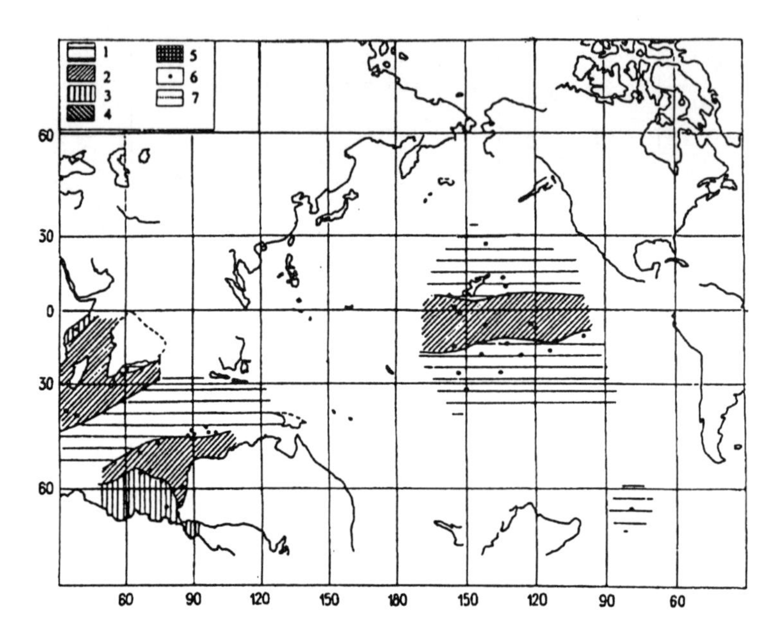

Fig. 3. Distribution of total organic carbon in sediments of Upper Cretaceous.
1- < 0.1%; 2 - 0.1-0.25; 3 - 0.25-0.5; 4 - 0.5-1.0; 5 - > 1%; 6 - sites; 7 - coastal line

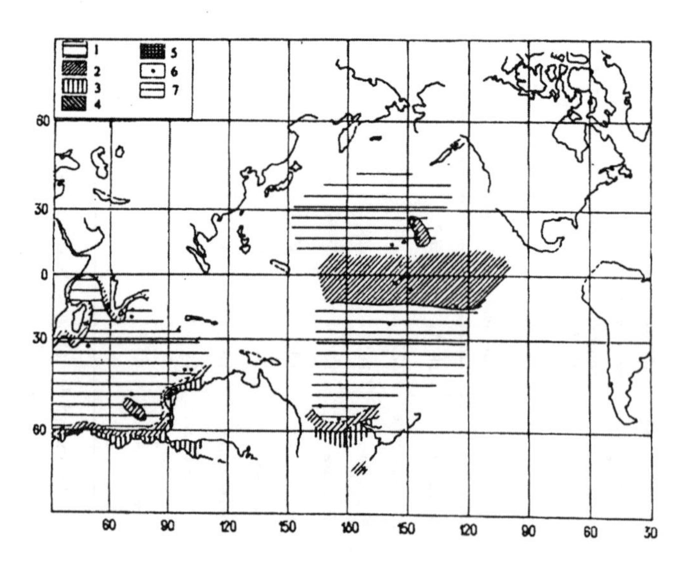

Fig. 4. Distribution of total organic carbon in sediments of Paleocene.
1- < 0.1%; 2 - 0.1-0.25; 3 - 0.25-0.5; 4 - 0.5-1.0; 5 - > 1%; 6 - sites; 7 - coastal line.

These sediments spread widely also in the Indian. In his central part the carbonate sediments contain low TOC concentration (< 0.1%) and were deposited on the vast elevated part of the floor. In the Paleocene the number of areas containing high TOC content was reduced abruptly. The Eocene (Fig. 5) was a time of global cooling with enhanced thermal gradients, provincialism in diatom assemblages, coastal upwellings along west margins of North America and New Zealand, that resulted in increasing TOC content in sediments of the ocean periphery. Terrigenous, siliceous and carbonate sediments, as a rule, containing increased TOC content were deposited in the marginal parts of the Indian Ocean. Thus, in the Eocene, in the Pacific and Indian Oceans the circumcontinental character of TOC distribution is clearly apparent. In addition, two dominant latitudinal zones of increased TOC content (north and equatorial) can be seen in the small, (but dominating by area), Pacific Ocean, as well as wide areas of low OM concentrations and low bioproductivity are in the regions of tropic anticyclone circulation. Particularly high TOC content is observed in the spreading area between Antarctica and Australia.

Characteristic events in the Oligocene (Fig. 6), such as continued cooling and enhancing of climatic contrast, rapid evolution of diatoms assemblages, considerable lowering of calcite compensation depth to 4-5 km and the enhancing of deep-water carbonate accumulation, regression and the enhancing of terrigenouos input, spreading of the Indian Ocean, increasing of depth of the North and South basins in the Pacific Oceans, did not result in any significant alterations of TOC distribution, nor in the average content in sediments. The near-continental zones of high TOC in the both oceans spread due to the enhancing of OM transfer from land. At the same time the areas of high TOC content (> 1%) decreased in both oceans. Carbonate sediments (nanoforaminiferal muds, nanoplankton muds and chalk) on elevated part of the seafloor in the Indian Ocean contain more TOC than some sediments in the Pacific Ocean. One of the causes was the calcite compensation depth, which from the Upper Cretaceous up to the present day,

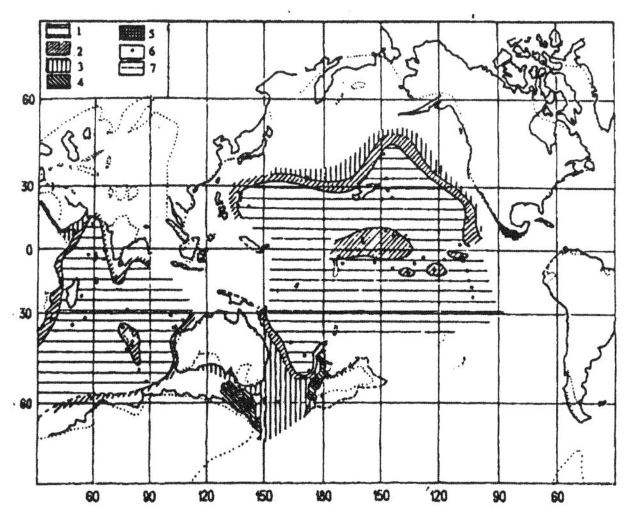

Fig. 5. Distribution of total organic carbon in sediments of Eocene. 1- < 0.1%; 2 - 0.1-0.25; 3 - 0.25-0.5; 4 - 0.5-1.0; 5 - > 1%; 6 - sites; 7 - coastal line.

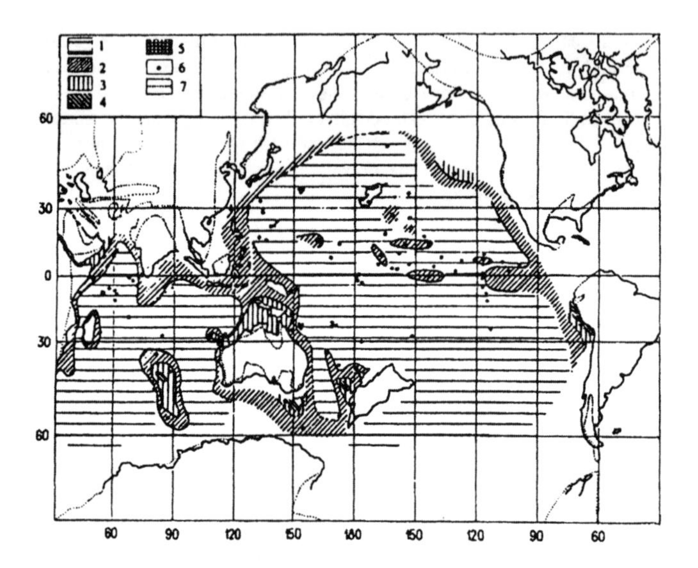

Fig. 6. Distribution of total organic carbon in sediments of Oligocene.
1 - < 0.1%; 2 - 0.1-0.25; 3 - 0.25-0.5; 4 - 0.5-1.0; 5 - > 1%; 6 - sites; 7 - coastal line.

was deeper in the Indian Ocean than in the Pacific Ocean. Fluxes of $CaCO_3$ and organic carbon to the ocean floor were higher in the areas of elevations than in basins.

In the Miocene (Fig. 7), in nearcontinental regions of the Pacific and Indian Oceans, occurrence of the deposits containing high TOC (1-2% and more 2%) increased, approximately by factor of 3, particularly after the cessation of the equatorial communication between the Pacific and Indian Oceans in the Middle Miocene [16]. The peculiarity of the Miocene was the enhancing of contrasts between climatic zones, vertical circulation, marginal upwellings and their influence on bioproduction in the ocean. For example, upwelling along the 2000 km equatorial-tropical region of the east part of the Pacific Ocean was responsible for 50% of "new" global bioproduction of the ocean [17]. High TOC content was observed in diverse types of sediments of the Okhotsk Sea and the Japan Sea (up to 3.0-8.5%), and in the near-continental parts of North and South America (up to 6.1-9.9%). The latter were deposited in conditions of active upwellings, high bioproductivity and the lack of oxygen. The glacial marine turbidites the edges of Antarctica also belong to the zone of increased TOC content (0.25-0.5%).

In the Pliocene (Fig. 8), patterns of TOC distribution in sediments resemble that described for sediments of the Holocene.

Thus, TOC distribution in the Pacific and Indian Oceans during the period from the Upper Cretaceous to the present day was zonal and reflected the existence of circumcontinental and latitudinal regularities of OM distribution. Gradients of TOC content, shelf-continental slope and connected with various morphostructures of the ocean floor (circumcontinentality) are stronger than latitudinal gradients at any epoch of the Cenosoic. Circumcontinental zonality of TOC distribution is not observed, as a rule, during the early history of the ocean. In new just formatted areas of the ocean accumulation of OM is similar to its accumulation in internal and rather small marginal seas [18]. Latitudinal zonality of TOC distribution is observed more clear in the Pacific Ocean through the existence of the equatorial band and high-latitudinal zones of higher TOC. Vertical zonality (TOC dependencies on depth) can be demonstrated by the existence of local zones of increased TOC content on elevated parts of the seafloor due to higher fluxes of OM deposited at the sediment-water interface.

The evolution of the TOC distribution is expressed through (i) rise of average TOC content in sediments (from the Upper Cretaceous to the Holocene) and the occurrence of its high (> 2%) values; (ii) graded enhancing and spreading of marginal upwellings and increasing of OM flux on the sediment-water interface; (iii) development of a circumcontinental character of TOC distribution.

Analysis of deep-water drilling data shows, that the regularities of TOC distribution in surficial sediments were controlled by complex sets of factors. Firstly, tectonic and climatic factors governing circulation of surface and subsurface water, the enhancement and spreading of coastal upwellings, resupply of nutrients to surface water, marine prroductivity and flux of OM to the seafloor, input of terrigenous OM from land mainly as dissolved and pariculate OM along with river run-off. Oxic and anoxic conditions in sediments corresponding to slow or rapid burial of OM, ratio of fluxes of terrigenous and planktonogenic OM, which have variable resistance to enzymatic hydrolysis and microbial destruction, resulted in varied peservation of OM in sediments.

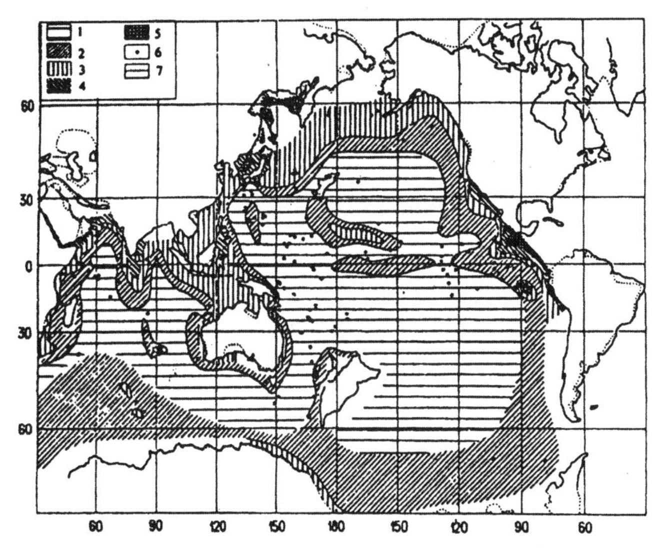

Fig. 7. Distribution of total organic carbon in sediments of Miocene.
1- < 0.1%; 2 - 0.1-0.25; 3 - 0.25-0.5; 4 - 0.5-1.0; 5 - > 1%; 6 - sites; 7 - coastal line.

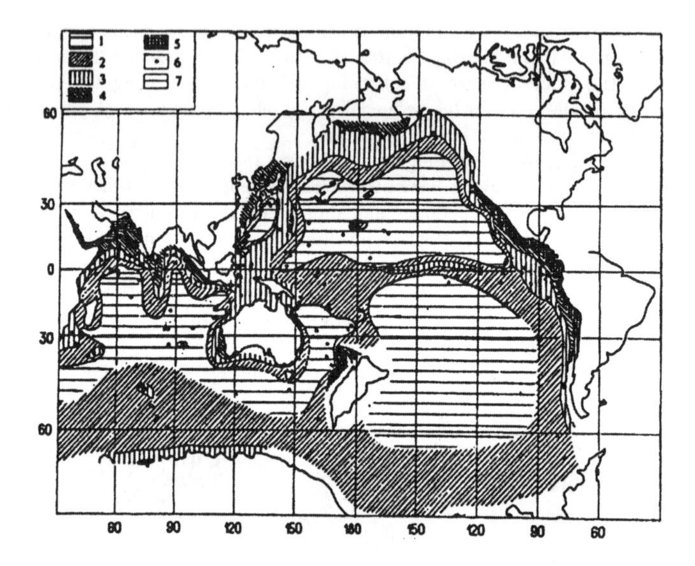

Fig.8. Distribution of total organic carbon in sediments of Pliocene.
1 - < 0.1%; 2 - 0.1-0.25; 3 - 0.25-0.5; 4 - 0.5-1.0; 5 - > 1%; 6 - sites; 7 - coastal line.

4. Burial rates of organic carbon in recent sediments of the ocean

The calculations of TOC balance in the ocean can be revealed by maps of TOC burial rates (mg cm^{-2} kyr^{-1}). The global map (Fig. 9) was based on the map of TOC distribution (Fig. 1), the maps of accumulation rates [19,20] and the maps of lithologic types of sediments [21] [22].

Basic patterns in the distribution of TOC burial rates are circumcontinentality, specific to each ocean, and latitudinal zonality observed in the Atlantic and Pacific Oceans. Circumcontinental zonality is shown as high values of TOC at the periphery of the ocean and decreasing with distance from land to central halistatic areas. Latitudinal zonality is shown through the existence of boreal, equatorial and notal areas of high values of TOC, separated by areas of low burial rates. Huge amounts of TOC are deposited in zones of river runoff in South and south-east Asia and West Africa, and also in some areas of marginal upwelling, high bioproductivity and high accumulation rates of sediments.

It is worthwhile noting that absolute amounts of TOC often do not correlate with concentrations. High accumulation rates can result in (i) rapid burial of OM, favourable for its preservation and high content due to anoxic conditions in sediments, (ii) dilution of OM by inorganic matter, an example of which is regions of the River Congo run-off river and Benguela upwelling (Fig. 10) [23]. Absolute amounts of TOC accumulating in the zone of avalanche sedimentation in the region of the Congo runoff exceed those in the Benguela upwelling area by one order of magnitude (230-6200 and 5-264 mg C cm^{-2} ky^{-1}), whereas percentage concentrations of TOC are equal (~ 2%) in these zones of high productivity. The area of the River Congo fan constitutes 0.0025% of the area of the Atlantic Ocean, nevertheless masses of TOC accumulated in this region constitutes about 0.6% of the amount of TOC accumulated in whole pelagic part of the Atlantic Ocean. It is necessary to stress that TOC accumulation on shelf and continental slope is insufficiently studied. The measurements of accumulation rates are especially necessary in these regions.

The map of OM fossilization (Fig. 11) was calculated as the ratio of absolute amounts of TOC to primary production. Two maps were used for calculation: the map of primary production, constructed in the P.P.Shirshov Institute of Oceanology [73, our modification] as a result of observations of chlorophyll concentrations by CZCS satellite radiometer during 1978-1986 (Nimbus-7). If "vessel effect" [62] in measurements of productivity is not taken into account, primary production of the World Ocean is estimated at $52 \cdot 10^{15}$ g yr^{-1}, and with the "vessel effect" $90 \cdot 10^{15}$ g yr^{-1}. The coefficient of fossilization of OM is 0.0004-0.83 and taking into account the "vessel effect" 0.0002-0.81%, (Fig.11). The average values are respectively 0.025 and 0.017%. The lowest coefficient of fossilization is found in deep-sea red clays. In our opinion the key factors are, (i) the large number of trophic chains in which OM is metabolised, (ii) very high activity of hydrolytic enzymes in surficial layer of sediments (our results), (iii) low accumulation rate and long time of exposition of OM in upper layer of sediments in active oxidative conditions.

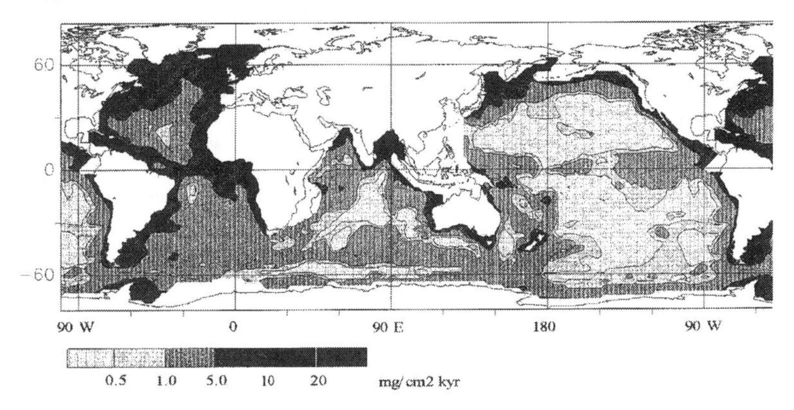

Fig.9. Distribution of burial rates of organic carbon in recent sediments of the World Ocean

The same map of primary production was used to construct the map of TOC fluxes reaching the benthic boundary layer. For this purpose we used the empirical relationship [24]

$$F_c = 33\, P_p\, Z^{-1} \quad (1)$$

obtained as result of analysis data of sediment traps, described particulate OM flux (F_c) as function of primary production (P_p) and depth (Z). In the pelagic part of the World Ocean (outside continental slope) total organic carbon reaching deep-sea floor is estimated at $460 \cdot 10^{12}$ g yr^{-1} and $730 \cdot 10^{12}$ g yr^{-1} without and with consideration of "vessel effect" respectively. Total organic carbon flux in the World Ocean can be estimated today approximately (due to the low 2° resolution of the map, the lack of an empirical relationship for depth less than 100 m, and underestimation of terrigenous OM missing by sediment traps). TOC flux reaching the benthic layer of sediments in the World Ocean is estimated at $3.1 \cdot 10^{15}$ g yr^{-1} and $4.4 \cdot 10^{15}$ g yr^{-1} without and with consideration of "vessel effect" respectively.

The map of burial coefficients (Fig. 12) was calculated as ratio of absolute masses of TOC to its fluxes to the seafloor. The range of variations of burial coefficients is rather wide (0.04-77%, averaging 2.4%). Maximum burial coefficients are observed in areas of high accumulation rates and anoxic conditions of sedimentation.

5. Fluxes and masses of organic carbon in the modern ocean

Starting from the 1970s the study of fluxes and masses of TOC in the ocean and other reservoirs of the Biosphere acquired a complex character. Along with estimations of single parameters of OM circulation, their joint estimations appeared [25] [26] [27] [1] [28]. However, it would be wrong to conclude that the problem of organic carbon cycle was not considered previously and that scientists have not tried to appreciate the balance between extremely complicated interactions of biological, geological and chemical processes, through which the matter and energy exchange in reservoirs of the Biosphere is recycled or dissipated at the Earth surface [29]. Sound generalisation were made from these earlier studies [30] [31]. Fig. 13 shows values of fluxes and masses of TOC in the modern ocean (most probable estimations, in our opinion, are given in parenthesis).

5.1. RIVER RUNOFF

The main terrigenous sources of TOC in the ocean are river runoff and eolian transfer. It has been shown, for example [5] [32], that the content of dissolved organic carbon (DOC) and particulate organic carbon (POC) in river runoff depends on the type of drainage area, climate, soil constitution, volume of runoff and varies in the range of two orders of magnitude: for DOC from < 1.0 to > 40 mg l^{-1} , and for POC from 0.3% up to 60% of dry weight, [33] [34]. From various estimates, total river input into the ocean varies in the range $(188\text{-}264) \cdot 10^{12}$ g DOC yr^{-1} and $(181\text{-}472) \cdot 10^{12}$ g POC yr^{-1}, giving a total of $(369\text{-}736) \cdot 10^{12}$ g TOC yr^{-1}. The input of river water is 41800 km^3 [35] and solid input $19288 \cdot 10^{12}$ g yr^{-1} [36], with concentrations estimated at 4.5-6.3 mg DOC l^{-1} and 4.3-11.3 mg POC l^{-1} (0.9-2.4% of dry weight of particulate matter) or 8.8-17.6 mg TOC l^{-1}.

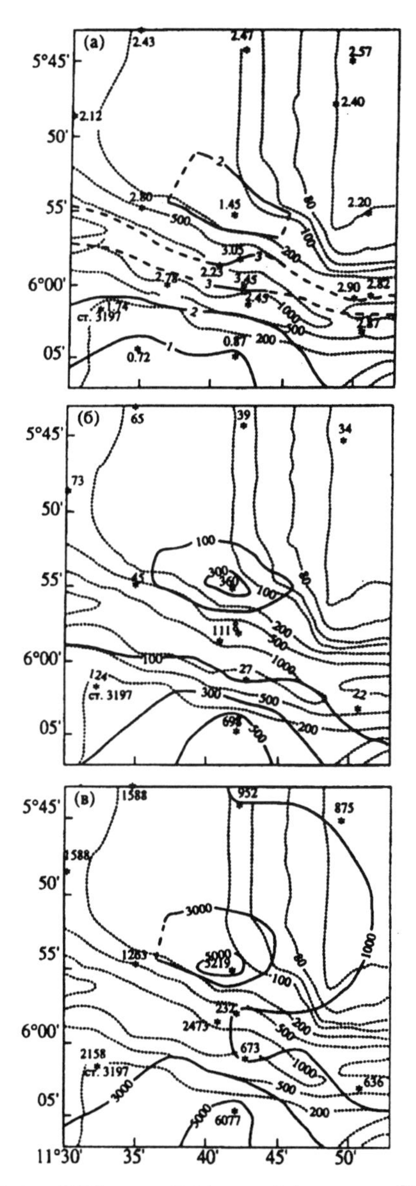

Fig. 10. Disrtibution of TOC contents, % (a), accumulation rates, g cm^{-2} kyr^{-1} (b) and burial rates of organic carbon, mg cm^{-2} kyr^{-1} (c) in the region of Congo river runoff.

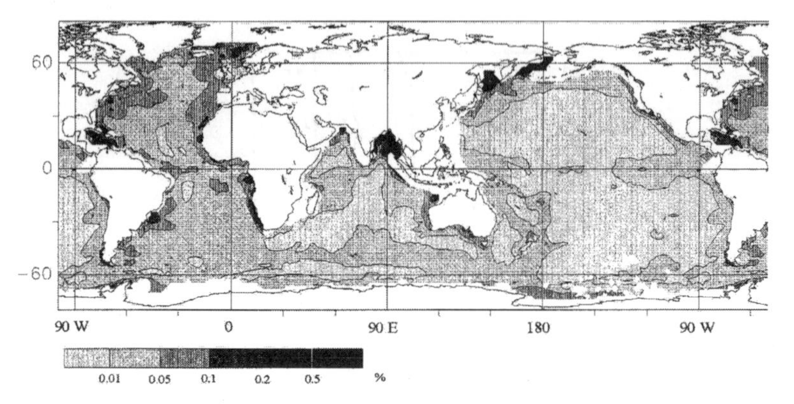

Fig. 11. Distribution of coefficients of organic matter fossilisation in the World Ocean.

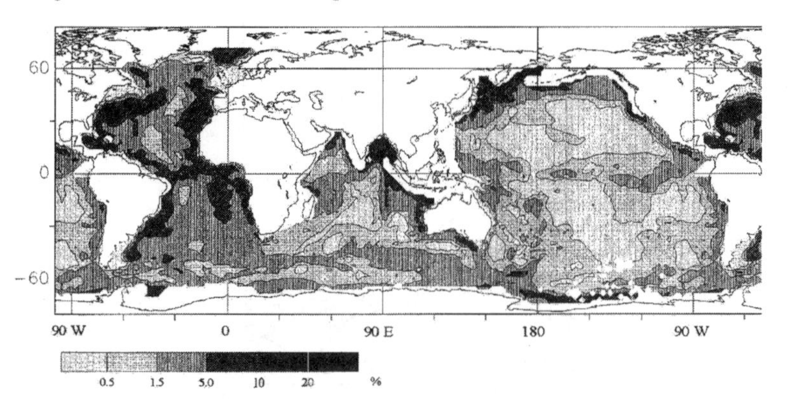

Fig. 12. Distribution of burial coefficients of organic matter in the World Ocean.

Analysis of field data and calculations show the most probable amount of the global river input at $210 \cdot 10^{12}$ g DOC yr^{-1} and $250 \cdot 10^{12}$ g POC yr^{-1}, corresponding to average values of 5 mg DOC l^{-1} and 6 mg POC l^{-1} (1.3% of dry weight of the solid river input). Thus, the shares of DOC and POC in river input are similar and the total input is estimated at $460 \cdot 10^{12}$ g TOC yr^{-1}, a little higher (15-38% $0.33\text{-}0.4 \cdot 10^{15}$ g TOC yr^{-1}) than published values [37] [38] [39].

Total river input of OM to the ocean constitutes 0.75% of primary production of the land ($60 \cdot 10^{15}$ g TOC yr^{-1}) and approximately the same amount of primary production of the ocean (0.75%, Fig. 13). However, river OM is preserved much better than planktonogenous one due to its resistance to biodegradation.

The main sources of river DOC are soil and the age of this OM is tens or hundreds of years. The sources of river POC are (i) erosion of soils (age hundreds years), (ii) erosion of sedimentary rocks (age millions and hundreds of millions years), (iii) biosynthesis of autochtonous OM (n x 0.01 years) [33] [34]. The proportion of river input from these sources depends on climate, conditions of wind erosion, soil cover, volume of river runoff, type of rivers. However, OM of soils almost always dominates in river runoff almost always (except for mountain rivers). Autochthonous POC usually constitutes about 1-2% of total POC in river runoff [25]. However, in eutrophic rivers autochthonous POC can reach 6 mg l^{-1} [34].

The average content of analytically measured biopolymers (aminoacids, aminosugars, carbohydrates, organic acids, lipids and others) in river runoff constitutes 30% of OM range 5-80%. The part of this OM (biopolymers that are not included in humic acids) belongs to the labile fraction of river runoff OM. This fraction metabolises rather rapidly in estuaries, and deltas and zones of mixing of fresh and sea waters. Enzymatic hydrolysis plays a very important role in destruction OM that will be discussed below. High concentrations of labile compounds (in % of POC) are observed in low solid runoff with high content of POC in particulate matter. However, particulate matter of maximum runoff is enriched by more stable degradation compounds. The remainding part of river OM (on average about 70%) is fulvic and humic acids, as well as insoluble OM, (e.g. kerogen in particulate matter), [3] [33]. Only a small fraction of stable, hardly-metabolising, humic and fulvic acids is adsorbed onto particulate matter [39].

5.2. EOLIAN TRANSFER

Eolian flux to the oceans can only be estimated approximately due to the scarcity of data. Gravitational sedimentation of dry insoluble terrigenous eolian matter is estimated in the World Ocean at $0.38 \cdot 10^{15}$g yr^{-1}, [40] [41]. In addition, $0.49 \cdot 10^{15}$ g yr^{-1} of eolian matter reaches the ocean surface along with atmospheric sediments. The average TOC content in eolian matter is about 20% [24 and new unpubl. data], and the total eolian flux is estimated at $0.87 \cdot 10^{15}$ g yr^{-1} x 20% = $174 \cdot 10^{12}$ g TOC yr^{-1}. Remarkable progress was achieved in the study of eolian matter in the SEAREX (The Sea-Air Exchange) program. This experiment used a wide range of lipid biomarkers of culutured plants (acids, alcohol, alkans). These data, as well as $\delta^{13}C$ (for fractions of OM) show that the largest fraction of TOC of eolian matter is of terrestrial origin. Eolian chemofossils are especially widely-spread in sediments located far from land and accumulated at low rate [42]. The distribution of matter of eolian origin in the ocean differs from the distribution of POC of the river runoff. The main mass of eolian matter is deposited in arid zones of the ocean, nevertheless a small fraction reaches high-latitudinal zones of the World Ocean. Eolian OM can constitute up to 50% of buried OM in pelagic deep-water basin of the ocean [43], but in nearcontinental regions of the ocean it is deluted by OM of different origin, and therefore is unlikely to be found using eolian biomarkers.

5.3. UNDERGROUND RUNOFF

Calculations performed during the last decade for some hydrological systems (480 studied areas) for all oceans show underground runoff, 1045 * 10^{12} g, constitutes about 42% of all river runoff, 2480 * 10^{12} g [44]. Input of dissolved matter through underground runoff is one of the main transport pathways of numerous chemical elementsto the ocean.

In recent years, the use of complex hydrologic-hydrogeologic methods has facilitated production of maps of underground runoff and estimates of the total volume of underground runoff, (excluding rivers) to the ocean. It constitutes about 2400 km^3 yr^{-1} (the Pacific Ocean 1300.3 km^3 yr^{-1}, the Atlantic Ocean 815.3 km^3 yr^{-1}, the Indian Ocean 219.4 km^3 yr^{-1}, the Arctic ocean 47.5 km^3 yr^{-1}) [44]. This value differs somewhat from that previously estimated by other hydrodynamic methods at 2100 km^3 yr^{-1}. Generalisations made from numerous data on chemical composition of underground water of the upper zone of hyperhenesis of the various regions of the Earth allows estimates of the average value of DOC from underground inputs at 5.86 mg l^{-1} [45]. As a result, underground runoff (without rivers) supplies to the World Ocean 2400 km^3 yr^{-1} x 5.86 mg l^{-1} = 14 *10^{12} g DOC yr^{-1}.

5.4. GLACIAL RUNOFF

The glacial runoff in the ocean is estimated at 3340 km^3 yr^{-1} [46] and its TOC content at 0.7 mg l^{-1}. It is not difficult to calculate that glacial runoff supplies to the ocean $2.3 \cdot 10^{12}$ g DOC yr^{-1}. The origin of this OM is unknown, a part is presumed to have an eolian genesis.

In addition to dissolved OM, ice carries into the ocean the products of decomposition of sedimentary, metamorphic and igneous rocks and, along with them, a certain amount of "ancient" OM (hundreds millions years old), which is resistant to degradation. Such OM predominates in some Antarctic sediments and is characteristically of rather specific composition [47]. Annually, 1500 mln. tons of solid glacial matter containing TOC about 0.4% or $6.0 \cdot 10^{12}$ g TOC yr^{-1} is carried out to polar regions of the ocean. Thus, about $8.3 \cdot 10^{12}$ g TOC yr^{-1} is carried into the ocean along with glacial runoff in dissolved and solid forms.

5.5. WAVE ABRASION

Wave abrasion is a small source of sedimentary matter in the ocean. From [48] about $500 \cdot 10^{12}$ g of clastic matter is supplied in the ocean along with products of wave abrasion. We assume, that the average TOC content in this matter constitutes 0.4% dry weight. It agrees with TOC content in sedimentary rocks including effusive ones (0.4%), as well as with relative area occupied by erosive sedimentary (80%) and granite (20%) rocks at the continents with TOC content 0.5% and 0.05% of dry weight respectively (averaging 0.4%) [25, 41]. The total mass of TOC supplied in the ocean as result of wave abrasion is estimated at $2 \cdot 10^{12}$ g TOC yr^{-1}. Particulate "old" OM, rather resistant to the biodegradation, predominates in its composition and is buried in sediments almost without alteration.

Thus, total terrigenous input of TOC in the ocean is estimated at about $658 \cdot 10^{12}$ g TOC yr^{-1} including $226 \cdot 10^{12}$ g DOC yr^{-1} and $432 \cdot 10^{12}$ g POC yr^{-1}. This input of TOC from the land is smaller by a factor of 100 than the production of phytoplankton. Nevertheless, compared with phytoplankton, it is important to note, that this matter is as resistant to enzymatic destruction and mineralization at thermodynamic conditions existing on the Earth surface.

5.6. PRIMARY PRODUCTION AND BIOMASS OF THE OCEAN

The estimates of primary production of phytoplankton in the ocean during the last 30-40 years show an increasing trend from $(15\text{-}25) \cdot 10^{15}$ g C yr^{-1} to $(60\text{-}103) \cdot 10^{15}$ g C yr^{-1} [49] [50] [51] and even up to $187 \cdot 10^{15}$ g C yr^{-1} [52]. Similar trend is observed in estimations of primary production of the land based on satellite mapping data, it is estimated today at $179.9 \pm 1.6 \cdot 10^{15}$ g C yr^{-1} [53]. The trend of phytoplankton production (as well as phytobentos) has increased significantly. It can be noted also that the estimates of TOC amount buried in the sediments also have been revised on the basis of new data and have had a tendency to increase. This relates entirely to estimates of biomasses of phytoplankton and phytobenthos.

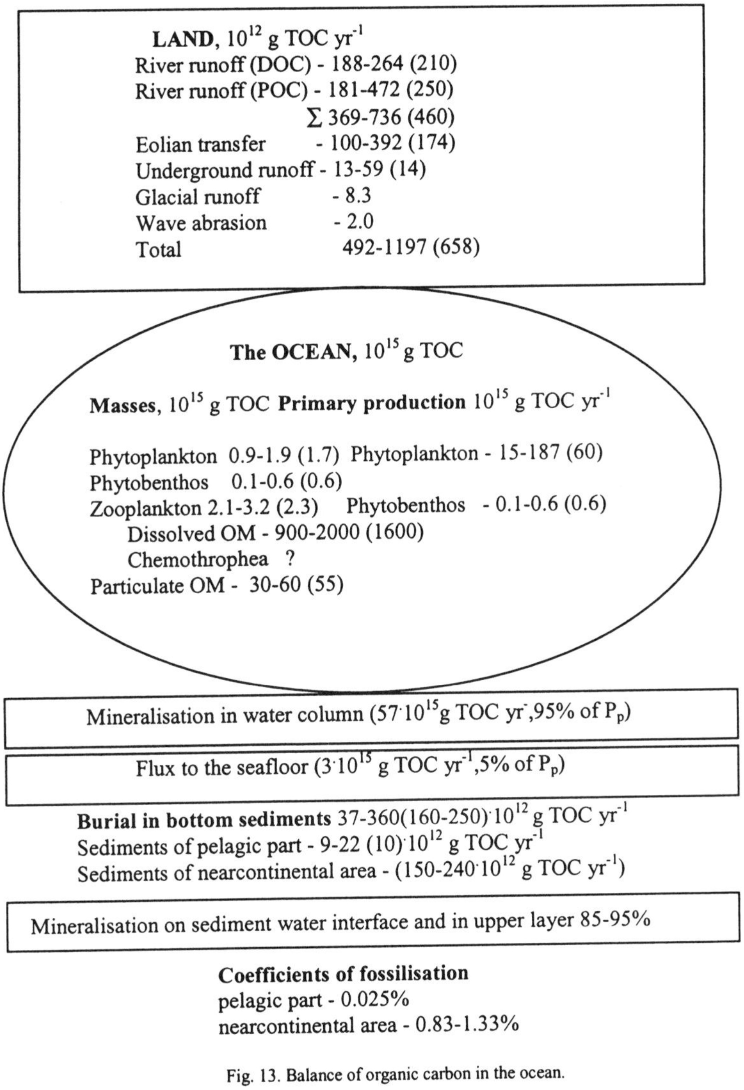

LAND, 10^{12} g TOC yr^{-1}
River runoff (DOC) - 188-264 (210)
River runoff (POC) - 181-472 (250)
Σ 369-736 (460)
Eolian transfer - 100-392 (174)
Underground runoff - 13-59 (14)
Glacial runoff - 8.3
Wave abrasion - 2.0
Total 492-1197 (658)

The OCEAN, 10^{15} g TOC

Masses, 10^{15} g TOC **Primary production** 10^{15} g TOC yr^{-1}

Phytoplankton 0.9-1.9 (1.7) Phytoplankton - 15-187 (60)
Phytobenthos 0.1-0.6 (0.6)
Zooplankton 2.1-3.2 (2.3) Phytobenthos - 0.1-0.6 (0.6)
Dissolved OM - 900-2000 (1600)
Chemothrophea ?
Particulate OM - 30-60 (55)

Mineralisation in water column ($57 \cdot 10^{15}$g TOC yr^{-},95% of P_p)

Flux to the seafloor ($3 \cdot 10^{15}$ g TOC yr^{-1},5% of P_p)

Burial in bottom sediments 37-360(160-250)$\cdot 10^{12}$ g TOC yr^{-1}
Sediments of pelagic part - 9-22 (10)$\cdot 10^{12}$ g TOC yr^{-1}
Sediments of nearcontinental area - (150-240$\cdot 10^{12}$ g TOC yr^{-1})

Mineralisation on sediment water interface and in upper layer 85-95%

Coefficients of fossilisation
pelagic part - 0.025%
nearcontinental area - 0.83-1.33%

Fig. 13. Balance of organic carbon in the ocean.

In the late 1960s the amount of phytoplankton and phytobentos in the oceans was estimated at $0.165 \cdot 10^{15}$ g OM and $0.03 \cdot 10^{15}$ g OM respectively, that corresponds to about $0.082 \cdot 10^{15}$ g C and $0.15 \cdot 10^{15}$ g C (averaging about $0.1 \cdot 10^{15}$ g C) [54]. In the 1980s to early 1990s, the mass of phytoplankton and phytobentos has been estimated at $(0.9\text{-}1.9) \cdot 10^{15}$ g C. Today the most probable estimates of biomass and production of organic carbon in the ocean are $1.7 \cdot 10^{15}$ g C yr^{-1} (biomass of phytoplankton and phytobenthos), $60 \cdot 10^{15}$ g C yr^{-1} (production of phytoplankton) and $0.6 \cdot 10^{15}$ g C yr^{-1} (production of phytobentos), for instance [49, 50,51]. The increased estimates of phytoplankton production can be explained by several causes, such as (i) procedures concerned mainly with "vessel effects" [53] and underestimation of production measurment by the radiocarbon method, (ii) underestimation of production mainly by picoplankton (0.2-2.0 µm) but also by nannoplankton (2-15µm), in the vast pelagic areas of the ocean that can be responsible for 70% of primary production, (iii) underestimation of production of organic carbon in marginal and nearcontinental areas of the ocean, which are still insufficiently investigated due to the complex dynamics and patchy character of the production processes, [54, 50]. Higher estimates of phytobenthos production earlier estimates are due to high biomass, high turnover rates and high rates of synthesis of OM and secretion of OM as dissolved products.

An important problem is primary production by chemolithothrophic bacteria in thermal zones of the ocean. The present estimation of the value of this production (< 1% of photosynthic production) probably needs revision [56] [51].

5.7. DISSOLVED ORGANIC MATTER

Dissolved OM significantly influences the chemistry and biology of the ocean and during more than 50 years has attracted the attention of investigators. The problem of accurate measurement of DOC in sea water has not been solved due to the extremely low concentrations (10^{-4} % or 1-2 ppm) and to contamination, (often about 10^4 times the DOC content). Today many methods have been rejected (for instance, the Menzel-Vaccaro method of persulphate oxidation), as well as data obtained by similar methods. Intercalibration of other methods, shows large variance in measured values, [57].

The most commonly used methods are (i) wet persulphate oxidation along with ultraviolet radiation of water for more complete oxidation of OM and (ii) dry high temperature incineration of OM with catalysts (platinum, oxides of cobalt and copper; TOC-500 and TOC-5000). The most common error with these methods are associated with the necessity for blank corrections (water, reagents), unstable action of catalysts and their ageing. Therefore we present estimates based on data obtained in Russia (and the former Soviet Union) using mainly the method of oxidation of dissolved OM by a mixture of sulphuric acid, potassium dichromate and silver dichromate at $130 \pm 1°$ with colourometric recording of evolved CO_2 [58] [59]. The method does not need blank corrections because incineration of contamination (in reagents, water, sulphuric acid, on the surface of reactor) is performed during analysis before burning of OM of sea water in the same reactor. The accuracy of DOC measurements using a new digital instrument (AACL) is $\pm$ 7% (1 mg l^{-1}), $\pm$ 5 % (2 mg l^{-1}). Unfortunately, the method has low productivity (1 analysis takes about 1.5 h).

The estimates based on 1000 of our DOC measurements show that DOC content varies from 0.8 to 1.9 mg l^{-1} at depths > 1000 m and constitutes 1.2 mg l^{-1} for the ocean on the whole. The volume of the ocean water is $1370 \cdot 10^6$ km^3 giving a total mass of ca. DOC amounts to about $1600 \cdot 10^{15}$ g. The mass of DOC in surface layer of the World Ocean (0-200 m, volume i.e ca. $68 \cdot 10^6$ km^3) accounts for about 0.3-0.4% of the total. If one assumes the average age of DOC is 1000 years the rate of annual withdrawal of DOC from the Ocean is about $1.6 \cdot 10^{15}$ g yr^{-1}. What fraction of this is resistant to degradation, is mineralised to CO_2 and is adsorbed on the particles or flocculates and reaches seafloor is unknown.

It is worthwhile noting that mechanisms to explain the great age of DOC in deep water of the ocean (up to 6000 years, [60]) are unknown. The flux of "ancient" OM along with flushed out pore water alone cannot be an explanation.

Recent calculations show that the accumulation rate of sedimentary water in the World Ocean is $3 \cdot 10^{15}$ g yr^{-1} on average and the annual value of the flushing out process during lithogenesis is $1.8 \cdot 10^{15}$ g yr^{-1}. Therefore, more than 60% of pore water buried in sediments returns to the maternal basin. [44]. The average DOC content in pore water from 150-200 m layer of sediments is 15 mg DOC l^{-1} and the total amount of DOC returning from sediments to the ocean is estimated at $1.8 \cdot 10^{12}$ l yr^{-1} x 15 mg l^{-1} = $0.027 \cdot 10^{12}$ g DOC yr^{-1}. Nevertheless, clarity in the issue will be reached only after measurements of the average age of flushed out pore water.

5.8. PARTICULATE ORGANIC MATTER

The content of POC in the ocean is estimated as an average of 0.04 mg l^{-1} with a total mass of 0.04 mg l^{-1} x 1370 km^3 = $55 \cdot 10^{15}$ g. Our estimations of the POC mass does not differ from published ones [28]. The higher value compared with earlier data (e.g [5]) is mainly due to better estimation of POC content in near-continental areas, particularly at the zones of river runoff and of high bioproductivity. The ratio POC/DOC is 1:32 in the World Ocean.

5.9. FLUX OF ORGANIC MATTER TO THE SEAFLOOR

The spatial variability of POC, the seasonal character of its flux to the seafloor and the interactions between authochthonous and allochthonous fluxes complicate the estimation of total OM settled on the sediment-water interface of the World Ocean. Earler, on the basis of estimates of the masses of DOC ($1800 \cdot 10^{15}$ g), POC ($30 \cdot 10^{15}$ g) and living OM ($2.8 \cdot 10^{15}$ g) and some assumptions (for instance that the average age of living OM is 1-2 years and the masses exist at dynamic equilibrium of synthesis and destruction), the residence time of POC in the ocean was estimated at 10-25 years, and $1\text{-}3 \cdot 10^{15}$ g POC yr^{-1} reaches the seafloor [5]. Based on results of analysis of the POC distribution in the water column at several dozen stations, located in various regions of the World Ocean it was estimated, that 5-10% of POC of euphotic layer or $(3\text{-}6) \cdot 10^{15}$ g POC yr^{-1} from primary production ($60 \cdot 10^{15}$ g C yr^{-1}) and OM input from land reaches seafloor. Based on sediment trap data and calculations of OM fluxes in the areas of shelf and continental slope, it was estimated, that in this area $1\text{-}1.5 \cdot 10^{15}$ g TOC yr^{-1} reaches the near-bottom layer [79].

The TOC flux to the seafloor based the map of TOC fluxes and using expression (1) amounts to $3.1 \cdot 10^{15}$ g TOC yr^{-1} for world oceans. In the water column of the open ocean outside the continental slope and continental rise $0.46 \cdot 10^{15}$ g TOC yr^{-1} reaches the seafloor. Therefore 95% of primary production mineralises in the water column and merely 5% reaches the sediment-water interface.

5.10. BURIAL OF ORGANIC MATTER OF SEDIMENTS

The absolute masses of TOC can be calculated directly, using the map of burial rate (Fig. 9), only for the water column of the open ocean (area about $290 \cdot 10^{6}$ km^{2}). The mass of TOC buried in sediments here amounts to merely $9.5 \cdot 10^{12}$ g TOC yr^{-1}.

The average mass of sedimentary matter deposited in thew water column (0.624 for terrigenous flux and $0.451 \cdot 10^{15}$ g cm^{-2} kyr^{-1} for biogenic flux [39]) and the average TOC in sediments of the ocean basins (0.38%, n=1686, our data) show that TOC flux makes up 4.2 mg cm^{-2} kyr^{-1} and the total mass of TOC buried in the pelagic part of the ocean amounts to $11.3 \cdot 10^{12}$ g TOC yr^{-1}. There is insufficient analytical data to calculate masses of TOC buried in the areas of continental slope and continental rise using the map and direct estimates.

Published data on TOC masses buried annually in the ocean sediments vary in the wide range $(37\text{-}360) \cdot 10^{12}$ g TOC yr^{-1},[14, 21]. Burial rates of $100 \cdot 10^{12}$ g TOC yr^{-1} calculated using accumulation rates of sediments and TOC content are underestimates because of the missing TOC masses deposited on the shelf and continental slope. Today the most probable estimation of TOC deposited in the sediments of the World Ocean amounts to $160\text{-}250 \cdot 10^{12}$ g TOC yr^{-1} [19, our data]. Estimates of TOC masses deposited on the shelf and the continental slope and the continental rise are extremely important. From [61] [2] 88-90% of TOC is deposited at the shelf, 4-6% at the continental slope and 5-6% in the deep sea basins. From our data and [62] 10% of TOC is deposited on the shelf, 85% on the continental slope and continental rise and 4-5% in the deep sea basins. New data collected in the Institute of Oceanology on TOC masses allow us to assert that the main deposition area for TOC is the continental slope and continental rise. Here 85% of TOC reaches ocean bottom along with fluxes of allochtonous and authochtonous OM.

The average TOC in dry matter from sediment traps is 5%, [63] [64] [65], and average TOC in surface sediments (0-5 cm) is 0.42% (estimated using the map on fig.1), it can be estimated roughly, that 92% of OM reaching sediment-water interface is mineralized at the surface and upper few centimetres of sediments. Similar estimates for the water column of the open ocean performed using the map of burial coefficients

(Fig. 12) accounts for 98% of OM. These approximate estimations show that the key zone for the transformation of OM is situated at the sediment water interface.

Coefficients of fossilization of OM in bottom sediments (ratio deposited TOC to organic carbon of primary production + allochtonous part) are estimated for basin at 0.02% or at 0.021% using the map (Fig. 11) and for near-continental area (shelf-slope-rise) at 0.8-1.3%. Therefore, the degree of closure of the OM cycle is the water column of the open ocean (99.98%) is significantly more than in main area of OM accumulation (99.2-98.7%). Such values of fossilization coefficients show very high ratios of synthesis-destruction of OM in the modern ocean and in general for the carbon cycle.

6. Hydrolytic enzyme processes of OM destruction

6.1. APPROACHES AND METHODS

The destruction of biopolymers and geopolymers to low molecular weight compounds is a key stage in the cycle of OM in the ocean. This process transforms part of the OM to a dissolved state. Then the low molecular weight fractions of OM can be utilized by microorganisms. Biopolymer destruction is primarily by means of hydrolytic enzyme reactions, the rates of which are 9-13 times greater than the rates of destruction of natural biopolymers by other chemical reactions [66]. Thus enzyme catalysis of the OM hydrolytic reactions is the dominant mechanism in biogeochemical cycling.

Studies of biogeochemical cycling have been done by many scientists in different countries, Russia, USA, Germany, France, etc. [67] [1] [68-76] [77] [78]. During the 1990s investigations of enzyme activity (EA) have been done by D. M. Carl and J. K. Christian (Hawaii University) [67] and Saliot et al. in the Arctic [77]. Our studies [71] show that EA occursnot only in surficial bottom sediments (0-20 cm) but also occurs at depths of 3 m.1. It has been shown [78] that marine suspended matter ("marine snow") acts as a permanent fermentor due to the sorption of enzymes, [65].

The development of enzyme studies has been limited by the absence of simple methods for the direct measurement of enzymes and their activity in water and bottom sediments without concentration, extraction, filtration and other procedures, due to their low concentrations. Therefore, direct measurement of enzymes and the estimation of their maximal potential activity (V_{max}) by addition of substrates peculiar to the enzymes are prteferred. It is desirable to use substrates which dissolve in sea water and to apply modified radioactive and chromophore techniques.

In the laboratory of ocean chemistry of P.P. Shirshov Institute of Oceanology the investigation of the EA role in OM destruction process has been done since 1989. The main efforts are development of sensitive and rapid methods allowing the detection EA in sea water and bottom sediments under field conditions without the complex preparation of the samples. Fast-soluble asocasein (Sigma, USA) are used for the determination of protease activity (PA) and modified by the M-Procionic-5CX dye starch - for determination of amylase activity (AA). As a result we developed new methods for: the determination of PA and AA in various ecosystem components (sea water, pore water, bottom sediments, products of biota metabolism) [67, 68, 70-73]; method of calculation of kinetic values of biopolymers EA in environment on the base of maximum activity (V_{max}) and the effective rate of polymer destruction as the reaction of the first order by substrate (v_1) [72]. High sensitivity test-systems (biosensors) have been developed for hydrolytic enzymes also have been developed [68].

Here the results of the OM biochemical transformation are represented in terms of enzyme hydrolyses of macromolecules (for assumptions and limitations of their classical postulates see [66]). In the case, when the substrate concentration in the sample is much higher than the enzyme concentration, the following equation is used :

$$v_1 = k_1 \cdot c_1,$$

where v_1 is the effective rate of the first order reaction by the substrate, $mg \cdot l^{-1} \cdot h^{-1}$; k_1 is the constant of the rate, h^{-1}; c is the initial substrate concentration, in $mg \cdot l^{-1}$. Since

$$k_1 = V_{max} \cdot K_m,$$

where K_m is Michael's constant, and the values in the right part of the equation are known, after calculation of k_1, the rate of the substrate loss in the studied samples (v_1) can also be determined.

6.2. THE REGIONS OF COMPLETE BIOGEOCHEMICAL INVESTIGATION

Biogeochemical investigations, including the determination of the biopolymers were carried out in 10 regions of the Ocean (Fig. 14), specifically on the transect in the Atlantic Ocean from the Congo river mouth to the Angola basin (6° S) comprised three detailed study areas 20x20 miles; and transect of Kunene river 17° S and also in the area of Benguela upwelling at sections on 23° S and 25° S.

An investigation in the Baltic Sea included the Riga and Finnish Bays. The studies were carried out on the longitude section from the north-western shelf through the deep-sea area to the south-eastern antycyclonic gyre in the Black Sea (Fig. 15).

In 1995-1996 the investigations of the Arctic`s ecosystem was fulfilled in the Barents, Kara, and Laptev Seas.

6.3. DISCUSSION OF ENZYME ACTIVITY RESULTS

EA was observed in only a few samples, the south-eastern area of the Atlantic Ocean, the Riga Bay, the Finnish Bay and in the Black, Barents, Kara and Laptev`s Seas. The maximum of EA was found in the shelf of the Black Sea. PA, AA and total hydrolytic activities were 300-600, 80-120, 300-700 enzyme units (e.u). respectively.

The general feature of the EA distribution into the Black Sea was the high activity in highly productive shallow shelf waters (Fig. 16). The tendency of activity to decrease with depth and the patchiness of the distribution at depth are related to the increased turbidity of these waters. In the Black Sea EA decreases with depth to a minimum of 1-3 e.u.

In the waters of the Arctic Seas the EA was 1.5 times lower than in waters of the Black Sea, Riga and Finnish Bays. In the Laptev`s Sea and Riga Bay high EA were detected not only into the photic layer, but also in the water near the bottom (1-5 m above the bottom).

The gradient of EA is: pore water < sea water < bottom sediments < material collected by sedimental traps, particulate matter < pellets, mucus lumps, strands..

The study of particulate matter shows that PA occurred in all samples. The near bottom layer of high turbidity was characterized by high values of EA. Particulate matter 1-3 m above the bottom has the highest PA (2-2.5 times that of PA at the surface layer of water). The high enzymatic destruction of OM near the bottom was connected with exocellular metabolism products input to the environment by micro- and macrobenthos, and pellet material which are enriched in labile biopolymers. High EA values were found also in sediment trap material. The results of the sediment material study are presented in Table 2 (Angola Basin, st. 3178 and Benguela upwelling area, (st. 3217).

TABLE 2. Fluxes of sediment matter, TOC and protease activity (PA)

	Angola basin			Benguela upwelling	
Depth, m	240	1005	4055	140	220
Sediment matter (mg m^{-2} day^{-1})	116.2	49.7	30.9	377.3	431.9
TOC, %	6.14	3.32	1.51	36.0	138.9
PA, e.u. l^{-1} h^{-1}	135	615	2740	194	4700

In sediment trap material, OM transformation ,estimated by PA, increased with depth. The sediment OM flux is accompanied by the high EA during OM destruction. This is due to microbial decomposition of dead plant and animal tissue and transformation of pellet material.

The EA study in bottom sediments has been carroed out in detail in the south-eastern part of the Atlantic Ocean in areas showed in the Fig. 14, and also in the Riga and Finnish Bays.

The distribution of rates of biopolymer destruction in the different sediment grain sizes is presented in Fig. 17. The rates decreased from the terrigenous clayish pelloid muds to calcareous clayish muds by a factor of ca. 2. The glauconitic sands in the Benguela upwelling area were characterized by high rates of biopolymer destruction. It should be noted that there is a correlation, usually negative, between the TOC content and OM enzyme destruction rates as a result of active work of enzyme systems hydrolyzing organic polymers.

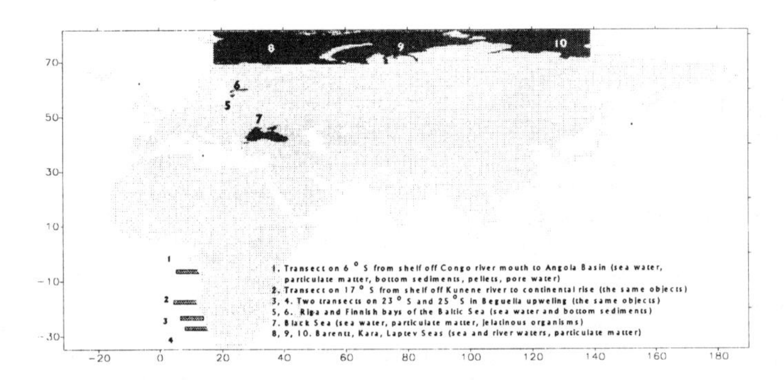

Fig. 14. Regions of determination of OM enzyme destruction.

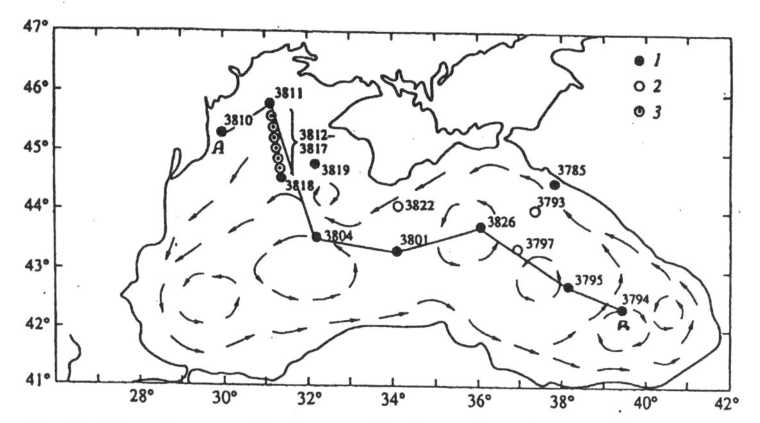

Fig. 15. Main surface current in autumn and location of the station (1) with complete series of experiments including enzyme activity, (2) with determination of the primary production and chlorophyll a content in the water column, and (3) with determination of chlorophyll a concentration in surface layer. Latitude section of hydrolytic activity distribution (A-B).

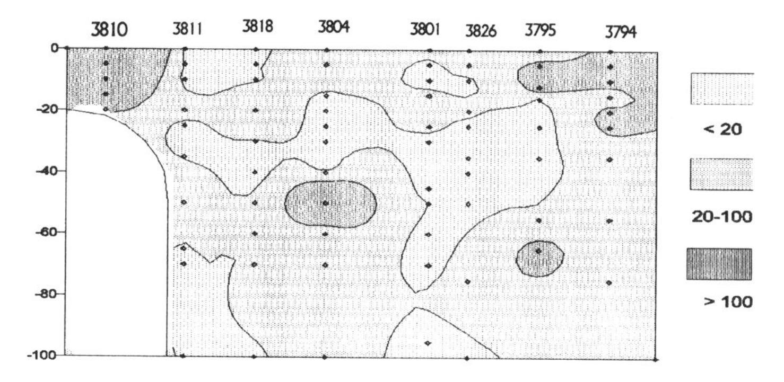

Fig. 16 Vertical distribution of hydrolytic enzyme activity (e.u./l) in Black Sea water (september 1992)

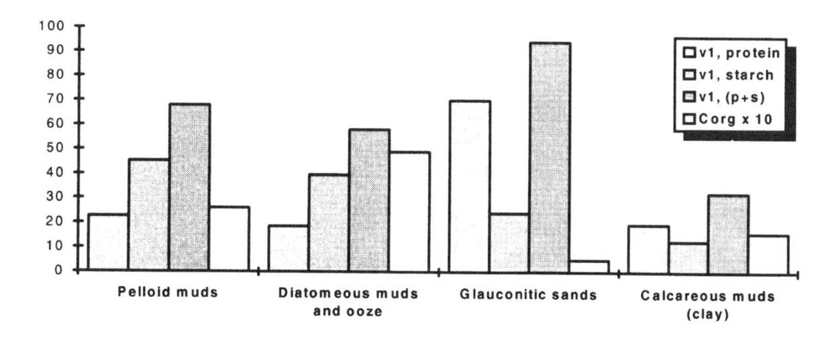

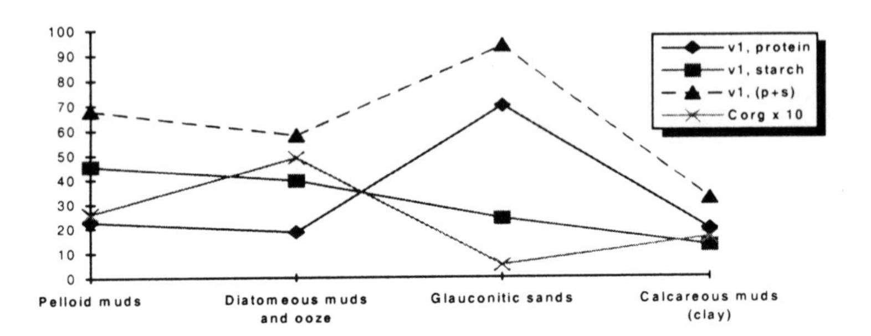

Fig. 17. Enzyme rate of organic matter destruction in various type of bottom sediments (average value)

Maximal values of biopolymer hydrolysis occurred in shelf sediments in the south-eastern part of the Atlantic Ocean (fig. 18). The average values of proteolysis, amylolysis and their total sum were respectively 170, 160, 350 $mg \cdot l^{-1} \cdot h^{-1}$. Moving from bottom sediments of the shelf to the continental rise the total EA sum decreases 7 fold (from 350 to 50 $mg \cdot l^{-1} \cdot h^{-1}$) for both proteolysis and amylolysis.

The organic carbon content of the bottom sediments of the shelf and continental slope was about 3.3% and 3.4% and decreased to 1% in bottom sediments of the continental rise and Angola basin. Calculations show that EA per unit OM in shelf sediments is 2-3 times that of the continental slope sediments. The high destruction rates were correlated with the pellet quantity in bottom sediments. The experimental study of pellet samples revealed that PA content was made up 10-4M per pellet which was 5 orders of magnitude higher then that of sea water.

The decrease in OM hydrolytic enzyme processes from shelf - slope - deep basins correlated with the decrease of total organic carbon and decrease in rates of its accumulation in bottom sediments, as shown in 1-3 parts of this paper.

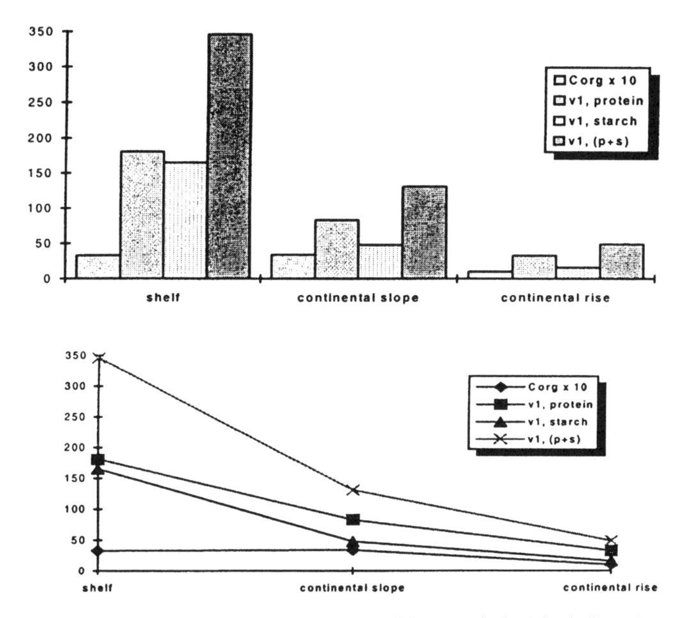

Fig. 18. Enzyme rate of organic matter destruction and TOC content in the Atlantic Ocean bottom sediments.

Acknowledgment

The research described in this publication was supported by Grant NE 1000 and NE 1300 from the International Science Foundation, Grant 97-05-64076 Russian Fond of Fundamental Research and Grant 97-02.06 from Minnauka, Russia.

7. References

1. Hedges, J.I. (1992) Global biogeochemical cycles: progress and problems, Marine Chem. **39**, 67-93.
2. Hedges, J.I. and Keil, R.G. (1995) Sedimentary organic matter preservation : an assessment and speculative synthesis, Marine chem. **49**, 81-115.
3. Ittekkot, V. and Arain, R. (1986) Nature of particulate organic matter in the river Indus, Pakistan, Geochim. Cosmochim Acta **50**, 1643-1653.
4. Berger,W.A., V.S.Smetacek and G.Wefer (eds.), *Productivity of the Ocean: Present and Past* (1989), John Wiley, Chichester, 473 p.
5. Romankevich, E.A. (1984) Geochemistry of Organic Matter in The Ocean, Springer - Verlag, Berlin, 334 P.
6. Romankevich, E.A. (1990) Organic matter in ocean lithogenesis, in P.P.Timofeev (ed.), *Accumulation and transformation of organic matter in recent and fossil sediments*, Nauka, •oscow, 3-20.
7. Premuzic, E.T., Benkovitz, C.M., Gafney, J.S. and Walsh J.J. (1982) The nature and distribution of organic matter in the surface sediments of world oceans and seas , Organic Geochemistry **4**, 63-77.
8. Initial Reports DSDP (1968-1983) Wash. (D.C.): US gov. Print. off. 1-96.
9. Proc. ODP, Initial Reports (1983-1995), College Station, Texas (Ocean Drilling Program) 100-143.
10. Ronov, A.B., Hain, V.E. and Baluhovsky A.N. (1989) *Atlas Lithologo-paleogeographic Maps of the World (The Mesozoic and Cainozoic of the Continents and the Oceans)*, Izd. AN SSSR, Leningrad.
11. Zonenshain, L.P., Savostin, L.A. and Sedova, A.P. (1984) *Global paleogeodynamic reconstruction for last 160 million. years,* Geotektonika, No. **3**, 3-16.
12. Zonenshain, L.P. and Kuzimin, M.I. (1993) *Paleogeodynamics*, Nauka, Moscow, 192 p.,
13. Ronov, A.B. (1993) Stratisphere or Sedimentary Cover of the Earth, Nauka, Moscow 143 p.
14. Geodekian, A.A. and Zabanbark, A. (1985) *Geology and distribution of hydrocarbon resources in the World ocean*, Nauka, Moscow, 192 p.
15. Trotsuk, V.Ya. (1992) Oil bearing rocks of aquatic sedimentary basins, Nedra, Moscow, 224 p.
16. Barron, J.A. and Baldauf, J.G. (1989) Tertiary cooling steps and paleoproductivity as reflected by diatoms and biosiliceous sediments, in W.H.Berger, V.S.Smetacek and G.Wefer (Eds.), *Productivity of the Ocean: Present and Past*, John Wiley, Chichester, 341-354.

17. Kemp, A.E.S. and Baldauf, J.G. (1993) Vest Neogen laminated diatom mat deposits from the eastern equatorial Pacific ocean, Nature **361**, 141-144.
18. Strakhov, N.M.(1960) Fundamentals of the Theory of Lithogenesis, 2, Izd. Akad. Nauk SSSR, Moscow, 460 p.
19. Lisitzin, A.P. (1991) *Processes of Terrigenous Sedimentation in the Sea and Oceans*, Nauka, Moscow, 271 p.
20. Emelyanov, E.M., Trimonis, E.S. and Kharin, G.S. (1989) *Paleoceanology of the Atlantic ocean*, Nedra, Leningrad, 247 p.
21. *Atlas of the Oceans. The Atlantic and Indian Oceans* (1977), Ministerstvo oboroni SSSR, VMF, 306 p.
22. *Atlas of the Oceans. The Pacific Ocean* (1974), Ministerstvo oboroni SSSR, VMF, 302 p.
23. Romankevich, E.A. (ed.), *Biogeochemistry of the boundary zones in the Atlantic ocean* (1994), Nauka, Moscow, 400 p.
24. Tseitlin, V.B. (1993) The relationship between primary production and vertical organic carbon flux in the ocean mesopelagial, Okeanologya **33**, 224-228.
25. Romankevich, E.A. (1977) *Geochemistry of Organic Matter in the Ocean*, Nauka, Moscow, 256 p.
26. *Global Carbon Cycle* (1977) B.Bolin, E.T.Degens, S.Kempe and P.Ketner (eds.), John Wiley, Chichester, 491 p.
27. Bolin, B and R.B.Cook (eds.)The Major Biogeochemical Cycles and Their Interactions (1983), John Wiley, Chichester.
28. Wollast, R., F.T.Mackenzie and L.Chou (eds.) *Interactions of C,N,P and S Biogeochemical cycles and Global Change* (1993), NATO ASI Series. 14, Springer-Verlag, Berlin.
29. Vernadsky, V.I. (1965) *Chemical Structure of the Biosphere of the Earth and her Environments*, Nauka, Moscow, 374 p.
30. Uspensky, V.A. (1956) *Balance of Carbon in the Biosphere Regarding to Carbon Distribution in the Earth Crust*, Gostoptekhizdat, Leningrad, 101 p.
31. Uspensky, V.A. (1970) Introduction in Geochemistry of Oil, Nedra, Leningrad, 309 p.
32. Degens, E.T. S.Kempe and J.E.Richey (eds.) *Biogeochemistry of Major World Rivers* (1990), SCOPE 42, John Wiley & Sons, Chichester, 356 p.
33. Artemyev, V.E. (1996) *Geochemistry of Organic Matter in River-Sea System*, Kluwer Academic Publishers, Dordrecht, 190 p.
34. Meybeck, M. (1993) Natural sources of C, N, P and S, in R.Wollast, F.T.Mackenzie and L.Chou (eds.), *Interaction of C, N, P and S Biogeochemical Cycles and Global Change* NATO ASI Series. 14, Springer-Verlag, Berlin, 163-193.
35. Livovich, M.I. (1974) *Global Water Resources and Their Future*, Mysl, Moscow, 448 p.
36. Vasiliev, V.P. (1987) Solid river runoff into the World ocean, Litologiya i poleznie iskopaemie, No. 6, 19-28.
37. Meybeck, M. (1982) Carbon, nitrogen, and phosphorus transport by world rivers, Am. J. Sci. **282**, 401-450.
38. Ittekkot, V. (1988) Global trends in the nature of organic matter in river suspensions, Nature **332**, 436-438.
39. Degens, E.T., Kempe, S. and Richey J.E. (1990) Summary: Biogeochemistry of major world rivers, in E.T.Degens, S.Kempe and J.E.Richey (eds.), *Biogeochemistry of Major World Rivers*, John Wiley, Chichester, 323-348.
40. Savenko, V.S. (1991) *Natural and Anthropogenous Sources of Pollution of Atmosphere*, Itogi nauki i tekhniki, Ser. okhrana prirodi, Viniti, Moscow, 207 p.
41. Savenko, V.S. (1995) Is the ocean source of carbon dioxide to atmosphere?, Geokhimiya,No. **11**, 1634-1642.
42. Prahl, F.G. and Muehlhausen L.A. (1989) Lipid biomarkers as geochemical tools for paleoceanographic study, in W.H.Berger, V.S.Smetacek and G.Wefer (Eds.*), Productivity of the Ocean: Present and Past*, John Wiley, Chichester, 271-289.
43. Zafirion, O.C., Gagosian, R.B., Peltzer, E.T., Alford, J.B. and Loder, T. (1985) Air-to-sea fluxes of lipids at Enewetalk Atoll, J.Geophys. Res. **90** (D1), 2409-2423.
44. Djamalov, R.G. (1991) *Underground water exchange between land and sea and its regularities*, Thesis, Leningrad Mountain Institute, Leningrad, 53 p.
45. Shvartsev, S.L. (1978) *Hydrogeology of zone of hyperhenesis*, Nedra, Moscow, 287 p.
46. Lisitzin, A.P. (1994) *Glacial Sedimentation in the World Ocean*, Nauka, Moscow, 448 p.
47. Daniushevskaya, A.I., Petrova, V.I., Yashin, D.S., Batova, G.I. and Artemyev, V.E. (1990) *Organic Matter of Bottom Sediments in Polar zones of the World Ocean*, Nedra, Leningrad, 280 p.
48. Lisitzin, A.P. (1974) *Sedimentation in the World Ocean*, Nauka, Moscow, 438 p.
49. Romankevich, E.A. (1990) Biogeochemical Problems of Living Matter of Present - Day Biosphere, in V. Ittekkot, S.Kempe, W.Michaelis and A.Spitry (eds.), *Facets of Modern Biogeochemistry*, Springer - Verlag, Berlin, 39-51.
50. Vinogradov, M.E., Shushkina, E.A., Kopelevitch, O.V. and Sheberstov S.V. (1996) Photosynthetic primary production in the Ocean, based on expedition's and satellite data, Okeanologiya **36**, 566-575.
51. Vinogradov, M.E., Tseitlin, V.B. and Sapojnikov, V.V. (1992) Primary production in the ocean, Jurnal obschey biologii **53**, 314-327.
52. Krupatkina, D.K., Berlak, B. and Maestriny, C. (1985) Leader of primary production is the ocean but land, Priroda, No. **4**, 56-62.
53. Myneni, R.B. and Los, S.O. (1995) Potential gross primary productivity of terrestrial vegetation from

1982-1990, Geophysical Research Letters **22**, 2617-2620.
54. Shushkina, E.A., Vedernikov, V.I., Kopylov, A.I., Mamaeva, T.I., Mikaelyan, A.S., Pasternak, A.F. and Sazhin, A.F. (1987) Alterations of plankton community in experimental bottles at phytoplankton and bacteria production measurements, Izv. AN SSSR, Ser. Biol., No.**1**, 42-54.
55. Bogorov, V.G. (1971) Regarding to amount of matter in living organisms of the World ocean, in N.B.Vossoevich (ed.), *Organic matter in recent and fossil sediments,* Nauka, Moscow, 12-15.
56. Lein, A.Yu., Galchenko, V.F., Pimenov, N.V. and Ivanov, M.V. (1993) Role of processes of bacterial chemosyntesis and methanotrophy in the ocean biochemistry, Geokhimiya, No.**2**, 252-268.
57. Hedges, J.L. and C.Lee (eds.) *Measurement of dissolved organic carbon and nitrogen in natural waters* (1993), , Proceeding of NSF/NOAA/DOE Workshop, Seattle, USA 15-19 July 1991, (Marine Chemistry 41, No.1-3, 290 p.)
58. Romankevich, E.A. and Ljutsarev, S.V. (1990) Dissolved organic carbon in the ocean, Marine Chem. **30**, 161-178.
59. Ljutsarev, S.V. and Chubarov, V.V. (1994) Methods of measurement of dissolved organic carbon content in natural waters, Optika atmosferi i okeana **7**, 479-491.
60. Williams, P.M. and Druffel E.R.M. (1987) Radiocarbon in dissolved organic carbon in the central north Pacific Ocean, Nature **330**, 246-248.
61. Berner, R.A. (1989) Biogeochemical cycles of carbon and sulphur and their effect on atmospheric oxygen over Phanerozoic time, Palaeogeogr. Palaeoclimatol. Palaeoecol. **73**, 97-122.
62. Gershanovich, D.E., Gorshcova, T.I. and Koniukhov A.I. (1974) *Organic matter of recent sediments of nearcontinental areas, in Organic matter of recent and fossil sediments and methods of its study*, Nauka, Moscow, 63-80.
63. Savenko, V.S. (1986) *Geochemical aspects of biosedimentation*, Dokladi AN SSSR 288, 1192-1196.
64. Wefer, G. (1989) Particle Flux in the Ocean: Effects of Episodic Production, in W.H.Berger, V.S.Smetacek and G.Wefer (Eds.), *Productivity of the Ocean: Present and Past*, John Wiley, Chichester, 139-154.
65. Lukashin, V.N., Shevchenko, E.A., Romankevich, E.A., Arashkevich, E.G., Borodkin, S.O., Korneeva, G.A., Oskina, N.S. and Pimenov, N.V. (1993) Fluxes of sediment matter in the south-east part of the Atlantic ocean, Dokladi AN SSSR **330**, 638-641.
66. Keleti, T. (1990) *Basic enzyme kinetics*, Academia Kiado, Budapest, 348 p.
67. Christian, J. R. and Karl, D. M. (1993) Palmer LTER: Bacterial exoprotease activity in the Antarctic Peninsula region during austral autumn 1993, Antarctic Journal U.S. **28**, 221-222.
68. Korneeva, G.A. (1966) Application of enzyme test-systems for monitoring of Black Sea water, Izv. Russ. Acad. Sci., Ser.Biol., No.**5**, 493-499.
69. Korneeva, G.A. and Artemyev, V.E. (1994) Hydrolytic enzyme process in Baltic bottom sediments, Dokl. Acad. Sci. **337**, 258-262.
70. Korneeva, G. A., Kharchenko, S. V. and Romankevich E. A. (1990) Studies of enzymatic hydrolysis of casein in sea water, Izv. Acad. Sci., USSR, Ser Biol., No.**6**, 821-827.
71. Korneeva, G. A. and Romankevich, E. A. (1994) Activity of hydrolytic enzymes in marine ecosystems, in E.A.Romankevich (ed.), *Biogeochemistry of the boundary zones in the Atlantic ocean*, Nauka, Moscow, 157-161.
72. Korneeva, G. A. and Romankevich, E. A. (1996) Dynamic characteristics of organic carbon transformation in bottom sediments, Izv. Russ. Acad. Sci., Ser Biol., No.**3**, 374-377.
73. Korneeva, G. A., Romankevich, E. A. and Artemyev, V. E. (1993) Processes of proteolysis and amylolysis in the bottom water of the Riga bay, Izv. Russ. Acad. Sci., Ser Biol., No.**2**, 280-286.
74. Korneeva, G. A., Romankevich, E. A. and Kharchenko, S. V. (1992) Enzymatic activity in components of the pelagic ecosystem of the south-earth Atlantic ocean, Izv. Russ. Acad. Sci., Ser Biol., No. **6**, 613-617.
75. Korneeva, G. A., Tropin, I. V. and Romanchevich, E. A. (1997) Processes of enzymatic hydrolysis of organic macromolecules in waters of different salinity and their correlation's with physiologically important and toxic metals on the Barents sea littoral, Oceanologiya **37**, 226-231.
76. Korneeva, G. A. and Vedernikov, V. I. (1994) Influence of ctenophore *Mnemiopsis* autointroduced in the Black sea on proteolytic processes of destruction of proteins and polysaccharides in sea water, Izv. Russ. Acad. Sci., Ser Biol., No. **1**, 127-131.
77. Saliot, A., Cauwet, G., Cahet, G., Mazaudier, D. and Daumas, R. (1996) Microbial activities in the Lena river delta and Laptev Sea, Marine chemistry **53**, 247-254.
78. Smith, D. C., Simon, M. and Alldredge, A. N. (1992) Intense hydrolytic enzyme activity on marine aggregates and implications for rapid particle dissolution, Nature **359**, 139-142.
79. Walsh, J.J. (1989) How Much Shelf Production Reaches the Deep Sea? in W.H.Berger, V.S.Smetacek and G.Wefer (Eds.), *Productivity of the Ocean: Present and Past*, John Wiley, Chichester, 175-191.

PRIMARY PRODUCTION AND DECOMPOSITION OF ORGANIC MATTER IN COASTAL AREAS OF THE NORTHERN ARAL SEA, WITH SPECIAL REFERENCE TO LAND-SEA INTERACTIONS

M.I. ORLOVA
Zoological Institute, Laboratory of Brackishwater Hydrobiology 199034. Universitetskaya EMB. B. 1 St.Petersburg Russia

Abstract
The biological systems of the Northern Aral Sea have adapted to new environmental conditions posed by a drastically lowered sea level. Based on field and experimental data collected between 1989 - 1996 and some earlier studies from the 1960-80s, new inferences on autochthonic organic matter production and cyling in planktonic and benthic communities are presented. The benthic system is shown to be of high importance in biosedimentation and utilization of produced organic matter. Due to natural seasonal and occasional man-made water level fluctuations caused by damming in Berg's Strait, the role of land-sea interactions in the supply and transformation of organic matter are shown to be important in some coastal areas.

1. Introduction

During the last decade many papers and reviews have been published containing descriptions of recent changes occurring in the Aral Sea and surrounding land areas [1,2]. These show that the Aral Sea is in an ecological crisis. Some papers even expressed the opinion that destruction of the whole Aral Sea ecosystem and transformation to a completely lifeless basin was imminent [3].

The relatively constant balance between water supply and evaporation, which was observed for the first half of the current century, was disturbed in the 1960s by bad water management in the catchment areas of the two contributing rivers (Amudarya and Syrdarya). As a result, from 1961 the water level has decreased. In the beginning this proceeded at a relatively low rate (ca 27 cm y^{-1} from 1961 to 1974), but between 1975-1985 the rates increased to as much as 71 cm y^{-1} (Smerdov, 1990, quoted in [4]). Moreover, from 1986 to 1990 the drop in water level was up to 88 cm y^{-1} [4]. The recession of the shoreline, the impossibility of navigation and fisheries, the desertification of dried bottoms, and induced salt and dust storms are all indicators of the sequence of the Aral Sea regression. As a consequence the majority of former USSR Asiatic republics adjacent to the basin have suffered great socioeconomical problems.

In relation to the Aral Sea ecosystem itself the complete destruction of the living associations has been considered a possibility for the last 30 years [5], and it has been suggested that the basin now is close to it's critical state [6]. However, the overwhelming majority of research projects on the recent state of living associations and their changes over the time [4] [5] [7] [8] etc.] have considered just data on species composition, abundance and biomass for animals' and plants' populations with no analyses of functional characteristics of the whole ecosystem. Papers analyzing connections between different trophic levels in the Aral Sea, the origin of organic matter and it's fate, are rare [9-12] for the period of the 1960s-19780s [13] [14].

For the years 1992-1996 a complex study of biological characteristics of the Northern Aral Sea was carried out the under the auspices of UNESCO-BMFT and by Russian projects (see acknowledgements). The study of the functional properties of the ecosystem at selected areas [15-18] was the main focus of these projects.

The objectives of this paper arde to consider the data collected in 1989-1996, on the rates of photosynthesis and decomposition, the structure of living associations, the content of organic matter in seston and bottom sediments in relation to stock, and the supply and cycling of organic matter in in relation to recent water-level oscillations in

J.S. Gray et al. (eds.), Biogeochemical Cycling and Sediment Ecology, 29–48.

coastal areas of the Northern Aral Sea.

Characterising the modern situation, it is suggested that there is a basis for anxiety for the socioeconomical problems of the Aral Sea region. Nevertheless, the water-level fluctuations (and other related parameters) in the Aral Sea are not unusual, (except for the man-mediated character of the last drop in level) in relation to global perspectives [19]. Moreover it was shown [19], that water-level fluctuations are the property of great saline lakes in other places of the world, both recently and in the geological past. In this respect the analysis of data on structure and functioning of the Aral Sea ecosystem could be used not only for better understanding of current situation in this very basin but also as a range of factorss describing common characteristics of saline lakes.

2. The Aral Sea

2.1. GENERAL CHARACTERISTICS

The Aral Sea is one of the largest inland seas in the world. Being at the Eastern margin of the Ustyurt Plateau it is stronlgy influenced by the Asian climate. Table 1 shows some general charateristics. The hydrology, water chemistry and biological characteristics of the Sea are completely dependent on drainage from two rivers, the Amudarya and Syrdarya (Figure 1). The interactions between the Aral Sea and adjacent land are also highly significant [9] [10] [20].

The Aral Sea was always considered as consisting of two unequal basins – the Northern smaller basin was named as "Small Aral", the larger Southern basin as "Larger Aral. Both basins were separated by island Barsakelmes. Since 1989 year due to progressive desiccating the Sea has been completely divided into two parts (Fig 1). Some characteristics of the Small Aral befóre and after the last man-mediated drop in water-level are shown in Table 2. The shoreline of the Small Aral varies greatly from area to area with the northern coast surrounded by clay cliffs, consisting mainly of Mesozoic oceanic sediment, whereas the Bolshoi Sary-Chaganack Gulf has exposed muddy and secondary sandy beaches, which have arisen following desertification. The eastern coast is a lowland area with small seasonal temporary pools, separated from or adjacent to open bays, which earlier formed large wetland areas. These wetlands completely disappeared when the Sea regressed during the 1970s and early 1980s. Recently, numerous small temporary pools were recorded in 1993 and 1994 at beaches near Bugun' settlement, Bolshoi Sary-Chaganack Gulf and Tastubeck Cape. This was probably due to damming in Berg's Strait, which raised the water level as much as 2-3 m. However, the dam is often damaged by floods and water level has not returned permanently to its former level.

TABLE 1. Northern Aral Sea Climate (after [20])

Characteristic	Value
Solar radiation per year (MJ m^{-2})	5860
Average temperature near the Sea (°C)	
in February	-10 -13
in July	25 - 26
Number of days with temperature below zero	120 - 150
Area of ice cover in February (%)	50 - 70
Number of days with ice covering	130 - 150
Humidity (%, average for a year)	65 - 70
Rainfall (mm/per year)	110 - 150

TABLE 2. Main characteristics of the Northern Aral Sea (after [20])

Characteristic	Before 1960s	1980s -1990s
Syrdarya inflow (km^3/year)	37	3 - 7
Volume of water (km^3)	80	28
Area (thousands of km^2)	6	3.5
Salinity (‰)	8-12	20-40
Greatest possible depth (m)	29.5	near 15

2.2. STUDY SITES

Since the beginning of the 1990s navigation of the Aral Sea is no longer possible. Almost all recent studies therefore, have been short-term and done as land expeditions or from helicopters. The main study sites for the work performed between 1989-1996 are shown in Fig.1 and cover estuarine areas, open sea areas (near Bugun' and Tastubeck cape) and semi-enclosed Gulfs. In 1994 the Bolshoi Sary-Chaganack Gulf with it's temporary embayments was included. This area dried out completely in 1991 and 1992 (N.V. Aladin pers. comm.). In 1993 and 1994 partial restoration occurred due to damming and by stopping the outflow of the Syrdarya river into the Large Aral.

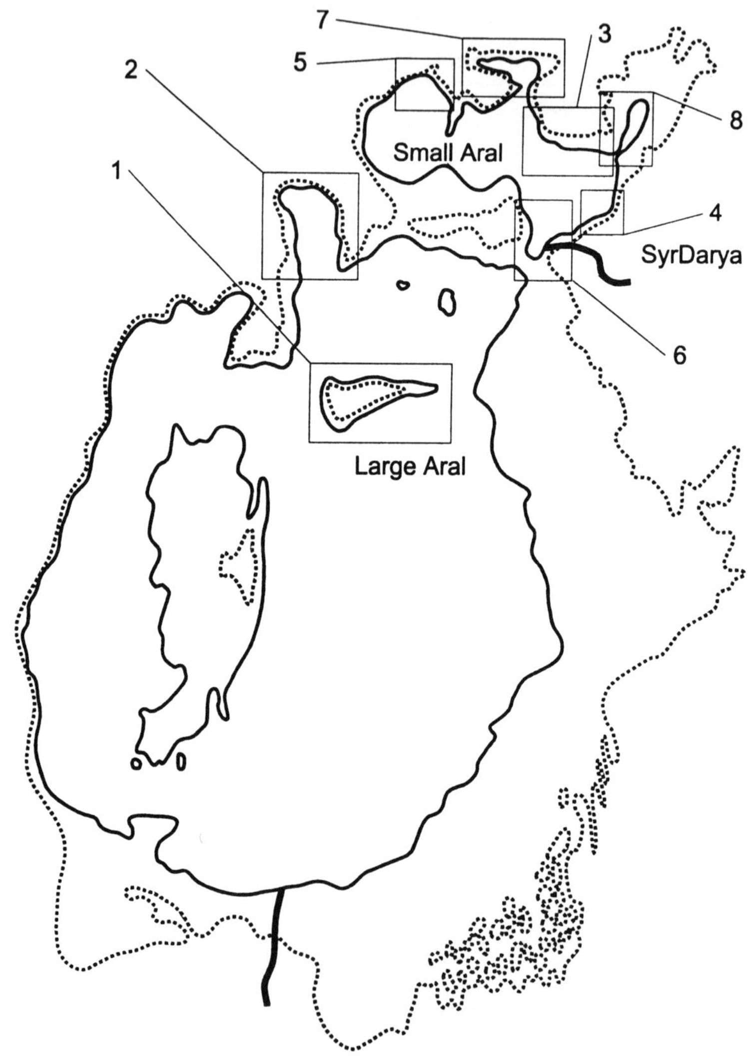

Figure 1. A map of the Aral Sea coastlines in 1960 (dotted line) and 1987 (solid line) with sites of investigations for 1989-1996 (rectangles). 1 - near I. Barsakelmes; 2 - Tshche-Bas Gulf; 3 - Tastubeck cape; 4 - near Bugun' settlement (Eastern coast); 5 - Shevchenko Gulf; 6 - Syrdarya pre-mouth area and Berg's strait; 7– Butakov Gulf; 8 -Bol. Sary-Chaganack Gulf.

3. Methods

3.1 SAMPLING

Samples of water and seston at stations with depths >1-1.5m were collected as a mixed sample from different depths using a "Hydrobios" water-sampler. In shallow waters and temporary pools samples were taken only from the surface. The samples were concentrated by filtration on "Whatman" GF/F fibreglass filters and the filters with particulate matter were dried and frozen before analysis.

Sediment sampling was performed using plastic tubes with Ø 5 cm to depths of 7-11 cm. From each station usually 3-8 cores were taken. In the majority of analyses only the 4 upper cm were used. The cores taken for the experiments on community metabolism (see below) were used afterwards for macrobenthos species and abundance determinations.

3.2. ANALYTICAL METHODS

Both particulate organic matter (POM) in seston and the fine fraction of bottom sediments (particle size <0.25 mm) were analysed for total organic matter by means oxidation by $K_2Cr_2O_7$ in concentrated H_2SO_4 at 140°C for 15 min [21]. Results are expressed as g C m^{-2} or as % organic matter per dry weight of bottom sediment. Salinity was measured by a refractometer. In all experiments rates of photosynthesis and decomposition were derived from oxygen measurements. Oxygen was determined by Winkler's titration or polarographically using a YSI-51 oxygenmeter.

3.3. EXPERIMENTAL METHODS

Estimates of community metabolism, such as rates of primary production (community photosynthesis) and decomposition (community respiration), were made using two approaches. The first based on diurnal oxyden-temperature curves [22] [23]. Measurements were performed *in situ*, in temporary pools and coastal shallow areas of the Sary-Chaganack Gulf. Computation of gross photosynthesis and respiration per day was done according to Odum [22] [24] and Ganning & Wulf [25].

The second approach was based on experiments in hermetically-sealed cylinders [26] [27] [28]. These experiments are similar to the light-dark bottle method [27]. The cylinders contained intact bottom sediment with living organisms (ca. 10 cm^2), which were covered by an intact column of upper-bottom water (volume 1 l). At shallow stations the cylinder sample usually covered the whole water column. At depths more than the length of the cylinder we have added to the result obtained a value for the remaining water column, estimated by the bottle method (see below). In cases when the length of seagrass was more then the length of the bottom sampler an additional series of experiments were performed in order to calculate the correction for photosynthesis and respiration of plants in these experiments. The cylinders, (n>5 of each kind per station + 4 controls, two dark and two light) were exposed *in situ* for one day or 4-6 h. Both daily values of photosynthesis and decomposition expressed in mg O_2 l^{-1} with further recalculation to organic carbon m^{-2}.

As mentioned above, the estimation of photosynthesis and decomposition rates in the water column have been done by oxygen modification of the light-dark bottle method [27]. For studies of the vertical distribution of photosynthesis, the bottles were placed at the levels from where samples were taken. For general evaluation of the photosynthesis at a station, mixture of waters taken from different levels were used and exposed in shallow water in fully illuminated conditions. Recalculation of the value for the whole water column has been done according to Bulion [29], multiplying experimental values by values of transparency. All experimental exposures lasted for 1 day. In cases where extremely high rates of photosynthesis were recorded the exposure time was reduced to 4-6 h. Due to technical reasons the experiments with cylinders were limited to depths < 3m for full illumination.

3.4. STATISTICAL METHODS

Values for the mass of organic matter, its supply and possible ways of utilization were expressed as averages with standard deviation ($\bar{x}$±SD). Cluster analyses, based on Euclidean distances with single linkage, (STATISTICS for Windows), was used to group areas having similar characteristics. The organic matter content of seston and bottom sediments and quantitative abundance of plankton and benthos were used as variables in the cluster analyses. The results show that the areas close to the mouth of the Syrdarya river and Bugun' settlement cluster together with high similarity. Sites from semiclosed bays formed a separate cluster. These results are considered below.

4. Environmental and biological characteristics and stock and sources of organic matter.

4.1. WATER LEVEL AND SALINITY

The drop in the level of the sea occurred in 1960-1980s and now two-thirds of the Small Aral has a depth of only between 2 and 5 m. Recently, however, the drying out of the Small Aral has slowed down and between 1992 and 1996 the water level has varied within limits of 2-3 m because of damming, mentioned above (section 2.1). Inter-annual changes in mean values of water level and salinity are given in Fig. 2.

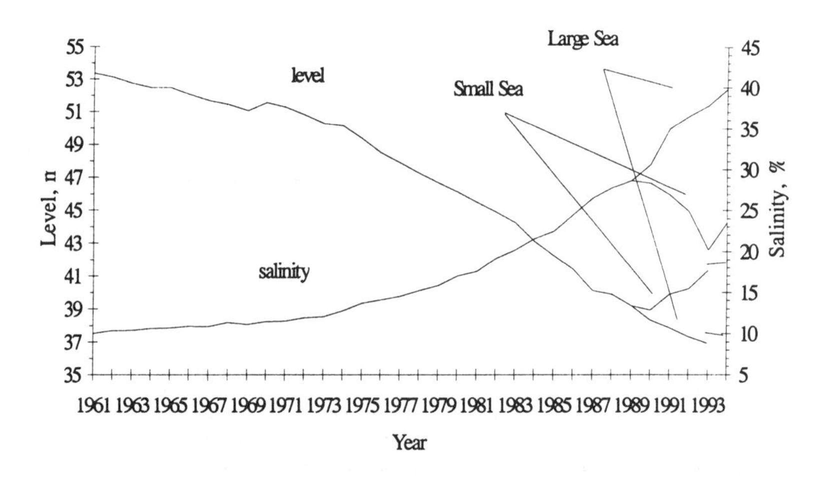

Figure 2. The Aral Sea level and salinity since 1961 till 1994 (from [30]).

Salinity is one of the key abiotic factors that has controlled the development of the biota in the Aral Sea over the last 30-35 years [5, 10]. Recently, minimum salinities were recorded in the Syrdarya area (Table 3). In other coastal areas salinity varied from 20‰ near Bugun' (1993) and at Sary-Chaganack (1994) to 36-41‰ at Tshce-Bas and Butakov Gulfs (1992).

TABLE 3. Some environmental characteristics of the Syrdarya area

Site	Date	Temp*	Transparency	Depth	Salinity	Organic matter Seston	Sediments**
		(°C)	(m)	(m)	(‰)	(mgC l^{-1})	%
Freshwater	May 92	20	0.3	0.6-0.7	1.8	0.74	0.52
	May 93	17	0.1	2(6-7)	1.8	3.42	-***
"Estuarine"	May 92	16	0.7	1.5	7	4.31	0.82
	May 93	17	0.65	1.4	18.5	21.86	3.07
Open Sea	May 92	16-20	Bottom	1.2-2	22-25	0.67-0.93	1.02-1.89
	May 93	16	Bottom	1.4	22	-	8.6

* surface water

** in upper 4 cm of bottom sediments

*** determination not possible from sample

4.2. TRANSPARENCY

Water transparency is a major factor affecting the development of the primary producers, (phytoplankton and phytobenthos). Transparency has decreased in parallel with depth (Fig. 2, Table 4). In shallow areas, where sediments that are easily resuspended dominate, such as clays and muds (see below), the transparancy decreases to only a few cm during windy weather. However, in calm weather the whole water column is illuminated. Table 4 shows that the ratio S/h has not changed over time. Thus the light conditions for phytoplankton and phytobenthos are still good.

TABLE 4. Mean transparency and depth for several periods in the main areas of the Northern Aral (for summers)

Period	Area	Transparency (S) (m)	Depth (h) (m)	S/h
*1942-1960	Central (deeper) part	10	20 - 28	0.4
	Main area (middle depths)	8	10- 15	0.6
*1961-1985	Central (deeper) part	8	15 - 18	0.5
	Main area (middle depths)	6	8 - 10	0.7
1992-1996	Central (deeper) part	5	8-10	0.6
	Main area (middle depths)	1.8 - 4**	2 - 5	0.8
	Syrdarya's delta area	0.7-2	1 - 2.5	0.5
	Marginal coastal zone	0.1 - 0.7**	<1 m	0.8

* after [20]

** in calm weather the water column was fully transparent

4.3. NUTRIENTS, PRODUCTIVITY AND AQUATIC COMMUNITIES OF THE ARAL SEA

In the past (before the 1960s), the concentration of nitrogen and soluble inorganic phosphorus was low (inorganic phosphorus only 4.2 $\mu g\ l^{-1}$) in waters of both the Amydarya and Syrdarya rivers, as their drainage basins start from glaciers. In summer in some areas the upper-bottom waters values were close to zero, [9]. Rapid development of agriculture and irrigation, and bad water management in catchment areas of the rivers, from 1961 to 1977, led to significant input of phosphorus from fertilizers leached from irrigated fields [31]. Recent observations indicate that the concentration of inorganic phosphorus is now rather constant and varies between 14-30 $\mu g\ l^{-1}$ [20] [32]. The dynamics of inorganic phosphorus are shown in Fig. 3.

In Karpevich's [10] survey in the 1960s the Aral Sea was characterised as a water-body of low productivity. She suggested that the low content of soluble nitrogen and especially phosphorus as well as the structure of first trophic level was one of the main reasons for the low productivity measured. At that time bottom macroalgae and submerged aquatic plants were predominant everywhere, (except for the deepest areas with muddy sediments). All sandy shallows and wetlands were surrounded by dense reeds and associated flora. Seaward

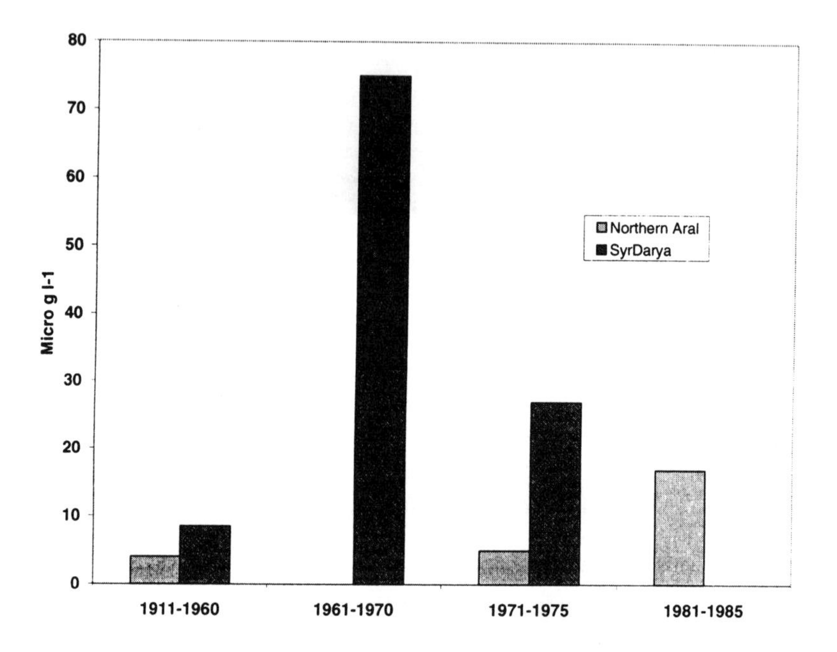

Figure 3. Dynamics of inorganic phosphorus content in water of the Northern Aral (after [20])

Charophyta covered any suitable bottoms. Her estimates of the biomass of phytobenthos for the whole Sea at that time are ca. 80-100 gm^{-2}. Phytoplankton, on the other hand, had a biomass of only 0.5-2.6 gm^{-3}. The low ratio between planktonic and benthic producers should lead to a slow turnover of matter, due to incorporation of an already limited stock of nutrients into long-lived benthic plants [9,10].

With the increase in dissolved inorganic phosphorus concentration (Fig. 3), changes in salinity (Fig. 2), the S/h ratio (Table 4) and the reestablishment of biota, an increase in planktonic primary production and of benthic producers has been noted [30]. By the early 1990s, the zooplankton and zoobenthos communities that previously existed had been completely replaced by a system consisting of very few species which are tolerant to environmental disturbances. These species-poor but abundant planktonic and benthic communities occur in all areas of the modern Aral Sea [5].

4. 4. THE ESTIMATION OF ORGANIC MATTER STOCK

4.4.1. General estimation and selected components of the stock.

Due to missing data, especially on dissolved organic matter it was not possible to assess the total organic matter stock. The total stock of particulate organic matter is, on average close to 1000 gCm^{-2} (Table 5) for the areas studied. Based on cluster analyses the areas fall into two groups, one the estuarine and open areas and the other the semi-enclosed gulfs. In both the highest proportion of organic matter is on the sea floor, (comprising both living organisms and bottom sediments). In the semi-enclosed gulfs there is a tendency for the living organisms to dominate over that from bottom sediments, the zoo- and phytobenthos exceeds 50% of the total organic matter on average (Table 5). The newly restored Bol Sary-Chaganack Gulf, in this group, has the highest dominance of biota. Here (June 1994), benthic plants alone comprised 75% of the total organic matter. In relation to phytobenthos biomass, the trophic structure of the latter Gulf is similar to that of the whole Aral Sea in the 1960s [10], although the species composition is completely different. One possible reason for this difference, is that samples were taken in summer sampling when the growth of seagrass was fastest.

TABLE 5. Some components of organic matter stock at selected areas (after [15, 16, 33-36])

Area, Time	Components	Value (gC/m^2)	% of total
Estuarine & Open areas n = 3	Particulated organic matter *(POM)	16.3±7.8	1.9±0.6
	**Zooplankton	3.3±2.8	1.2±0.8
	***Bottom sediments	600±218	75.2±2.2
	Zoo +phytobenthos	183.7±78.7	22.4±3.2
	Total	**803.6±303.6**	
Semiclosed gulfs n = 3	*(POM)	5.1±1.6	0.5±0.1
	**Zooplankton	0.1±0.06	<0.01
	***Bottom sediments	339.6±127	34.8±13.8
	Zoo +phytobenthos	684.2±249	65±13.9
	Total	**1029.1±174.2**	

* - includes detritus, bacterioplankton, phytoplankton and small representatives of zooplankton

**- Crustacea,

*** - without macrozoobenthos and macrophytobenthos

The higher abundance of zoobenthos in semiclosed gulfs, represented mainly by seston-feeders and deposit-feeders is probably due to the accessiblity of food and more sheltered and constant environmental conditions compared with open areas. There is also a relatively high biomass of phytobenthos, represented by *Ruppia sp* and *Zostera sp* with their epiphytic algae. Clearly light conditions are better then in open areas and with sufficient levels of nutrients, both leached from soft muddy bottoms and released by the abundant benthic organisms, so far phytobenthos biomass is high.

In the water column, the abundance of phytoplankton has also increased to up to 6.8 $g.m^{-2}$ [17] [37] against 0.5 - 2.6 $g.m^{-2}$ earlier [10]. The increase of biomass was accompanied by absence of freshwater species and dominance of tolerant and halophylic species (Fig. 4), except for the Syrdarya area [37]. An increase in bacterioplankton biomass was also recorded [38] [39]. No trends were noted for an increase in zooplankton filter-feeder biomass [33] [34].

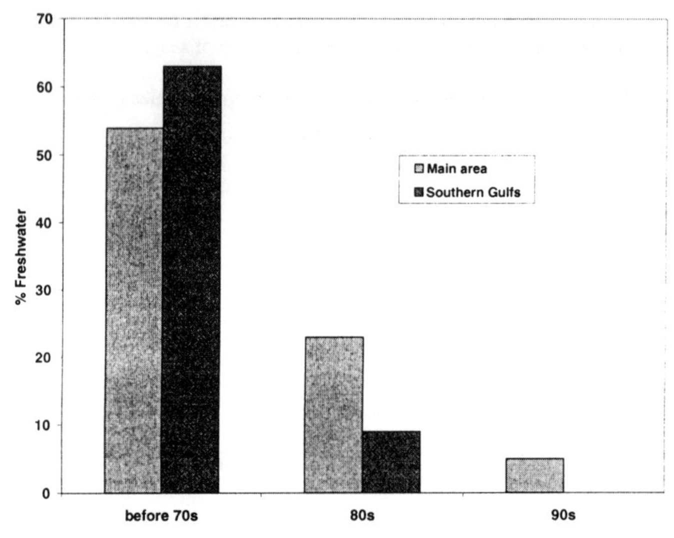

Figure 4. Percent of freshwater and oligohaline algae groups in phytoplankton of the Aral Sea (after [37] [40-43])

Unfortunately, there is no information on the amounts of seston and POM that occurred earlier in the Aral Sea. I believe that there has been enhancement of the organic matter, at least in coastal non-estuarine areas. This is due to the redistribution of sediments, mass mortality of freshwater aquatic plants during the period of increasing salinity, the lowering of the water level and as a consequence the increased frequency of bottom sediment resuspension, and the increased exchange between land and sea. Probably, the increase of macrozoobenthos abundance especially that of the suspension-feeders (*Cerastoderma isthmicum)* [35], can also be considered as indirect evidence of enhanced suspended organic matter in recent years.

It is likely that in the estuarine Syrdarya area particulate matter load was greatly reduced due to the almost ten-fold decrease of annual river input since the 1960s (Table 2).

4.4.2. Bottom sediments in modern conditions. The role of bottom sediments in organic matter stock.

At the main area of the central depression silt-clay sediments occur, often just below shorelines, so that many beaches are swampy. Sandy soils now are situated above the shoreline, whilst before the water level drop they were commonplace to depths between 10-12 m. Formerly these sand graded into muddy sands and muddy clays with increased depths, (Fig. 5.). In some areas, reflooding of dried-out sediments leads to sub-littoral secondary sands (e.g. the Bol. Sary-Chaganack area).

The content of organic matter in sediments changes with depth and in relation to the shoreline (Fig. 6). Usually at stations close to the shoreline (depth 0.5 - 1 m) sediments have low concentrations of organic matter (< 1% of dry weight). Up to 2-3 m depth (Shevtchenko Gulf) or even to 6-8 m (Tastubeck cape area) sediments become more enriched by organic matter, up to 6%. In the eutrophicated Chesapeake Bay [44] organic matter content varied between 1.5 –2% at depths up to 3 m. So far high concentrations of organic matter in sediments of the new coastal zone have been found.

At several stations (Fig. 7) the vertical distribution of organic matter in the upper 10-15 was not homogenous along sections. Sediments with mud, usually had lower values in the surface layers, while from depths of 2-4 cm the content of organic matter increased. Samples, collected from secondary sands from shallow waters had low values along the whole core.

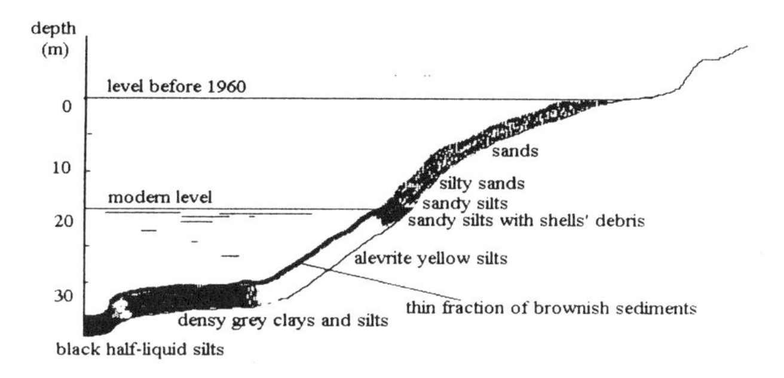

Figure 5. The distribution of main types of bottom sediment in the Northern Aral Sea (after [20])

More detailed sedimentological studies are needed before the patterns in the spatial (Fig. 6) and vertical (Fig. 7) distribution of organic matter content in the majority of areas investigated can be explained. However, the role of resuspension processes in shallow waters and active biotic consumption at surface layers by such abundant deposit/seston feeder as *Syndosmya segmentum* (Bivalvia), which is dominant at the main area are probably highly important. The increase in the total weight of seston and POM towards land may be considered as another indirect confirmation of the importance of resuspension. The heterogeneity in deeper sediment layers may be due to bioturbation. The main burrowing organism *Nereis diversicolor* (Polychaeta), occupies all types of sediments right up to the shoreline and temporary pools.

The temporal and spatial distribution of organic matter in bottom sediments of Syrdarya area (Table 3) are most likely due to dam-building and repairs, man-induced disturbance of river-bed upstream and inter-annual changes in river input [16].

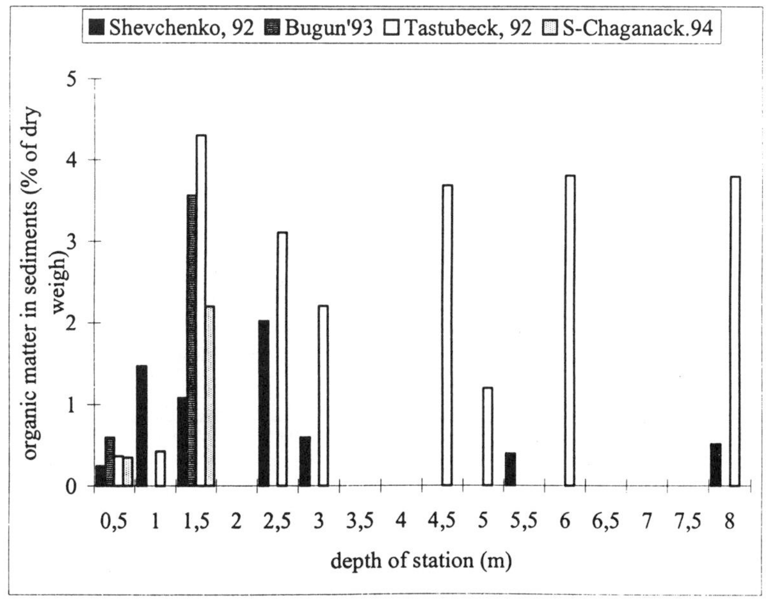

Figure 6. Organic matter content in bottom sediments.

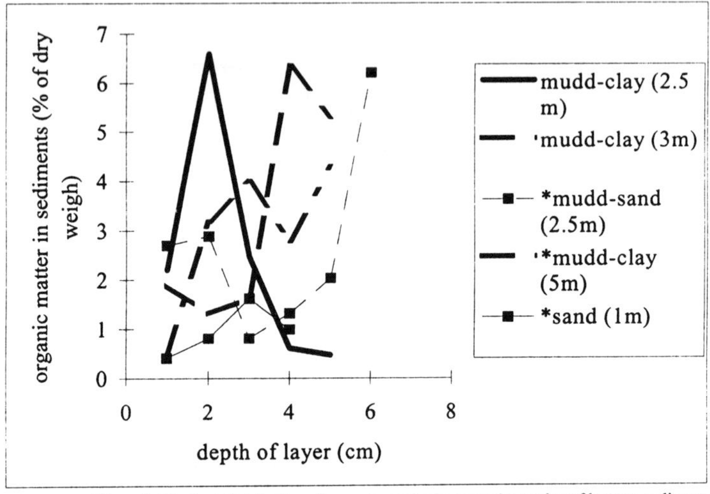

Figure 7. Vertical distribution of organic matter in several samples of bottom sediments.

4.5. THE ESTIMATION OF SOME SOURCES OF ORGANIC MATTER SUPPLEMENT

Among allochthonic sources of organic matter supply, the Syrdarya river input is highly important, throughout the history of the Aral Lake [9] [10] [20]. The main autochthonic source is primary production in the water column and on the seabed. In recent years bottom sediments (section 3.4.2.) have become a significant source for supply and cycling of organic matter in coastal areas.

In the past the exposed wetlands along the eastern coast were regarded as crucial for the development of the Aral Sea ecosystem [45]. The oscillations and then periodical stabilisation of the water level over the last 3-5 years have led to restoration of the belt of temporary basins around lowland bays. This is likely to be of significance for organic matter cycling in adjacent coastal areas.

4.5.1.Syrgarya contribution.

Based on the annual water input and the content of organic matter in the Syrdarya river water the approximate daily supply per m^{-2} of the Northern Aral in 1992 was ca. 34 $mgCm^{-2}$, and in 1993 182 $mgCm^{-2}$.

4.5.2. Primary production and decomposition in water column

Direct measurements of the rate of photosynthesis and indirect estimates of primary production for the 1960s [9-12] give extremely low rates of primary production of the planktonic system, values typical for oligotrophic waters [9], for all areas of the Small and Large Aral. At the mouths of the rivers more typically mesotrophic conditions occur, (Fig. 8).

In the water column daily values of decomposition [9] [12] exceeded total values of primary production by factors of 1.5-2. The ratio A/D (where A-is the value of total photosynthesis in the water column $1m^{-2}$; D - decomposition, m^{-2}) is the simplest estimate of the balance between autochthonic supply and decomposition of organic matter in the ecosystem. The system is in balance when the ratio is close to 1. The dominance of decomposition over photosynthesis is evidence that pelagic associations existed previously and were based on allochthonous organic matter with some contribution from the decomposition of phytobenthos. However, recently a balance has been established between rates of primary production and decomposition in many areas. At some shoreline sites the ratio significantly exceeds 1 (Fig. 9) due to the relatively high rate of photosynthesis in the water (Fig. 10).

4.5.3. Primary production and decomposition rates at bottoms

In the 1960s it was established from data on the distribution and abundance, that the phytobenthos dominated the primary production contributing 80-90% of the total, [10] [12].

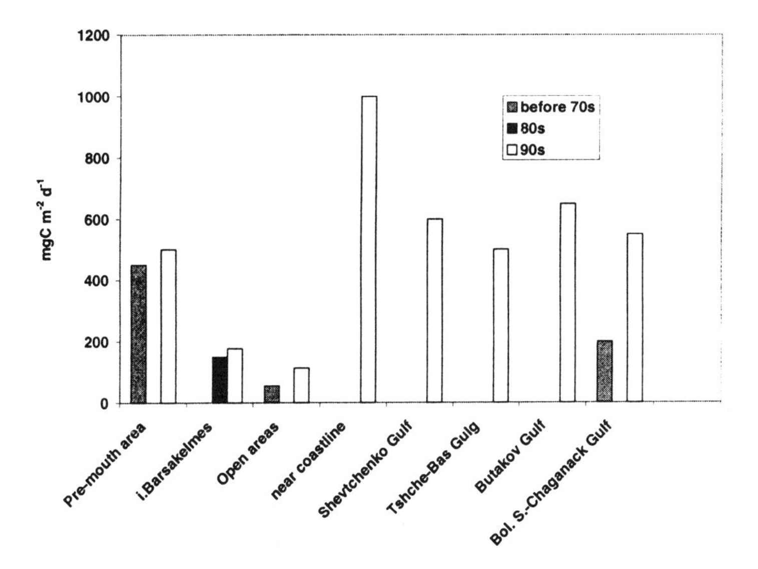

Fig 8. Rates of planktonic primary production (mean values) in the Northern Aral (from [5, 9, 12-14, 16] and by expeditions in 1992-1994 years).

In our experiments (Fig. 11) it was shown that even where full illumination occurred, i.e. the most favourable conditions for development of bottom plants), the relative contribution of benthos, both in primary production and decomposition was significantly lower than in the 1960s. In the majority of areas the gross photosynthesis on the bottom was close to 50% of the total value. Only in some areas, such as the newly restored Bol. Sary-Chaganack Gulf, (where 30% of the bottom is densely covered by seagrass, the depth is no more 2 m and the phytoplankton biomass does not exceed 1 mg l^{-1} [46]), the contribution of benthic communities is extremely high both in the main area and near the shoreline.

5. Some suggestions about fate of organic matter supplement.

Of the autochthonic sources the Syrdarya area and the area near Bugun', were chosen for more detailed analyses. It was shown earlier that the Syrdarya River is the main allochthonic contributor of organic matter supply. For other areas, however, the contribution from land is probably significant, but has not been possible to estimate this during this study.

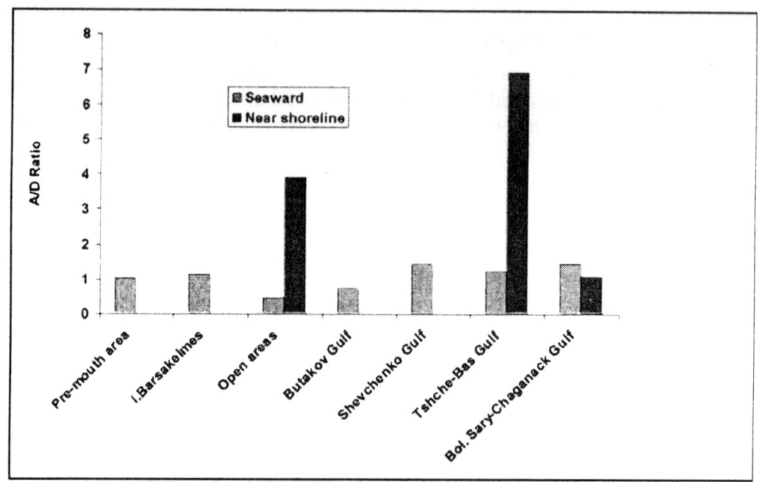

Figure 9. Mean values for A/D ratio (from [13] expeditions in 1992-1994).

In relation to the selected area (Table 6) the Syrdarya river contribution is estimated at one tenth of the total primary production (in the water and on the bottom). Thus the autochthonic sources are far more significant then the allochthonic. In spring the balance between pelagic and bottom production of organic matter is almost equal. In summer when rapid growth and development of seagrass beds occurs the autochthonous contribution to the organic matter supply is even higher.

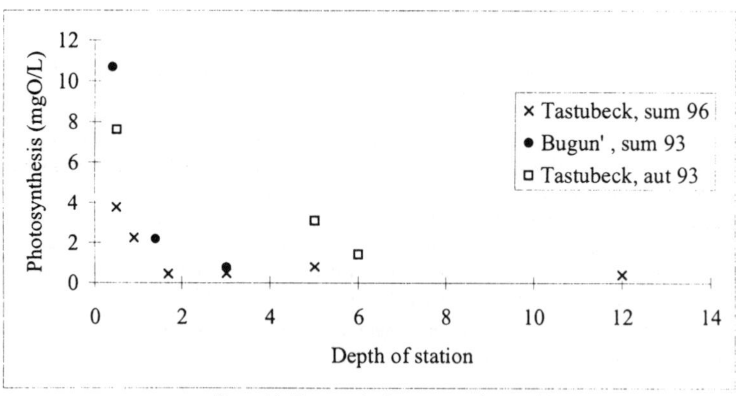

Figure 10. Photosynthesis rates at stations and depths

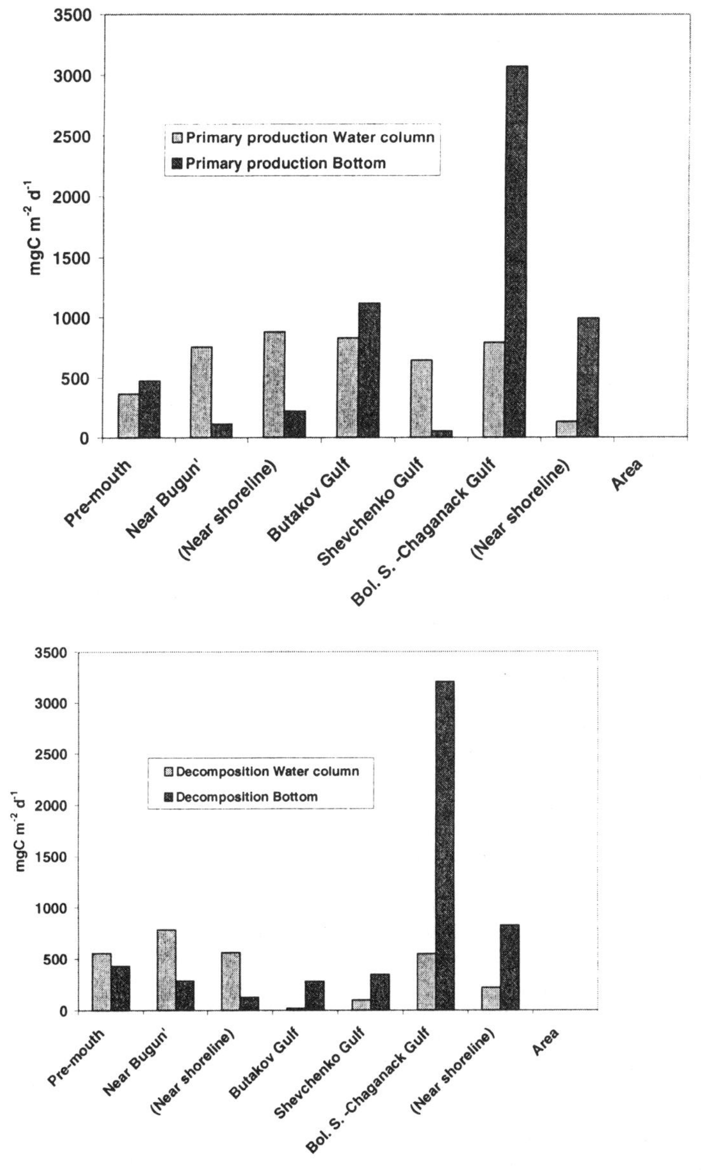

Figure 11. Community metabolism at some coastal areas (two first charts in mg Cm^{-2})

TABLE 6. Supply routes and biotic utilization and transformation of organic matter at selected coastal areas (daily values) for the pre-mouth area + Bugun' area, spring (1992-1993)

Area	Characteristics	Water ($mgCm^{-2}$)	Bottom ($mgCm^{-2}$)	Total ($mgCm^{-2}$)
1.	SUPPLY			1005±141
1.1.	*River Syrdarya input	97±41		
1.2.	**Primary production	498±128	410±228	
2.	UTILISATION			
2.1.	*Biosedimentation	351±149	625±538	975±469
2.2.	***Assimilation	30±19	79±26	108±32
3.	RETURN			
	*(pseudofaeces+ faeces)	341±238	502±332	843±278
4.	DECOMPOSITION			
	**(Aerobic decomposition)	577±137	382±142	960±119
5.	BALANCE	1.01±0.12	1.01±0.40	1.05±0.09
5.1.	(Supply/ Decomposition)			
5.2.	Primary production /decomposition	0.86±0.1	1.07±0.46	0.94±0.09

*by calculations
**by experiments
***Zooplankton and Macrozoobenthos

In relation to organic matter cycling primary production from the water column is completely utilised by suspension feeders both planktonic and benthic. Phytobenthos production is incorporated mainly into seagrass beds and to a lesser degree stored in microalgae. It is possible that peryphytic and microbenthic microalgae are consumed directly by the common Aral Sea grazer *Caspiohydrobia spp* (Gastropoda).

Filtration and biosedimentation of organic matter by suspension feeders was almost twice the value of photosynthesis measured in the water. It is likely the remaining part of their diet is made-up of detritus present as seston and resuspended material from the surface of bottom sediments.

Nevertheless the calculated requirements for growth and respiration (assimilation in Table 6) of all planktonic and benthic consumers could be completely recovered by primary production, if one accepts Bulion's suggestion that extracellular (dissolved) production contributes half the total photosynthesis [47].

Animals return organic matter as faeces and pseudofaeces (Table 6), of which the latter is fresh particulate matter (from primary production) which passes through consumers with minimal changes and then may be secondarily used in the nutrition of surface deposit-feeders, burrowing deposit feeders and seston-feeders. The macrozoobenthos associations in most areas are simple systems consisting only from four main functional groups represented by four species (taxa). The bivalve mollusc *C. isthmicum* is the only obligate filtering suspension feeder and consumes directly from the pelagic pool of organic matter. Thus it represents the benthic functional group mostly dependent on primary production. Where suspended material is available *Syndosmya segmentum (Abra ovata)* also can behave as a suspension feeder [18], however, it usually is a surface deposit feeder consuming sedimented matter, (resulting from both biotic and abiotic sedimentation) and microphytobenthos. Both bivalve species usually contribute more then 50% of the total biomass of the zoobenthos. *Caspiohydrobia spp* is a microalgal grazer, as mentioned above. *N. diversicolor* is the only burrowing deposit feeder and predator and to a lesser extent directly dependent on primary production and sedimentation processes.

The consumers in sediments are of equal in organic matter cycling as pelagic organisms. Thus there is strong benthic-pelagic coupling in the areas studied. Similar processes probably occur in the Small Aral Sea [35]. It is likely therefore, that the zoobenthos is a prerequisite for rapid organic matter cycling, ensuring faster rates of primary producition in the water column and rapid material exchange between water and sea-bed via benthic-pelagic coupling.

6. Community metabolism and accumulation of organic matter in temporary pools

From Fig. 11 the rates of primary production and decomposition on the seabed of the Gulf are extremely high and exceed those for other areas. Caution must be exercised in this interpretation since only one season was sampled, (the end of June). The rates are also high in temporary pools above the shoreline (Table 7). However, in their late developmental stages decomposition processes dominate strongly over production. In my opinion this is caused by an increase in salinity (from 20 to 43‰), rapid oscillations in temperature and oxygen saturation etc.

TABLE 7. Community metabolism in Bol. Sary-Chaganack Gulf and adjacent temporary pools (June, 1994)

Area	Value for:	Primary production $mgCm^{-2}$	Decomposition $mgCm^{-2}$	A/D
Gulf	Water column	359±186	281±51	1.2±0.6
Pools		394±264	605±214.4	0.7±0.4
Gulf	Total	2465±1785	2377±1810	1.1±0.01
Pools		3158±1380	4206±967	0.7±0.03

The results suggest that during in the last stage of pool succession evaporation leads to rapid rates of accumulation of organic matter from autochthonic and allochthonic processes, from increased community metabolism and from mass mortality of living organisms (Figure 12,b). Accumulation of organic matter is probably faster in the belt of temporary embayments than in the most productive areas of the Gulf proper.

These temporary pools occur from early spring until the early autumn and then disappear as the shoreline retreats, contributing to organic deposits and the seasonal development of biological systems in the Bolshoi Sary-Chaganack Gulf.

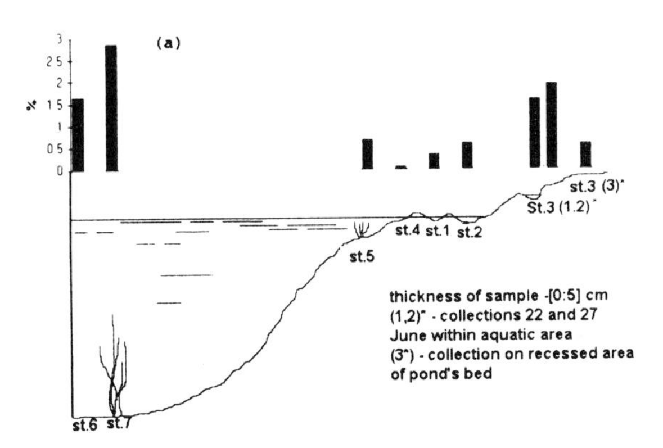

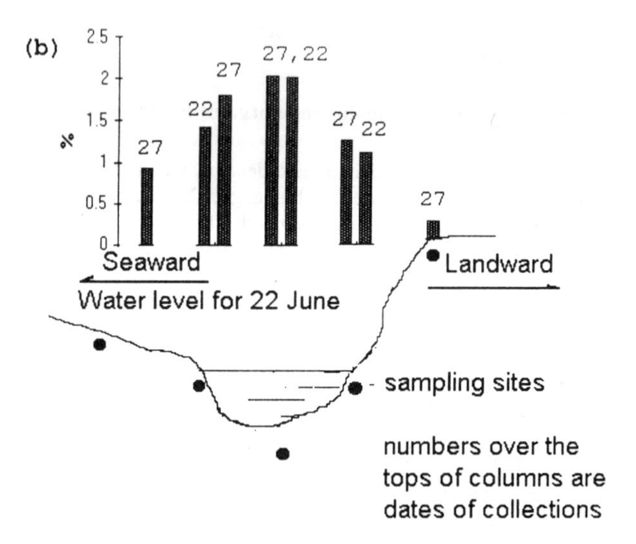

Figure 12. Content of organic matter in bottom sediments 22 - 27 at Bol Sary-Chaganack Gulf (a) with temporary pools (b)

7. Acknowledgements

The present study has been carried out under the financial support of UNESCO and BMFT project 59/RAS/40-SC 213.056.5 and project 96-04-48 114 Russian Foundation for Basic Research. The author expresses her thanks to Ms. Olga Rusakova and Dr. Andrei Filippov for their help in both field and experimental work; to Mrs. Lubov Zhakova for identification and for the quantitative study of macrophytobenthos, to Dr. Nikolai Aladin for information about the history of the Aral Sea basin and possible ways of conserving the Northern Aral Sea; to Mr. David Piriulin, for personal communications about the evolution of the coastline and desiccated areas. The special gratitude of the author is addressed to the editor of the volume, Prof John Gray who has edited this paper and solved technical and stylistic problems in the preparation of the manuscript.

8. References

1. Micklin P. (1991) The water management crisis in Soviet Central Asia, *The Carl Beck Papers in Russian and East European Studies*. No **95**, University of Pittsburg Center for Russian and East European Studues, 120 p.
2. Glantz M.H., Rubinstein A.R., Zohn I. (1992) Tragedy in the Aral Sea basin: Looking back to plan ahead, *Proceeding of seminar. Central Asia: it's strategic importance and future prospects*, Villanjva University, Pennsylvania, 1-42
3. Ellis W.S. (1991) The Aral Sea. Soviet Sea lies dying, *National Geographic*, February, 73-92
4. Aladin N.V., Plotnikov I.S., (1995) On the problem of possible conservation and rehabilitation of the Small Aral Sea, *Proceedings of the Zoological institute of Russian Academy of Science*, **262**, 3-17 *(in Russian)*
5. Aladin N.V., Kotov S.V.(1989) Natural state of the ecosystem of the Aral Sea and it's changes during anthropogenous impact, *Proceedings of the Zoological institute of Academy of Science of the USSR*, **199**, 4-25. (in Russian)
6. Kusnetzov L.A., (1993) Preface, *Proceedings of the Zoological institute of Russian Academy of Science*, **250**, 3-5 (*in Russian*)
7. Aladin N.V. (1989) Zooplankton and zoobenthos of coastal area near i. Barsakelmes (The Aral Sea), *Proceedings of the Zoological institute of Academy of Science of the USSR*, **199**, 110-114 (*in Russian)*
8. Andreev N.I., Andreeva S.I., Filippov A.A. (1990) Zoobenthos of the Aral Sea in conditions of progressive salinization, *Proceedings of the Zoological institute of Academy of Science of the USSR*, **223**, 24-31 (*in Russian)*
9. Novozhilova, M.N. (1973) *The microbiology of the Aral Sea*, Nauka, Alma-Ata(in Russian)

10. Karpevich A.F. (1975) *The theory and practice of acclimatization of aquatic organisms*, Pischevaya promishlennosct', Moscow (in Russian)
11. YablonskayaE.L, Lukonina N.K. (1962), To the question of the productivity of the Aral Sea, *Oceanology*, **2**, 299-304 (in Russian)
12. Yablonskaya E. A. (1964) To the question of importance of phytoplankton and phytobenthos in food chains of organisms of Aral Sea, in *The Stock of Marine Plants and Animals and Their Use*. Pischevaya promishlennos, Moskow, p 71-91 (in Russian)
13. Dobrinin E. G., Koroliova N. G., Burkova T. M. (1990) The estimation of the Aral Sea ecological state near the Isle Barsakelmes, *Proceedings of the Zoological institute of Academy of Science of the USSR* , **223**, 31-35 (in Russian).
14. Dobrinin E. G., Koroliova N. G. (1991) Production and microbiological processes in Butakov Bay (The Aral Sea), *Proceedings of the Zoological institute of Academy of Science of the USSR*, **237**, 49-59 (in Russian)
15. Orlova M.I. (1993) Materials for the general evaluation of production and decomposition processes in the coastal zone of the Northern part of the Aral Sea, *Proceedings of the Zoological institute of Academy of Science of the USSR*, **195**, 47-64 (in Russian)
16. Orlova M.I. (1996) Primary production and decomposition in the coastal zone of the Northern part of the Aral Sea near the delta of the Syrdarya, *International Journal of Salt Lake Research*, **5**,**1**, 17-33
17. Orlova M. I., Rusakova O. M. (1995) Structural and functional characteristics of phytoplanktonic community in the district of Tastubeck Cape in September 1993, *Proceedings of the Zoological institute of Academy of Science of the USSR*, **262**:208-230 (in Russian)
18. Orlova M.I., Komendantov A. Yu.(1995) To the ecology of dominating species of bottom macroinvertebrates of the Aral Sea, *Proceeding of the Zoological institute of Russian Academy of Science*,**262**, 174-189 (in Russian)
19. Williams W.D. (1995) The Aral Sea: A limnological perspective, *Proceedings of the Zoological institute of Russian Academy of Science*, **262**, 237-247 *(in Russian)*
20. *Seas of the USSR. Hygrometeorology and hydrochemistry of Seas ot the USSR. V.7, Aral Sea*. (1990), Hydrometeoisdat, Leningrad (in Russian)
21. *Methods of chemical analyse in hydrobiological studies* (1978).Edition of Far East Scientific Centre, Vladivostok, (in Russian)
22. Odum H. T. (1956) Primary production in flowing waters, *Limnology and Oceanography* **1**, 102-117
23. Odum H. T. (1957) Primary production measurements in eleven Florida springs and a marine turtle grass community, *Limnology and Oceanography*, **2**, 85-97
24. Odum H.T., Hoskins C.M. (1958) Comparative studies on the metabolism of marine waters, *Publishing of institute of Marine Science*, **5**, 16-46
25. Ganning B, Wulf F (1970) Measurements of community metabolism in some Baltic brackish water rockpools by means of diel oxygen curves. *Oikos*, **21**, 292-298
26. Assman A. V. (1953) The contribution of algal foulings in to organic matter in Glubokoye Lake. *Proceedings of All Union Hydrobiological Society*, **5**, 138-157 (in Russian)
27. Vinberg G. G. (1960) *Primary production in water bodies*, Visheisha shkola, Minsk (in Russian)
28. Elmgren R, Ganning B. (1977) Ecological studies of two shallow brackish water ecosystems. *Contribution from the Asko Laboratory University of Stockholm*, Sweden **6**: 55 p
29. Buljon V.V. (1979) Primary production of plankton, in G.G. Vinberg (ed.), *General basis for the study of aquatic ecosystems*, Nauka, Leningrad, pp.187-199 (in Russian)
30. Aladin N.V., Potts W.T.W., Filippov A.A., Orlova M.I., Plotnikov I.S. (in press) Hydrobiological and palaeolimnological characteristics of the ecosystem of the Northern Part of the Aral Sea in it's modern State, *Chemistry and Environment*
31. Alekin O.A., Liakhin Yu.I.(1984) *Chemistry of the Ocean*, Nauka, Moscow (in Russian)
32. Zizarin A.G. (1991) Modern state of elements of hydrological regime of the Aral Sea, *Proc. of Geological inst of RAS*, **183**, 71-91 (in Russian)
33. Plotnikov I. S. (1993) Zooplankton of the Aral Sea in 1992, *Proceeding of the Zoological institute of Russian Academy of Science*, **250**, 46-52 (in Russian)
34. Plotnikov I.S. (1995) Zooplankton of the Aral Sea (Small Aral Sea) in conditions of stabilization, *Proceeding of the Zoological institute of Russian Academy of Science*, **262**, 167-174 (in Russian)
35. Filippov A. A. (1995) Macrozoobenthos of coastal zone of the Northern Aral in modern conditions of polyhalinity: abundance, biomass and spatial distribution, *Proceeding of the Zoological institute of Russian Academy of Science*,**262**, 103-167 (in Russian)
36. Zhakova L.V. (1995) Notes to content, distribution and biomass of aquatic plants and filamentous algae in Bolshoi Sary-Chaganack Gulf (Aral Sea), *Proceeding of the Zoological institute of Russian Academy of Science*, **262**, 231-237 (in Russian)
37. Rusakova O. M (1995) Concise characteristics of the Aral Sea phytoplankton qualitative composition in spring and autumn 1992, *Proceeding of the Zoological institute of Russian Academy of Science*, **262**, 195-202 (in Russian)
38. Sulalina A. V., Smurov A.O. (1993) The state of bacterioplankton of the Aral Sea in autumn, 1991, *Proceeding of the Zoological institute of Russian Academy of Science*, **250**, 104-108 (in Russian)
39. Sulalina A.V., Smurov A. O., (1993) The State of bacterioplankton of the Aral Sea in autumn 1992, *Proceeding of the Zoological institute of Russian Academy of Science*, **250**, 108-114 (in Russian)
40. Pichkily L.O. (1970) Content and dynamic of phytoplankton of the Aral Sea. *Manuscript of Doct. thesis*, Leningrad, (in Russian)

41. Pichkily L.O. (1981) *Phytoplankton of the Aral Sea in conditions of anthropogenous impact (1957-1980 years),* Naukova dumka, Kiev, (in Russian)
42. Elmuratov A. E. (1981) *Phytoplankton of the Southern part of the Aral sea*, Fan, Tashkent (in Russian)
43. Elmuratov A. E. (1988) Content and distribution of phytoplankton on the South of the Aral Sea in conditions of changed regime, in *The structure of aquatic living associations in low portion of Amudarya*, Fan, Tashkent, pp. 25-34 (in Russian)
44. Safianov G.A. (1987) *Estuaries*, Misl', Moskow
45. Khusainova N.Z. (1958) Biological base of acclimatization of fishes and invertebrates in the Aral Sea, in *Acclimatization of fishes and invertebrates in water bodies of the USSR*, Moskow: pp. 100-144. (in Russian)
46. Filippov A. A., Orlova M. I, Rusakova O. M, Plotnikov I. S, Smurov A. O, Zhakova L. V (in press) The state of the ecosystem of Sarychagahak Bay (The Northern Aral) in june 1994 year. *Hydrobiological Journal* (in Russian)
47. Bul'on V.V. (1993) *Primary production in Lymnetic systems*, Nauka, Leningrad (in Russian)

BIOGEOCHEMISTRY OF WATER AND SEDIMENT IN THE OB AND YENISEY ESTUARIES

VYACHESLAV V.GORDEEV
P.P.Shirshov Institute of Oceanology
Russian Academy of Sciences
36, Nakchimovsky prospect
117851 Moscow
Russian Federation

Abstract

In August-September 1993 hydrochemical, hydrooptical, geochemical and biological researches were carried out in the Ob and Yenisey estuaries and adjacent regions of the Kara Sea (49-th cruise of the R/V "Dmitry Mendeleev"). The results of this expedition allow comparison of the two large estuarine systems in relation to the role of organic matter and biota in the behaviour of some elements in water, suspended matter and bottom sediments. At relatively similar hydrochemical and hydrophical conditions (the same salt wedge estuarine type, similar temperature (T) and salinity (S‰) distributions) there are significant differences in the level of dissolved carbon and nutrient concentrations (higher in the Ob estuary), and in photosynthetic activity and primary production (higher in the Yenisey estuary). The results show that biotic factors play an important role in trace metal behaviour in the water column (for both dissolved and particulate forms of elements). The influence of biotic factors is more marked in the Yenisey estuary due to higher photosynthetic activity and primary production of phytoplancton.

1. Introduction

In Russsia three main approaches or directions in the studies of oceanic geochemistry have been developed during the last decades: physico-chemical, sedimentary (mechanical and chemical differentiation), and biogeochemical.

The Physico-chemical approach was developed actively after the brilliant lecture by Prof. L.Sillen, Sweden at the First International Oceanographic Congress in New-York, 1959. He considered not the real ocean, but a theoretical solution with content similar to sea-water content. This solution was free of dissolved organic carbon, biota, organic detritus etc. This concept was developed later by Garrels, McKenzie, Lerman, and other.

The Sedimentary approach was developed by V.M.Goldschmidt, L.V.Pustovalov and N.M.Strakhov. Academician Strakhov wrote: «In the ocean geochemical process is first of all (on 90-93%) the physical process, and more exactly the mechanical one, that is the process of mechanical transportation and fractionation of hard phases entering from shore, allochtonous, to a very small degree (6-9%) this physical process is complicated by biogenic (carbonates Ca + Mg, $C_{org.}$, SiO_2) and to absolutely insignificant degree - by physico-chemical one: by coagulation of Fe and Mn colloids and sorption of trace elements on them» [10]. This concept is in contradiction with the biogeochemical idea by V.I.Vernadsky and with the law of environmental zonality by V.V.Dokuchaev because the role of organisms, is in Strakhov's opinion, in oceanic sedimentogenesis is quite insignificant - it is only 6-9%, as was mentioned, versus 99% that is following from Vernadsky concept.

The Biogeochemical approach originated from the works of V.V.Dokuchaev and especially of V.I.Vernadsky. Academician Vernadsky laid down the principles of

J.S. Gray et al. (eds.), Biogeochemical Cycling and Sediment Ecology, 49–68.

biogeochemistry in his works «Biosphere» (1926) and «Essays of geochemistry» (1934). He pointed that «live organisms are not a secondary but a principal factor of chemical elements migration on the Earth (including the ocean)». The important contribution in development of this concept in the Soviet Union was made by A.P.Vinogradov, B.B.Polynov and A.I.Perelman. The scientific school under the leadership of academician A.P.Lisitzin recognizes the biogeochemical concept in oceanic geochemistry to be the basic one. The main approaches to the problem and the results of multiannual studies were published in the collective monography «Biogeochemistry of the Ocean» (1983).

The key functions of living organisms in the ocean are the following: 1) concentration organisms in their bodies *(biogenic accumulation or bioassimilation)*, associated with the accumulation of many chemical elements; 2) the biogeochemical functions associated with growth, reproduction, and transportation of organisms *(biogeochemical differentiation and biotransport)*.

All these processes have significant influences on transport, distribution and transformation of chemical elements, especially in the areas of sharp gradients of environmental parameters such as river-sea mixing zones, or estuarine zones. We have tried to develop the biogeochemical concept in the study of the Arctic Ocean and its coastal seas and estuaries of the great Siberian rivers. These studies started in the framework of the Russian-French-Netherlands Scientific Programme on Arctic and Siberian Aquatorium (SPASIBA, 1989-1995). There were three expeditions - to the Lena delta and the Laptev Sea (1989 and 1991) and a large expedition to the Ob and Yenisey estuaries and the Kara Sea (1993).

The objective of this chapter is to show the role of organic material and biota in the distribution and transformation of chemical elements in water and suspended particulate matter and bottom sediments in the Ob and Yenisey estuaries and adjacent area of the Kara Sea. It is interesting to compare these two estuaries since both estuaries are situated in the same climatic (ice) zone, but they differ greatly in their concentrations of organic carbon, nutrients and primary production.

2. Materials and Methods

The results were obtained during the 49-th cruise of R/V «Dmitry Mendeleev» to the Kara Sea in August-October 1993, in the SPASIBA programme [7]. Scientists from Russia, France, Norway and Germany took part in this expedition. Hydrochemical, optical, geochemical and biological studies were carried out along two transects across the Ob and Yenisey estuaries up to 76°N (Figure 1).

River- and sea-water samples were sampled using an acid-cleaned Teflon pump, fitted with PTFE tubing and connectors and collected using a portable laminar-flow clean bench. Deeper samples (>10m) were collected by Teflon Go-Flo bottles. Water samples were filtered on board ship through Nuclepore filters (0.4 μm pore size). Few samples of suspended matter (SM) were collected by centrifugation of large volumes of water (a few tons).

Bottom sediment sampling was carried out by grab, gravity tube and box corer. Determinations of dissolved trace metals were made using two methods - one from France [2] and the other from Russia [5]. The French group analysed Cu, Ni, Pb, Cd and Fe in truly dissolved, colloidal and «dissolved» fractions. The Russian group made determinations of labile forms of Cu, Zn, Pb and Cd. Standard reference samples of river water, suspended matter (SM) and bottom sediments were used to control reliability and reproducibility of analyses.

Chemical element determinations in water, suspended matter and bottom sediments were done by flame and flameless AAS, ICP and INAA in laboratories of Russia, France and Belgium.

3. The main characteristics of the Ob and Yenisey estuaries

The Yenisey and the Ob Rivers are the largest in the Arctic basin (together with the

Lena River) in terms of both catchment area and water discharge (Table 1). Both river estuaries are orientated from south to north. The Ob estuary is a marine bay with length about 800 km, width 35-95 km and depth 13-24 m. The mouth area of the Yenisey River includes the delta and the Yenisey bay with total extent of 350 km^2. Both estuaries are belonged to the same salt wedge estuary type.

TABLE 1. Physical characteristics of the Ob and Yenisey estuaries

Characteristic	Ob estuary	Yenisey estuary
Watershead square, km^2	$2.54 \cdot 10^6$	$2.59 \cdot 10^6$
Mean river discharge, $m^3 \cdot s^{-1}$	12800	20000
Maximum river discharge, $m^3 \cdot s^{-1}$	43000	158000
Mean concentration of SPM, $mg \cdot l^{-1}$	38	10
Mean SPM discharge, $t \cdot a^{-1}$	$16.5 \cdot 10^6$	$5.9 \cdot 10^6$
Mean tidal range, m	0.5	1.0
Maximum depth, m	24	30
Estuarine type	salt wedge	salt wedge

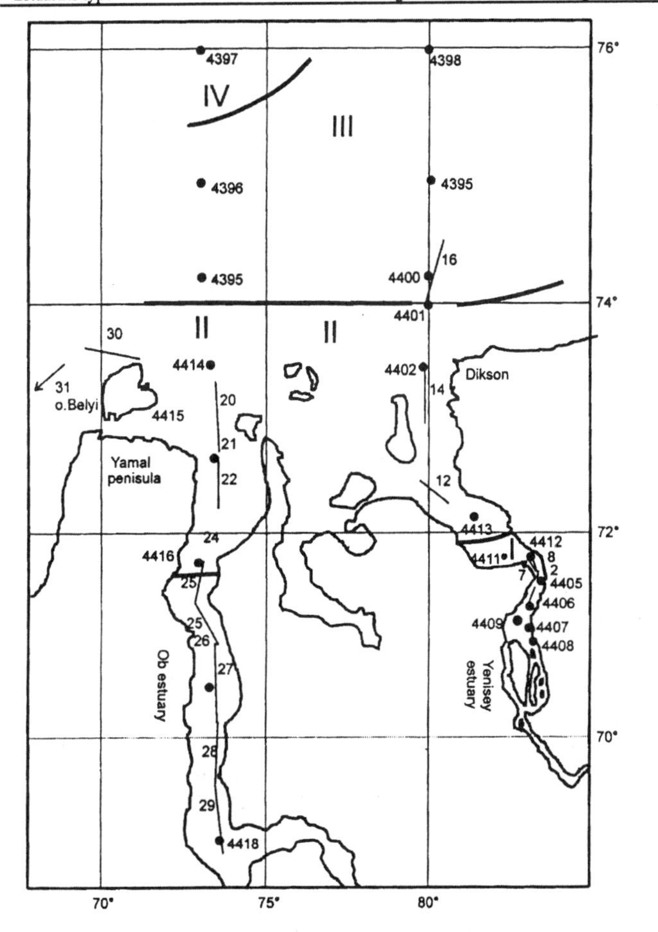

Figure 1. Map of sampling locations of suspended matter (bars) and bottom sediment stations, and borders of the facies zones I-IV (see text).

3.1. SALINITY AND TEMPERATURE

Salinity and temperature distributions in the Ob and Yenisey estuaries are shown in Figures 2 and 3.

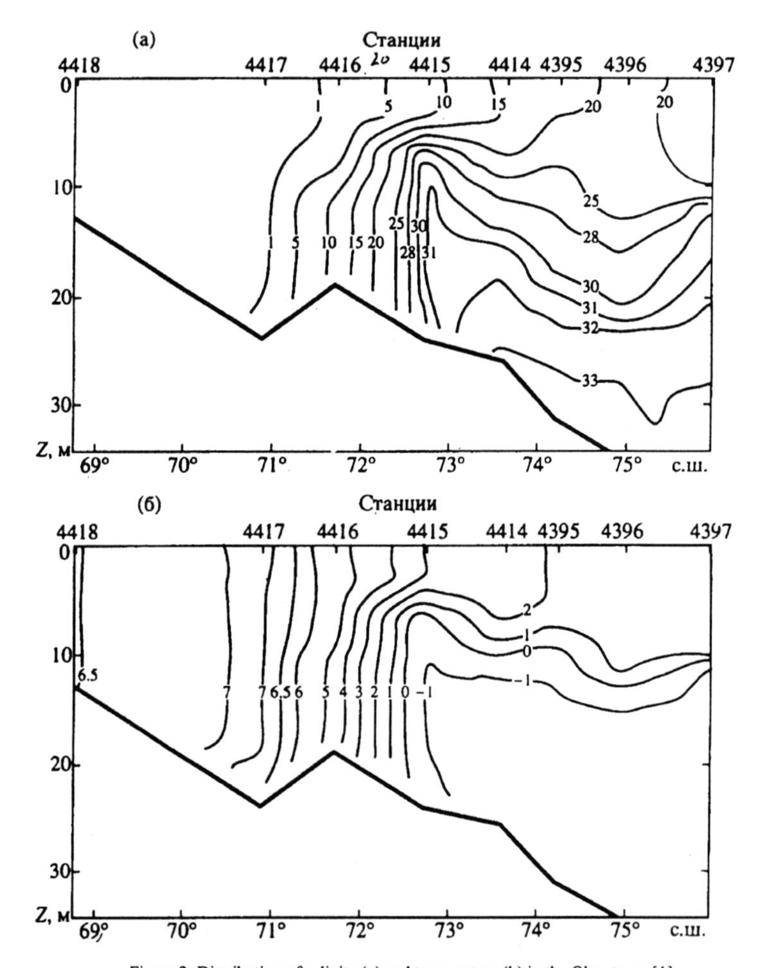

Figure 2. Distribution of salinity (a) and temperature (b) in the Ob estuary [1].

Both transects for salinity are characterised by freshening of surface waters, under the influence of river disharge, and by increasing salinity with depth. There are two frontal zones in each transect due not only to salinity and temperature gradients but also to hydrochemical parameters (T°, $O_{2\ diss.}$, NO_3, PO_4, SiO_2, Alk, pH, Eh) and the concentration of suspended matter. The differences between the two estuarine zones are:

1) the frontal zones in the Ob estuary are not as clear as in the Yenisey estuary;
2) the invasion of saline waters into the Ob estuary is significantly weaker. In the Yenisey estuary saline waters penetrate up to 71°30 N due to a southward near-bottom current, and also to tide generated waves and wind-driven waves.

The temperature distributions in both estuaries are very similar to the salinity distributions.

3.2. NUTRIENTS

The distribution of dissolved nitrates and orthophosphates is presented in Figure 4 and 5. There are common features in nutrient behaviour in the two estuaries. The extreme values of almost all parameters were found in near-bottom waters of the frontal zone 1 in the Ob estuary (min $O_{2diss.}$, max NO_3, NH_4, PO_4). The bulk of terrestrial organic matter is deposited and oxidized in the sea bed, nutrient regeneration occurs, and the so-called «biogenic trap» is formed here.

The frontal zone 2 (S = 15-20‰) is also characterised by sharp gradients of parameters and nutrients. Concentrations of NO_3 and PO_4 are relatively stable in surface waters, but decrease strongly after crossing the frontal zone. The reason for this is an active consumption of nutrients by phytoplankton, high nutrient input and high transparency of the water column, (due to sedimentation of the bulk of SM), which results in a phytoplankton bloom and high primary production (PP).

Regeneration of organic matter deposited down to near-bottom waters results in high concentration of nutrients with maximum at S = 30-32‰.

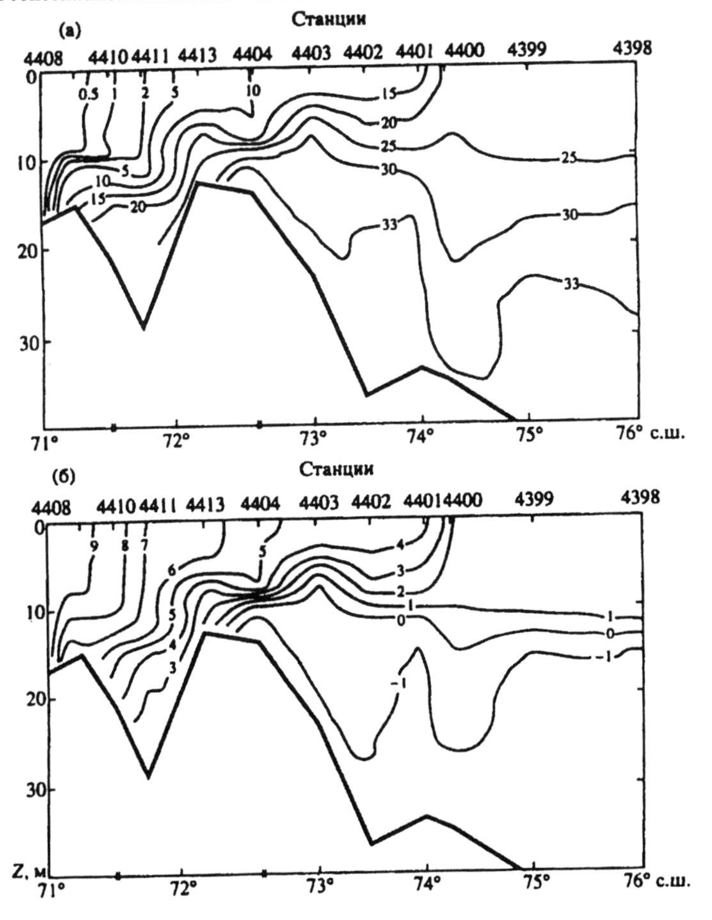

Figure 3. Distribution of salinity (a) and temperature (b) in the Yenisey estuary [1].

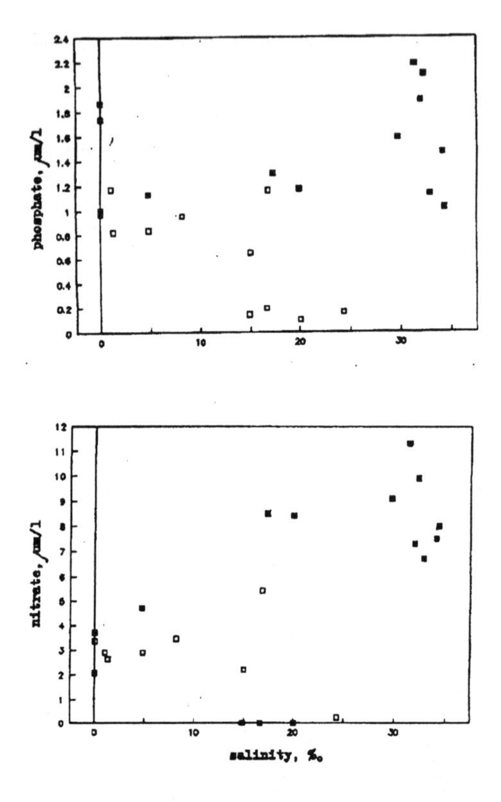

Figure 4. Distribution of phosphate (a) and nitrate (b) in the Ob estuary (data from Stunzhas and Makkaveev).

In the Yenisey estuary (Figure 5) concentrations of nitrates were permanently low. Nitrates were probably the limiting factor for PP in this estuary. Low nitrate concentrations and high concentrations of ammonia demonstrate the active consumption of phytoplankton by zooplankton. In the Ob estuary a high NO_3 concentration and low variability along the transect indicate lower nutrient consumption. The phosphate concentrations in the Yenisey estuary show some increase with salinity (Figure 5, a). The probable explanation is the increasing rate of PO_4 turnover as a result of higher PP and the intensity of OM degradation in near-bottom waters of the Yenisey estuary.

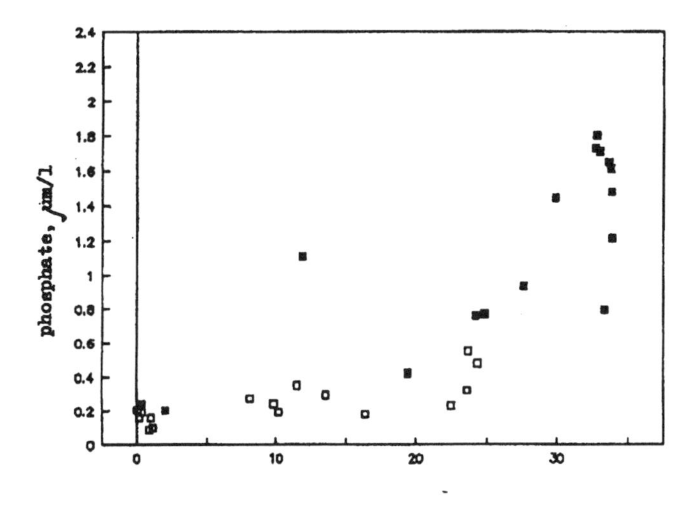

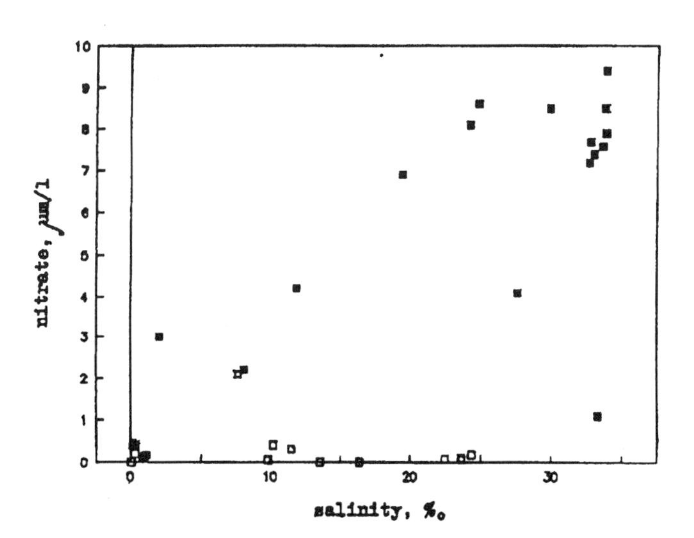

Figure 5. Distribution of phosphate (a) and nitrate (b) in the Yenisey estuary (data by Stunzhas and Makkaveev).

TABLE 2. Concentrations of nutrients and organic carbon in the Ob and Yenisey rivers (in μM) [2,8]

Component	Ob river	Yenisey river
NO_3	2.1-3.2	0.08-0.19
PO_4	1.3-1.5	0.07-0.22
$DOC_{diss.}$	600-800	300-350
$DOC_{colloid.}$	250-360	100-130

A comparison of mean concentrations of NO_3, PO_4 and dissolved and colloidal C_{org} (Table 2) demonstrates higher values in the Ob estuary.

3.3. PRIMARY PRODUCTION (PP)

Primary production in the Ob and Yenisey estuaries was determined during the 49-th cruise of R/V «Dmitry Mendeleev». In the Ob estuary PP in the water column ranged between 25 and 63 mgC $m^{-2}d^{-1}$, in the Yenisey estuary - 100-300 mgC $m^{-2}d^{-1}$, i.e. 4-5 times more.

Low PP in the Ob estuary was connected with lower water transparency, low physiological activity of phytoplankton, and unfavourable weather conditions during the period of measurements [11]. The high content of nutrients in the Ob estuarysuggests they are unlikely to limit PP. The main abiotic factors of PP limitation in the Yenisey Bay were nutrient deficiency and low transparency of water column.

Thus the two large estuaries that are situated in the same climatic zone, and of the same estuarine type (salt wedge), with similar hydrology, have significantly different levels of DOC, nutrients and PP (in the Ob estuary DOC and nutrients are higher, while in the Yenisey estuary are higher physiological phytoplankton activity, rates of nutrient turnover and PP). Further we will consider the differences in behaviour of dissolved and particulate metals in water column and in bottom sediments in the two estuaries.

3.4. DISSOLVED METALS

It is interesting to see, firstly, in what degree the metals interact with organics and nutrients, and secondly, what whether or not metals in the mixing zone behave conservatively or nonconservatively.

The comparison of dissolved and colloidal metal concentrations in the lower course of the Ob and Yenisey Rivers shows that (Table 2): 1) metal concentrations in the Yenisey River (except Cd) are overall lower than in the Ob River. This may be not only related to the lower weathering rate, but also to the higher biological productivity in the Yenisey River [2]; 2) the colloidal fraction in the dissolved form for all the metals except Pb is higher in the Yenisey waters. This leads to a closer association of metals with organic material in the Yenisey waters. Let us consider the behaviour of dissolved and colloidal trace elements in the mixing zones. (Figures 6 and 7).

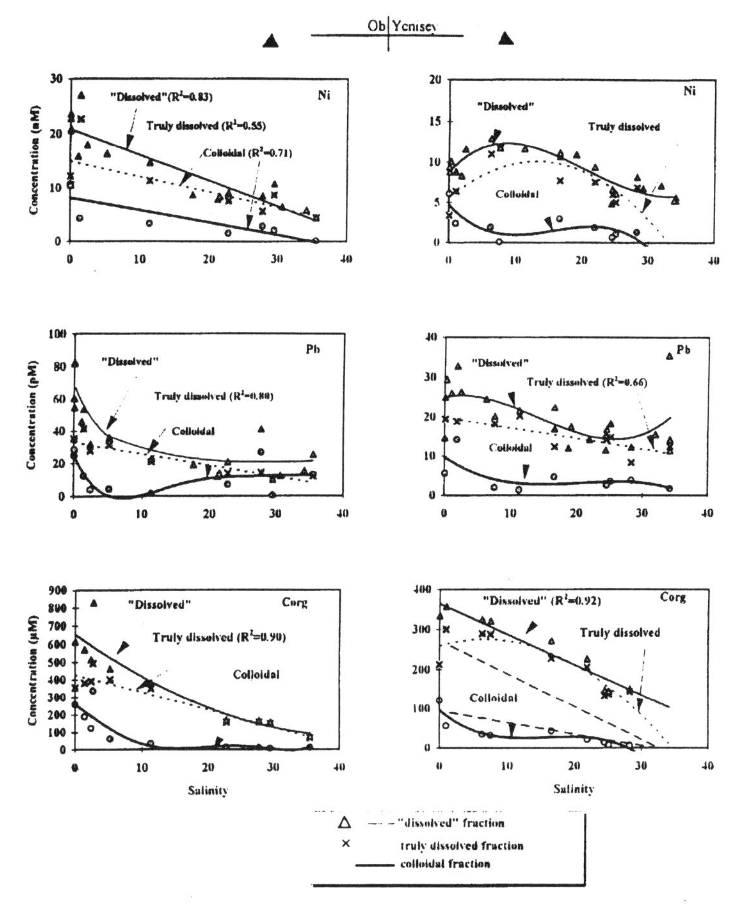

Figure 6. Cu, Cd and Fe in «dissolved» fraction, truly dissolved fraction, and colloidal fraction as a function of salinity in the Ob and Yenisey estuaries [2].

Copper shows conservative behaviour in both estuaries. The copper conservativity has been hypothesized to be due to the organic character of colloidal Cu preventing its flocculation. A good correlation between colloidal Cu and organic colloidal carbon in both estuaries provides some direct evidence for such a mechanism (Figure 8) [2].

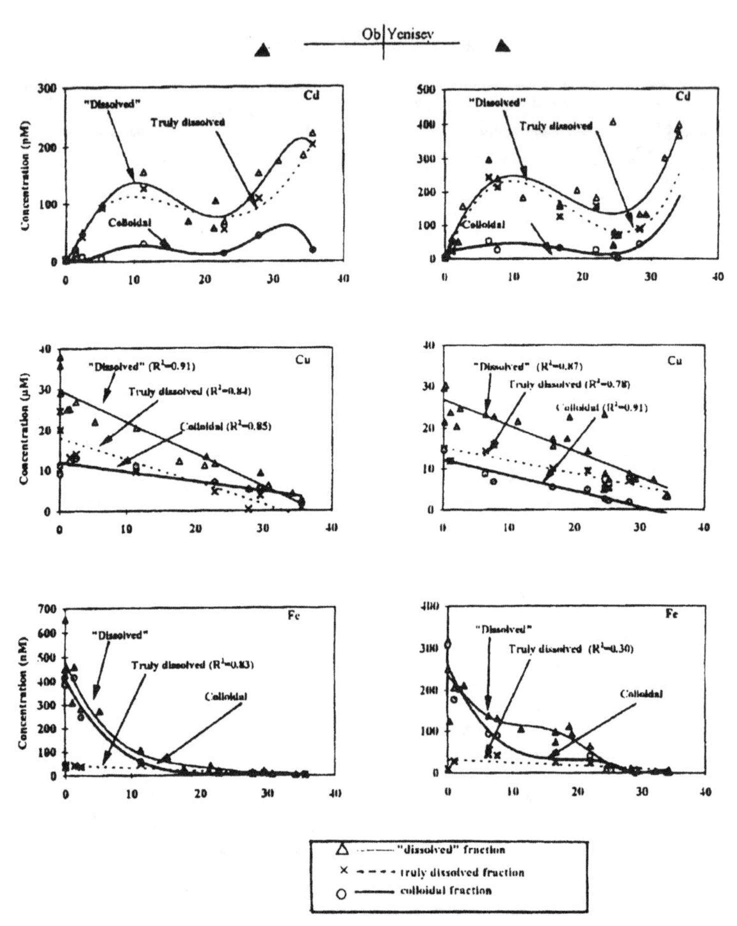

Figure 7. Ni, Pb and $C_{org.}$ in «dissolved» fraction, truly dissolved fraction, and colloidal fraction as a function of salinity in the Ob and Yenisey estuaries [2].

TABLE 3. Dissolved and colloidal fractions of trace metals in the Ob and Yenisey river waters [2]

Element	**Ob river**		**Yenisey river**	
	Dissolved, nM	Colloidal % from diss	Dissolved, nM	Colloidal % from diss
Cd	0.005-0.008	50-57	0.011-0.016	76
Pb	0.055-0.083	42-52	0.025-0.029	22
Cu	29-38	37	22-29	46
Ni	21-24	50	8.8-9.4	60
Fe	470	89-92	260	97

Nickel is conservative in the Ob estuary and nonconservative in the Yenisey estuary. Some excess at S<10‰ is a result of a biological regeneration process. At lower biological activity it does not appear in the Ob estuary. Additional support for this hypothesis is provided by the close relationship between nitrate and truly dissolved Ni (Figure 9). But the correlation is absent between NO_3 and colloidal nickel which suggests that additional Ni has the same origin (regenerated from OM) as nitrate whereas colloidal Ni is primarily associated with organic matter ($Ni_{col.}$ - $C_{org.col.}$ R^2 = +0.84 in the Ob and +0.81 in the Yenisey).

Cadmium has a complex distributional trend with salinity and a patchy occurrence in both estuaries (Figure 6). The concentration increase is probably related to the well-recognized desorption of particulate Cd. But there is an alternative explanation, since a direct correlation occurs between $Cd_{diss.}$ and nutrients (Figure 10) and suggests that maximum Cd may be also associated with nitrate and phosphate regeneration [2].

Iron and lead show removal in both estuaries, less significant for Pb than for Fe. Truly dissolved Fe and Pb are conservative indicating that the flocculation of colloidal material is responsible for removal of dissolved Fe and Pb. The nature of the colloidal fraction for iron is not clear and may be organic, inorganic or both. Dai and Martin [2] consider that it is probably related to humic substances, which are constituted of macromolecules and removed in estuaries. The significant removal of Fe and Pb in the Ob estuary is a result of higher content of colloids in this river.

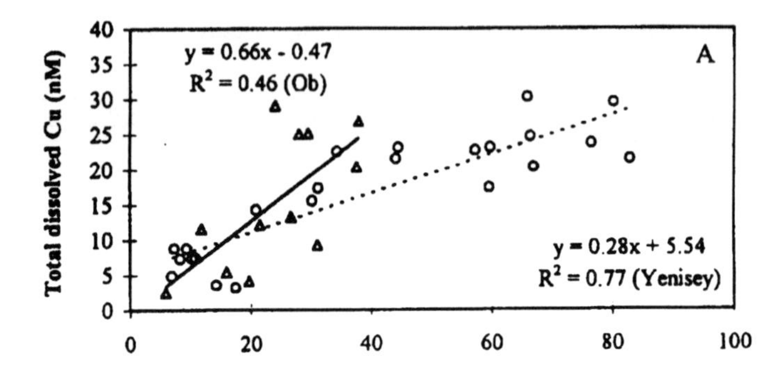

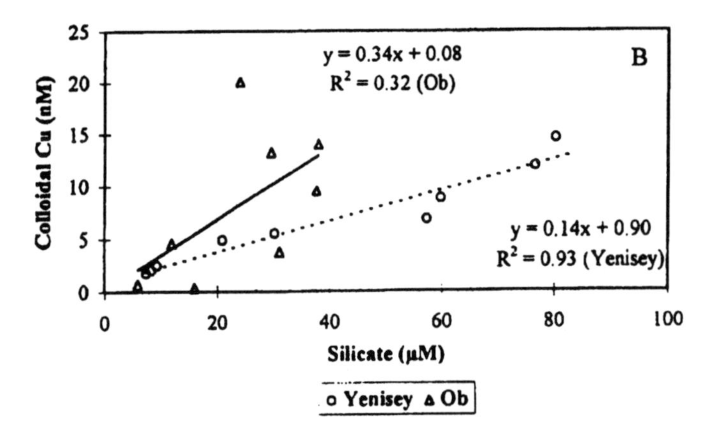

Figure 8. Linear relationship between organic carbon and copper in the colloidal fraction [2].

3.5. SUSPENDED MATERIAL

Two sets of suspended material samples were collected during the expedition. POC in centrifuged samples show that SM of the Yenisey River was richer in organics in comparison with the Ob SM (4.7% against 2.9%). POC increased with salinity in both estuaries, in the Yenisey Bay up to 12.5% (within a salinity range 12-22‰), in the Ob estuary up to 4-5% (salinity 10-13‰, but there are no SM samples from higher salinities). Unfortunately, wec were unable to evaluate the correlations between POC and metal concentrations in centrifuged samples due to the small number of samples.

In order to evaluate the influence of biota on the geochemistry of elements in SM the method of Sc-normalization was used. The particles of river SM entering into the river-sea mixing zone may undergo processes of sorption-desorption, partial dissolution among others. Flocculation transforms part of the dissolved metal fraction, especially in colloidal form, into particulate form. The calculations show however, that at typical concentrations of dissolved metals and SM in water, even for the element most liable to flocculation in mixing zone, iron, this process does not increase the Fe content in SM more than 10-20%. As a result of photosynthesis the suspended OM is regenerated, which increases the content of metals in suspended matter by one to several orders of magnitude by accumulation of dissolved metals from water.

Let us consider the distribution of three Sc-normalized elements which characterised three groups of elements in accordance with a degree of their concentration in SM.

Group 1. Lantanum, an element from the REE group, is a typical element. It remains unaltered with salinity when compared to its content in river SM. All other REE, Fe, Hf, Th and Zr show similar behaviour in the mixing zone (Figure 11).

Group 2. Distribution of Sc-normalized chromium, a representative of this group, is shown in Figure 12. Cr is stable up to salinities of 10-15‰. One can see a significant increase in the ratio (5-7 times) at higher salinity. We explain this increase by a higher photosynthetic activity of phytoplankton as a result of sedimentation of the major part of river SM and higher transparency of waters. Ba, Co and Cs show similar behaviour. It is worth noting also that the enrichment factor at salinities of 15-20‰ (frontal zone 2) is either similar in the two estuaries (Cr, Ba), or higher in the Yenisey SM (Co, Cs) (Table 4).

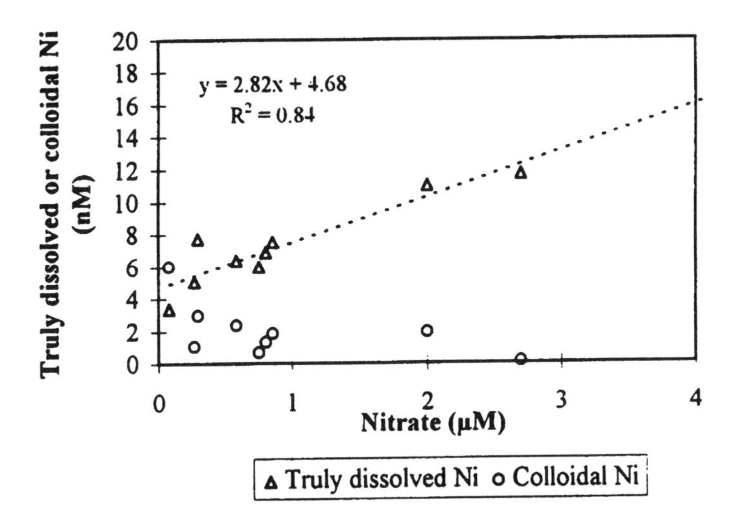

Figure 9. Truly dissolved and colloidal nickel in the Yenisey estuary as a function of nitrate concentration [2].

Table 4. Enrichment factor of elements in suspended matter of the Ob and Yenisey estuaries *

Element	Enrichment factor		Comments
	Ob estuary	Yenisey estuary	
Fe, Hf, Th, Zr, La, Ce, Nd, Sm, Eu, Tb, Yb, Lu	< 2	< 2	These elements remain unaltered when compared to their av content in river SPM
Cr	4-9	5-8	The elements accumulate in SPM at salinity 15-20‰ (frontal zone 2)
Ba	3-8	4-9	
Cs	2-2.7	3-7.5	
Co	3-5	4-9	
Br	36	130-150	Increasing is visible at S=5-10‰ (frontal zone 1) and especially at S=15-20‰ (frontal zone 2)

* Enrichment factor = $(Me/Sc)_{est.susp} / (Me/Sc)_{av.river\ susp.}$

Group 3. A very high enrichment factor was shown by bromium (up to 36 in the Ob estuary and 130-156 in the Yenisey estuary, Figure 13). A significant increase of the ratio at salinities of 5-12‰ for this element is obvious, that is immediately after the frontal zone 1. At the same time there are the samples without any enrichment even at high salinity, the distribution is not homogeneous and we need better data to find a pattern.

Available data show that the biogenic factor plays an important role in the geochemistry of many elements in the river-sea-water mixing zone. We see also that despite higher nutrient concentrations in the Ob estuary these processes appear to be more intensive in the Yenisey estuary due to higher phytoplankton activity and higher PP.

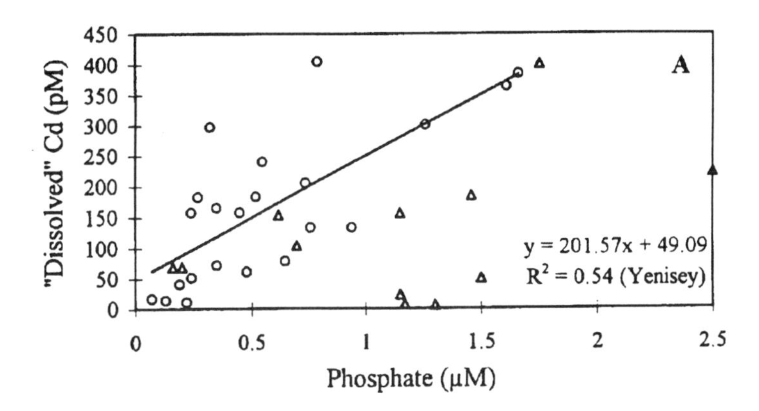

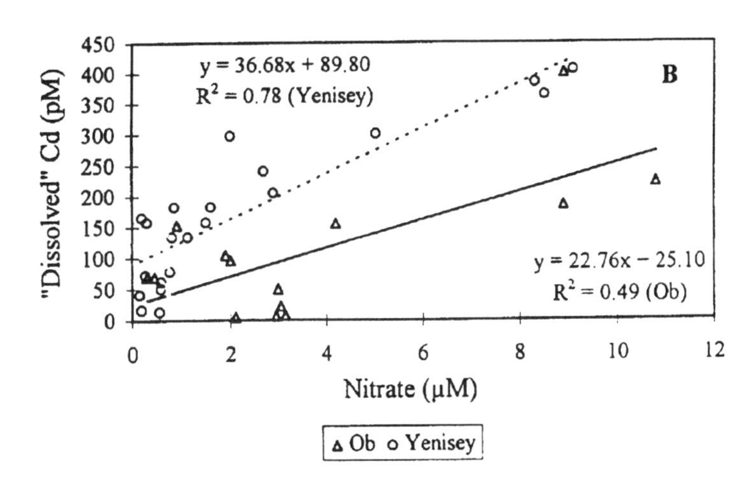

Figure 10. «Dissolved» cadmium as a function of phosphate (A) and nitrate (B) in the Ob and Yenisey estuaries [2].

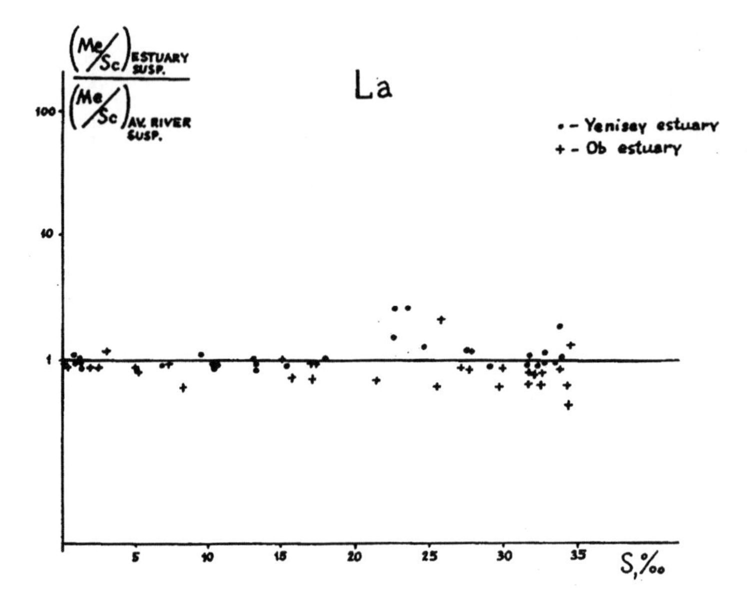

Figure 11. Sc-normalized La in estuarine suspended matter with salinity.

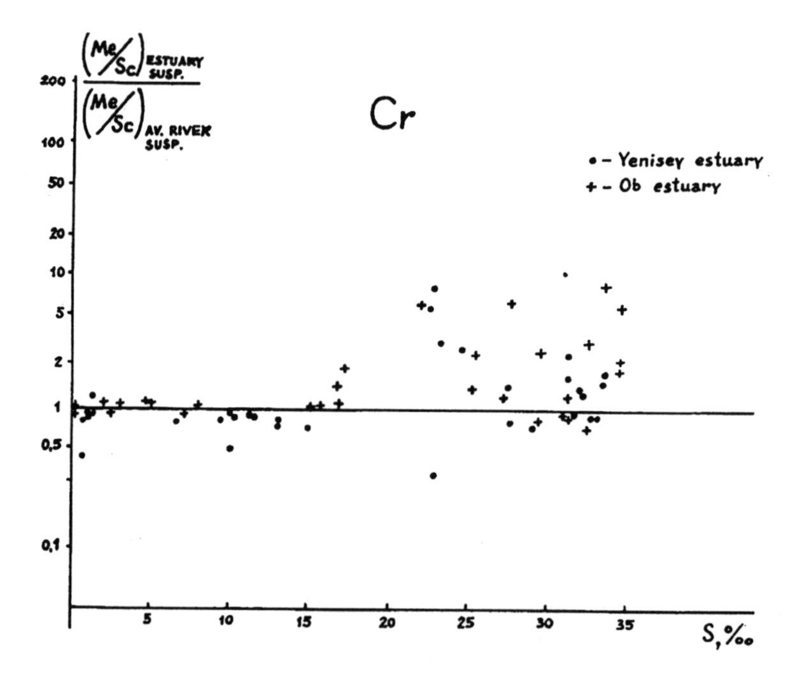

Figure 12. Sc-normalized Cr in estuarine suspended matter with salinity.

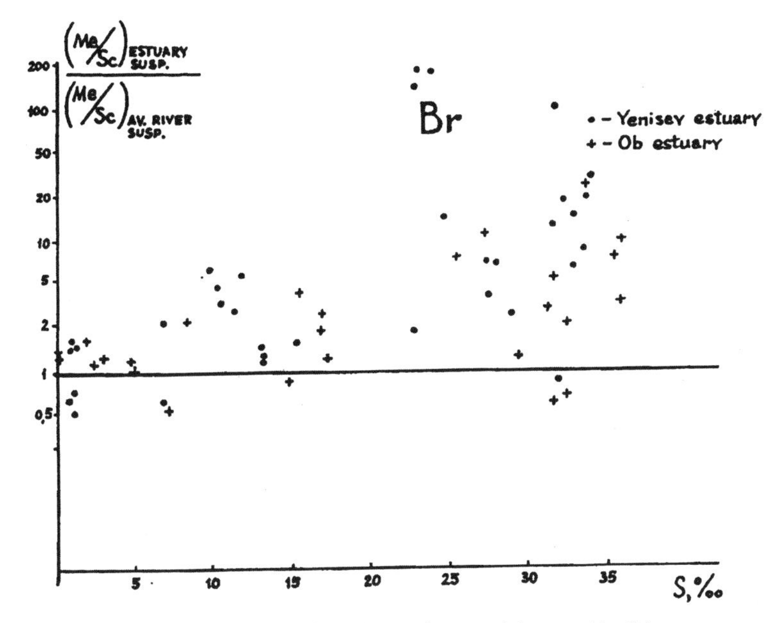

Figure 13. Sc-normalized Br in estuarine suspended matter with salinity.

3.6. BOTTOM SEDIMENTS

Based on a lithological study, four zones were selected in two estuaries with different deposit-structure types of the Upper Quaternary (Figure 1) [6]. The river deposits proper in the Ob and Yenisey occur in zone I. The Yenisey sediments were represented by silts and sands of various grain size. Plant detritus is very abundant in the deposits. The upper 2-3 cm are oxidized sediments, whereas in the Ob sediments the oxic layer and plant detritus are absent.

Zone II is the vast zone where river and sea-water mix. In this geochemical barrier zone the interactive biogeochemical processes take place, that lead to flocculation and sedimentation of huge masses of material. Muds prevail in this zone.

Zone III is interpreted as a zone of mainly bottom erosion and transit of sediments. In September 1993, westward currents dominated here with velocities up to 60 $cm \cdot s^{-1}$. Silts and sands occurred everywhere. The last zone IV, is a zone of intensive accumulation of fine clayey muds.

The distribution of major and trace elements in the surface of bottom sediments corresponds indentically to these zones [4]. Variations of some metals and C_{org} in the Ob estuarine bottom sediments are presented in Figure 14.

The content of elements in sediments depends largely on their grain-size composition. Only Si and Ca are concentrated in the coarse fraction, while the majority of metals (Fe, Cu, Pb, Ni and other) and C_{org} are concentrated in the finer fractions. The same picture is observed in the estuarine bottom sediments of the Yenisey [4].

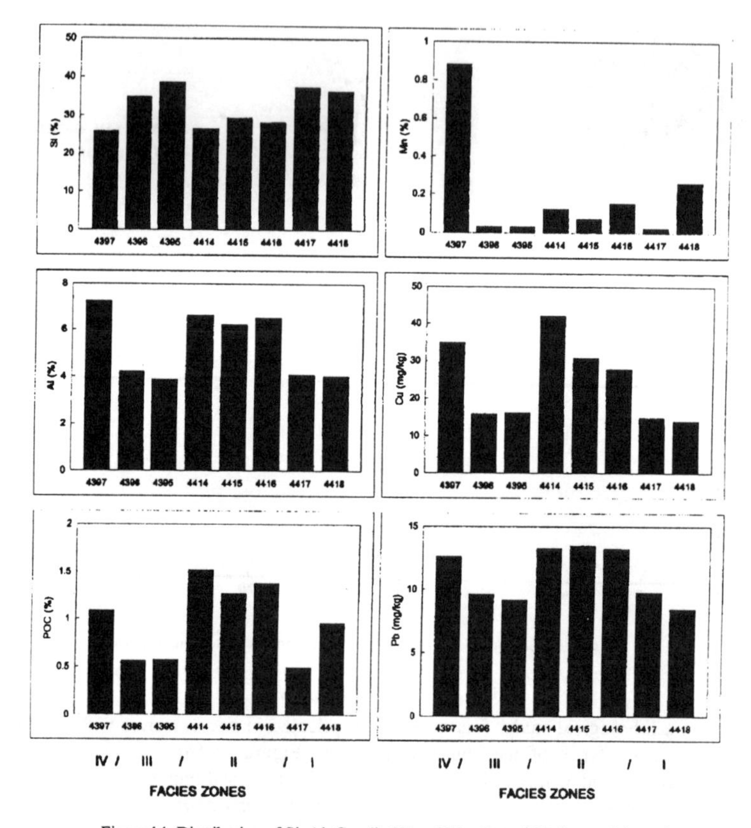

Figure 14. Distribution of Si, Al, $C_{org.}$ (in %) and Mn, Cu and Pb (in ppm) in surface bottom sediments of different facies zones in the Ob estuary and adjaceut Kara Sea [4].

To evaluate the interrelations between metals and organic matter a study was carried out to determine the forms of elements in bottom sediments by the method of successive chemical leaching [3]. The results show that for Fe, Co, Cr, Pb and Ni the refractory fraction contributes 70-100% of the total. Only for Cu and in lesser degree for Zn the associated fraction organic plays an important role.

The distribution of different elemental forms in sediments depends first of all on the lithological type of sediments. The available data do not show any significant difference in behaviour of the organic-associated fraction of metals in sediments of the two arctic estuaries. No appreciable difference was found in metal - C_{org} correlations between the sediments of the Ob and Yenisey estuaries (Table 5).

TABLE 5. Correlation coefficients between metals and organic carbon in surface bottom sediments in the Ob and Yenisey estuaries

Element	Ob estuary (n = 20)	Yenisey estuary (n = 29)
Fe	0.77	0.71
Co	0.45	0.66
Cr	0.64	0.63
Ni	0.76	0.69
Cu	0.83	0.73
Pb	0.82	0.53
Zn	0.71	0.71
Si	-0.83	-0.66

Thus, even if some differences in the geochemical behaviour of some metals in water and suspended matter were found depending on the level of primary production, in estuarine bottom sediments no visible differences were observed using these approaches.

4. Conclusion

Comparisons were made between the two large estuarine systems the Ob and Yenisey estuaries, to determine the role of biota in geochemical behaviour of some metals in water, particulate material and bottom sediments. The results show that biogenic factors play an important role in the water column (dissolved and suspended elements). This influence is more visible in the Yenisey estuary due to higher photosynthetic activity and primary production, even under lower concentrations of C_{org} and nutrients compared with the Ob estuary.

The severe climatic conditions of the Arctic did not prevent the biogeochemical processes of metal transformation in mixing zone in periods of ice cover. We could not find any difference in level of interaction of metals with organics in the bottom sediments of the two estuaries. Probably, these differences will be revealed in the study of interstitial waters.

6. Acknowledgments

This work was supported by the Russian Fund for Basic Research, Grant N 97-05-64576.

7. References

1. Burenkov, V.I. and Vasilkov, A.P. (1994) Influence of river discharge on spatial
2. distribution of hydrological characteristics in the Kara Sea, *Oceanology* 34, 652- 661, (in Russian).
3. Dai, V.-H. and Martin, J.-M. (1995) First data on the trace metal level and behaviour in two Arctic river/estuarine systems (Ob and Yenisey) and the adjacent Kara Sea (Russia), *Earth and Planetary Science Letters* 131, 127-141.
4. Demina, L.L. and Polytova, N.V. (1998) On the speciation of certain heavy metals in bottom sediments in the estuarine zone of the Ob and Yenisey rivers, *Marine Pollution Bulletin* (in press).
5. Gordeev, V.V., Paucot, H., Wollast, R. and Aibulatov, N.A. (1998) Distribution of chemical elements in grain-size spectrum of suspended matter and bottom sediments of the Ob and Yenisey estuaries, *Estuarine and Coastal Shelf Science* (in press).
6. Kravtsov, V.A., Gordeev, V.V. and Pashkina, V.I. (1994) Labile dissolved forms of heavy metals in the Kara Sea-waters, *Oceanology*, 34, 673-680 (in Russian).
7. Levitan, M.A., Khusid, T.A., Kuptsov, V.M., Politova, N.V. and Pavlova, G.A. (1994) Types of Upper Quaternary cross-sections in the Kara Sea, *Oceanology*, 34, 776-788 (in Russian).
8. Lisitzin, A.P. and Vinogradov, M.E. (1994) The international high-latitude expedition to the Kara Sea (the 49-th cruise of the R/V «Dmitry Mendeleev»), *Oceanology*, 34, 643-651 (in Russian).
9. Makkaveev, P.N. and Stunzhas, P.A. (1994) Hydrochemical characteristics of the Kara Sea based on results of the 49-th cruise of R/V «Dmitry Mendeleev», *Oceanology*, 34, 662-667 (in Russian).
10. Monin, A.S. and Lisitzin, A.P. (eds.) (1983) *Biogeochemistry of the Ocean*, Science Publisher, Moscow, 368 pp. (in Russian).
11. Strakhov, M.N. (1976) *The problems of geochemistry of recent oceanic lithogenesis*, Science Publisher, Moscow (in Russian).
12. Vedernikov, V.I., Demidov, A.B. and Sudbin, A.I. (1994) Primary production and chlorophyll in the Kara Sea in September, 1993, *Oceanology*, 34, 693-703 (in Russian).

13. Vernadsky, V.I. (1926) *Biosphere*, Scientific Chemical - Technological Publisher House, Leningrad (in Russian).
14. Vernadsky, V.I. (1934) *The essey of geochemistry*, Leningrad, (in Russian).

THE CONTINENTAL-OCEAN BOUNDARY AS A MARGINAL FILTER IN THE WORLD OCEANS

A.P.LISITZIN
P.P.Shirshov Institute of Oceanology Russian Academy of Scieces
36, Nakhimovsky, 117851 Moscow, Rrussia

Abstract

On global scales a marginal filter is found as a rather narrow belt where mixing of fresh fluvial water and saline seawater occurs. The two types of waters have quite different compositions of suspended matter, dissolved material and biota. The processes that take place in a marginal filter are unique. There is not merely mixing, but a combination of a variety of physical, chemical and biological transformations which may lead to up to 90-95% of suspended particles and 20-40% of dissolved substances of the fluvial water masses sedimenting.

Five successive stages of filter operation have been identified: three abiotic and two biotic. Of special significance is the formation of natural sorbents with flocculation and coagulation of dissolved (colloidal) forms of organic matter, iron and some other elements. Of sorbents argilleous (claylike) materials are of major importance (15.2 Gt); organic matter (OM) ranks second, (360 x 10^6 t as suspended matter and 80-90 10^6 t as floccules), while Fe ranks third (1.28 10^6 t of newly formed and 170 10^6 t of decrystallized oxyhydrates per year). Al and Mn play less significant roles.

A substantial part of dissolved elemental forms is captured (i.e. transferred into suspended particles) by phytoplankton and afterwards removed from the water column by filtering organisms, (zooplankton and benthos).

Thus the marginal filter governs the concentration of substances derived from land (including pollutants), and in certain areas (deposition centres), determines the geochemistry of many elements, including the carbon cycle and contaminants. A transect through the marginal filter shows that on the landward side the main form of most elements is as suspended material and seaward of the filter as dissolved material.

1. Introduction

The mixing zone of the two most widespread of Earth's water types, river water and seawater, up to now has been studied largely from the standpoints of morphology, hydrodynamics and hydrology. To date, proper recognition has not been given to the role of the mixing zone of fluvial water and seawater as a global-scale filter for both particulate and dissolved substances, (including contaminants), arriving from land. In recent years some aspects of this large and complex system have been explored, especially the relationships between living and dead matter coming from continents and from the ocean. Complex interactions occur in the marginal filter between salts and suspended particulate (including colloidal) systems, and with gases and biological systems.

On a global scale the marginal filter is a rather narrow belt (from hundreds of kilometres for large rivers to hundreds of metres for streams) where river- and sea-waters are mixed. This belt is extensive near the mouths of large rivers in humid climatic zones with substantial river flow and is comparably smaller in small rivers, and is not found in arid zones. At the filter chemical transformations of river water occur accompanied by removal of most the suspended material, colloids and a considerable proportion of soluble elements. This then is the principal accumulating area of, not only mineral substances, but also of organic carbon. Therefore study of the

J.S. Gray et al. (eds.), Biogeochemical Cycling and Sediment Ecology, 69–103.

marginal filter is of particular significance in understanding the global carbon cycle. According to current beliefs, annual sedimentation amounts here to 360 $.10^6$ t of suspended organic C (out of a total input by rivers to the ocean of $392*10^6$ t) and 80-90 10^6 t of dissolved organic carbon [1]. According to estimates, up to15% of anthropogenic carbon dioxide is accumulated here, i.e. marginal filters affect the climate through the carbon dioxide cycle.

Marginal filters are contemporary parts of avalanche sedimentation provinces (the first global level, [2] [3] [4] which change with time due, first of all, to variations in level of the World Ocean. Nowadays the level is rising at an average rate of 1 to 2 mm a year. Marginal filters reached the present position about six thousand years ago, when the present-day level of ocean had been established. During glaciations, (the last of which occured 18 thousand years ago) the position of marginal filters was drastically altered. The level dropped by 120 to 140 m and has caused displacement of river mouth and estuaries towards the outer portion of a shelf and often, onto the continental slope. Loose sedimentary material accumulated when sea levels were high was deposited from marginal filters and under the effect of waves and currents moved to the foot of the continental slope (the second level of avalanche sedimentation). At active borders, material was moved to the bottoms of deep-sea grooves (the third global level). So, one should search here for residues of marginal filters of interglacial and more ancient periods.

It is important to note that simultaneously with effects of various sorbents, biological processing and rapid removal of even the smallest suspended particulate and dissolved pollutants, substantial dilution occurs in bottom sediments by the enormous amounts of inert mineral matter already present.

Thus the processes taking place in these outlying areas of continents are of global importance. The bulk of the planet's sedimentary matter is concentrated here, (about 92-93% of substances being discharged from rivers to ocean), and among these substances the majority of anthropogenic contaminants. Up to 90% of the total flux of riverine organic matter is accumulated, and oil, gas, coal and combustible shale deposits have been arisen in the past. Hence, this is the main area for the accumulation of both matter and energy and is extremely important for understanding the global exchange and cycles of virtually all elements, but especially carbon. Thus more studies of the marginal filters of oceans are urgently needed.

The processes peculiar to the outer areas of continents are best seen in estuaries and riverine deltas. The key processes are not simply the mixing of river- and sea-water (EMR - River and EMS - Sea), but also precipitation of sedimentary matter, and chemical transformations. Of these chemical reactions the key processes are sorption and desorption, flocculation, coagulation, co-sedimentation and biological transfer of elements and contaminants pollutants in dissolved form and suspended particulate matter. Furthermore, removal of products by by filter-feeding organisms is highly important here. The process begins with the gravitational (mechanical) and physico-chemical (colloidal) differentiation and is completed by biological processes such as extraction of dissolved forms, their transfer by phytoplankton into suspended particulate matter and, at last, removal of suspended matter through biofiltration by zooplankton.

Based on data derived from numerous investigations, particularly in the Arctic, Amazon and rivers of the Baltic and Black Sea basins, the working of the marginal filter has been deduced for estuaries and deltas woldwide. The magnitude of the processes has been determined from the end members of the mixing process: soil substances, weathering of the crust in catchment areas, the material being delivered by rivers (suspended particles, colloids, solutions, organism) and substances in seawater (suspended particles, colloids, solutions, organisms).

The figures for the masses involved in the processes taking place within marginal filters are immense. Rivers annually deliver 35550km^3 of water, 18.5 Gt y^{-1} of suspended particulate matter and 3.6 Gt y^{-1} of dissolved elements. Among the latter, as far as is known today, colloids are of especial significance, i.e. organic matter and Fe, Mn and Al. The rates of the processes occuring there are also enormous. Suspended particulate matter content reaches 1 to 10 g l^{-1} or more (compared to 0.1 mg l^{-1} for the pelagial), vertical fluxes of particulate matter (more than 1000 mg $m^{-2}d^{-1}$ in comparison with 20 to 30 mg $m^{-2}d^{-1}$ for open seas), sedimentation rates exceeding 100 mm per thousand of years (in some locations more than 2000 mm $1000y^{-1}$), against the average of 1 to 10 mm$1000y^{-1}$), and absolute masses of substance (in excess of 5 g $cm^{-2}1000y^{-1}$). The marginal filter is thus the largest sedimentary, sorption and filtration systemsof the Earth. The main objective of this paper is to show mechanisms of matter transformation in these decisively important regions.

2. Materials and Methods

Geological and geochemical investigations of the river-sea exchanges began more than 20 years ago at the Physico-Chemical Research Laboratory in the Institute of Oceanology, the Academy of Sciences of Russia. Initially, river-mouths were studied and concentrated on those of the basins of the Black, Baltic and Caspian Seas, [5] [6]. Later investigations began into seas of the western Arctic and Subarctic [1] and of the Far East, (by the Pacific Institute of Oceanology, [7] [8]. These studies were done as a single programme using common methods that allow comparisons to be made. Research was performed also on both particulate matter and bottom sediments and on elements in suspended matter and in the pelagial of the World Oceans, [7] [9] [10] [11] [12].

Comparisons of the data obtained lead to the formulation of a concept of three vertical levels of "avalanche" sedimentation in the World Ocean. It is in these areas that the bulk of the planet's sedimentary matter are found, [2] [13].

Based on an analysis of the large database for the oceanic pelagial and on data on fluvial flux, the important conclusion has been drawn that 93% of sedimentary riverine matter does not reach depths exceeding 3000m. The material is deposited in peripheral oceanic areas such as river mouths and shelves, [10] [14). Subsequent investigations in estuaries of the largest rivers have confirmed that deposition is in narrow belts, (Table 1). In addition important new data has been obained for understanding the proceses occuring there.

TABLE 1. Loss of sedimentary matter in marginal filters of rivers

Area	(% of input to a river mouth)
World Ocean pelagial environment	93
Estuaries of the rivers	
Amazon	95
Mississippi	90
Zair	95
Saint Lawrence	93
Kura (Caucasus)	90-95

The conclusion, which is fundamental for lithology and geochemistry, is that deposition near their mouths of 90-95% of sedimentary matter delivered by rivers, is based on two independent data sets, pelagic sediment deposit formation and estuarine deposition.

Data from the mouth of the Amazon River were obtained by the RV "Professor Shtokman" cruise in 1983 [15], (Fig 1) and from other expeditions

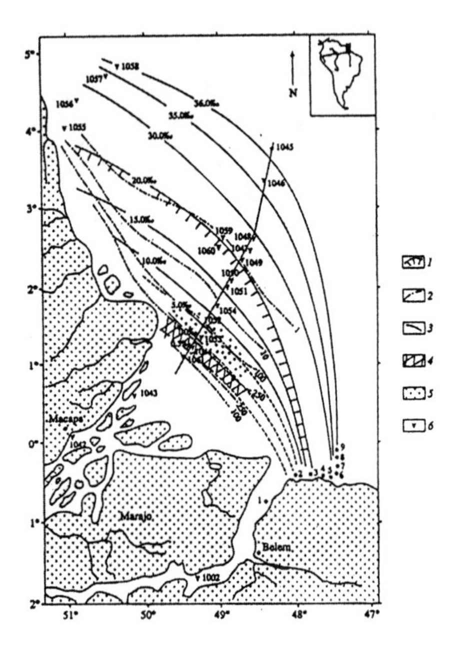

Fig.1. Marginal filter of the Amazon River (equatorial zone) 1) outer boundary of marginal filter; 2) isolines of suspended particle content on surface (mg/l); 3)salinity; 4) maximum concentrations of suspended particles (more than 250 mg/l) - "silt plug"; 5) maximum plankton contents - "biological plug"; 6) stations of the expedition on the research vessel "Professor Shtokman" in 1983 [15]. Data from subsequent expeditions have been taken into account.

to the mouths of the Congo, Cunene and Orange rivers by the RV "Vitiaz" in 1990. A new research era has been started with organization of the French-Russian Scientific Programme on Arctic and Siberian Aquatorium (SPASIBA), in the mauth of the Lena River, one of the largest arctic rivers, and in the Laptev Sea. This expedition undertook research in summers of 1989 and 1991 as part of the Joint Ocean Flux Study (JGOFS).

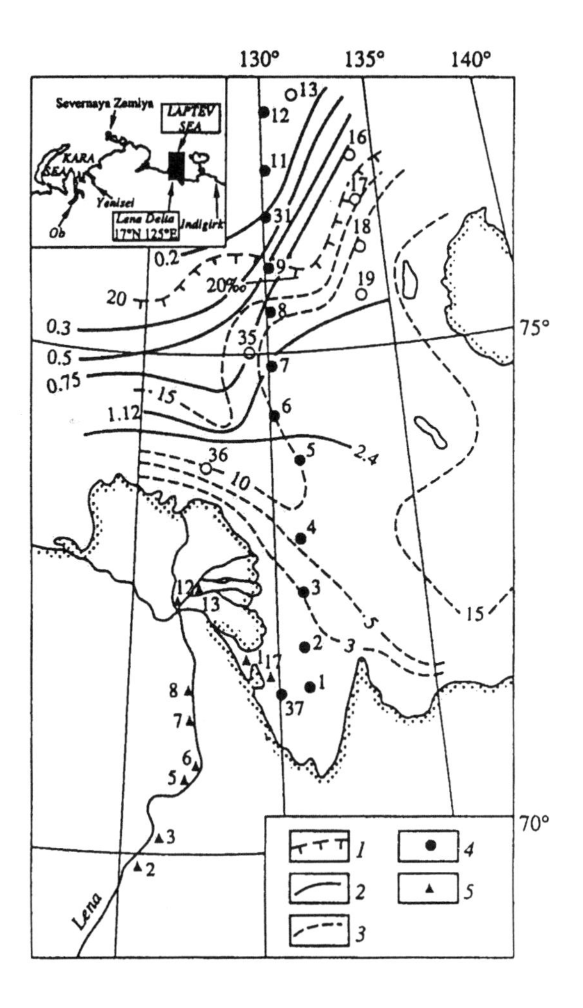

Fig.2. Marginal filter of the Lena River (Arctic) 1)outer boundary of marginal filter; 2) isolines of suspended particle content on surface (mg/l) [16]; 4) stations on the SPASIBA-2 expedition transect in the Laptev Sea; 5) the same in the lower course and delta of the Lena River.

In 1993 the large international expedition to the Kara Sea was conducted by the RV "Dmitrii Mendeleev" (SPASIBA-3) in which 92 researchers took part (Fig.3). The estuaries of two large Siberian rivers the Ob and Yenisey were studied. Unlike the earlier work, the prime focus was given to studying fluxes of suspended particulate material by means of numerous sediment traps along river-sea transects, and bythe use of isotopic and biological methods. In particular the biological parts of the filter were emphasised (bacterio-, phyto- and zooplankton, and benthos). Special studies were done on organic matter, forms of elements and diagenesis processes, including the carbonate system, gases, etc.. This article is based on the data derived from these expeditions and takes into account material obtained by other researchers.

In order to illustrate the principal concepts we base our analyses on two zones with marked differences in environmental conditions, a cold part of the humid zone, (the Lena, Yenisey and Ob rivers in Siberia) and the tropical zone (Amazon). In both these study areas a single programme was followed with common methods and done by the same teams of scientists.

3. Results from the Amazon and Arctic Rivers

It is well-known that climatic conditions within a river catchment area together with geological conditions governs the key parameters of fluvial flow. I now compare results of investigations done on the Amazon, as well as on the three large rivers flowing in cold areas of the temperate zone (Lena, Yenisey and Ob in Siberia). Dependence of the amounts of suspended particulate matter on salinity is clearly seen on maps of suspended material and salinity distributions for the upper water layer of the Amazon (Fig.1), Lena (Fig.2), Yenisey and Ob (Fig.3) rivers.

High suspended matter content, (100 to 500 mg l^{-1}), was found in the Amazon River. The particulate matter content of the surface layers decreased rapidly with distance from the river mouth and with higher salinity. When the salinity was 20‰ the suspended particle content was only 1 mg l^{-1} and with a further increase of alinity the value usually found in the ocean, 0.1 mg l^{-1} was found.

The second specific feature observed was the appearance of high suspended particulate matter concentration not in the river mouth, but at some distance away (the "silt plug") where concentration of 250 mg l^{-1} or more are found. These values are two to three times higher than in the river. The same zone has maximum sedimentation rates of 20 000-100 000 B. (B = Bubnoff units where 1 Bubnoff unit = 1 mm/1000 years). In the pre-mouth offshore area of the Lena River isolines of both salinity and suspended matter content are similar. However, initial particulate matter content of the river water is much lower than in the Amazon River, (three - to five-fold), so particulate matter content of 1 mg l^{-1} is found at15‰.

The estuaries of the Yenisey and Ob rivers, as seen from Fig.3 merge into a large common estuary stretching almost to Novaya Zemlya Island. Salinity of the upper water layer varies substantially from year to year.

In Fig.3 the average salinity over many years is plotted. (It should be noted that Fig. 3 differs from the salinity distribution pattern in the year and season of our observations, September 1993). Nevertheless, the suspended matter content of 1 mg l^{-1} is close to the 20‰ isohaline. So this isoline is a convenient first approximation to the outer boundary of the marginal filter. The vertical transect is shown in Fig. 4.

As seen from Fig.4 in the area where suspended particulate matter content decreases markedly, (some 200 km offshore), salinity rises to 20-30‰. A "silt plug" 70 to 100 km wide is very clearly seen. Profiles of the distribution of both dissolved and particulate organic substances (Fig.4 c, d) show a "colloidal plug", resulting from coagulation of colloids and their transition to suspended particles. The dissolved organic matter content maximum gradually gives way seaward to a maximum of particulate organic mass. The amounts are approximately five times those in the river mouth. The chlorophyll distribution pattern shows a phytoplankton bloom, situated offshore from the "silt and colloidal plugs". This suggests that the position of the bloom is governed by the location of the silt and colloidal plugs which produce a sharp decrease in water turbidity.

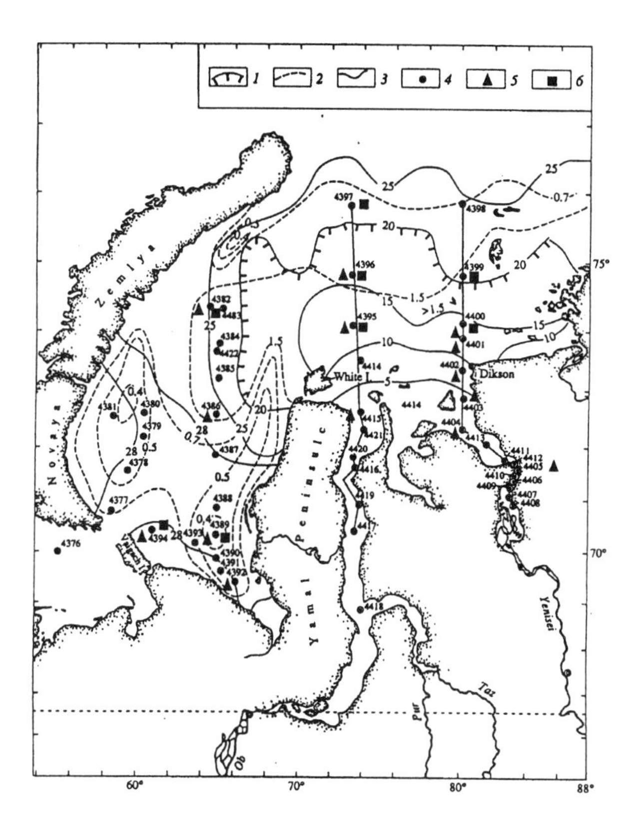

Fig.3. Marginal filters of the Yenisey and Ob rivers (Arctica) 1) outer boundaries of marginal filters; 2) isolines of suspended particle contents according to data of the SPASIBA-3 expedition [17]; 3) average multiannual salinity [16] 4) stations with studying suspended particulate matter on surface and vertical profile, by means of hydrooptical investigations, STD as well as the set of biological and chemical investigations [17]; 5) stations where suspended matter fluxes are measured through ^{234}Th [18].

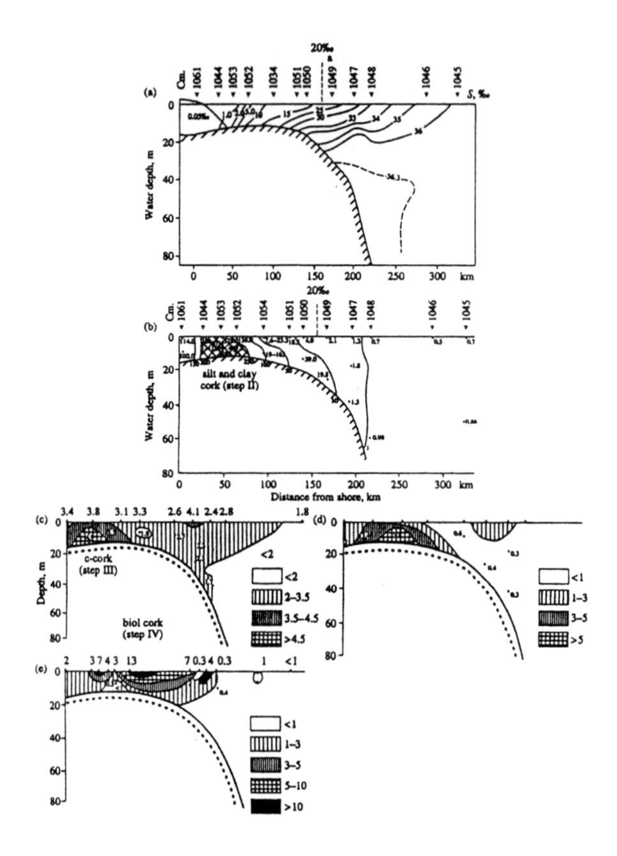

Fig. 4. Transect across marginal filter of the Amazon River; see location of the cross section in Fig.1 [15]. a - salinity distribution (level of high water); b - distribution of particulate matter according to data of membrane ultrafiltration (mg l^{-1}) and hydrooptical research. "Silt plug" is identified, with suspended particle content higher than in the river and adjacent part of the ocean (more than 250 mg l^{-1}) : losses of mineral particulate matter (step I+II) - 95%; c - distribution of dissolved organic carbon (DOC) (mg l^{-1}): losses in the filter, up to 80% (step III of the filter); d - distribution of particulate organic carbon (POC) (mg l^{-1}): losses in the filter, up to 70-80%; e - distribution of chlorophyll as an indicator of phytoplankton part of filter (mg l^{-1}). Step IV is identified.

In rivers of the Arctic (Fig. 5), a surface fresh-water wedge over a lower saline wedge are shown. Initial river-water turbidity is much lower than in the Amazon River. Therefore, no "silt plug" formation was observed in this season, (September). The filter appears to be more extended than that of the Amazon River. In this Arctic the 20‰ isohaline is 550 km from the river mouth, compared with 170 km for the Amazon River. Nonetheless, the 20‰ salinity front is closely related to the particulate matter concentration isolines and seaward values typical for the open ocean (0.1 mg l^{-1}) are found. During spring-summer freshets one can expect the appearance of an even more dramatic difference in the "silt plug".

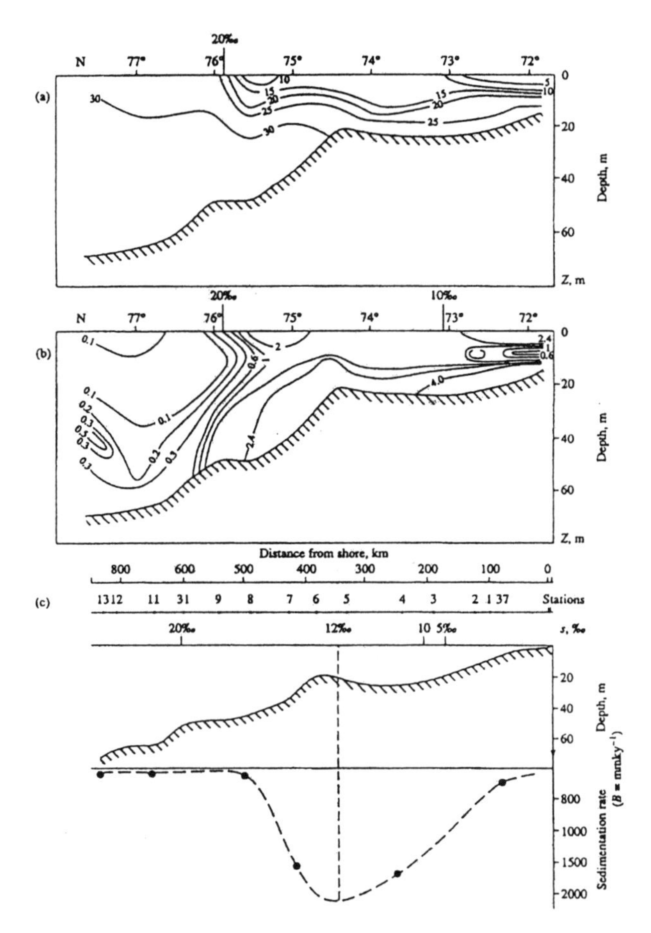

Fig.5. Transects across marginal filter of the Lena River (see location in Fig.2) a - salinity distribution ‰ ; b - particulate matter distribution (mg l^{-1}); outer boundary of the filter conforms to 20% salinity [16]; c - embodiment of marginal filter in bottom sediments: in the Bubnov units (1B = 1 mm 1000 y^{-1}). Depository centre of the filter corresponds to salinity of some 12%, is situated 350 km from shoreline. The marginal filter basin is abound 400 km long [18]

Fig.5 c shows a comparison of the quantitative distribution of suspended particulate matter in the Lena River estuary with the amounts of particulate matter deposited on the bottom. The area of the "avalanche-type" sedimentation, with rates of up to 2000 B are shown. The location of the marginal filter is indicated in the bottom sediments by an enormous (some 400 km long) sedimentary lens of the sedimentary-rock basin with a deposition centre situated 350 km offshore from the river mouth. The deposition centre is spatially related to the 12‰ isohaline location. This is based on the sediment thickness. The maximum sedimentation from the marginal filter increases from the 3-5‰ isohaline to a maximum at the 12‰ (2000 B), and thereafter declines again to values of 10-20 B, quite usual for a shelf, (with salinities of 15-20‰). Yet sedimentation rates measured for the marginal filter of the Lena River are not exceptional. In the Gironde estuary using two independent methods (historical data and ^{210}Pb assay) sedimentation rates were found to be within the range of 2.94-2.29 cm y^{-1}, i.e. 22900-29400 B [19].

The interrelationship between particulate matter and salinity is clearly seen in transects across the estuaries of the riversYenisey (Fig.6 & 8). The transition to values typical for the open sea for these two rivers is shown by the 20‰ isohaline. It is remarkable that in the same area of the estuary the dissolved organic matter content is an order of magnitiude lower (Fig.6 c), due to the transfomation from solutions to floccules, "estuarine snow".

In this process heavy chemical elements in solutions are removed from the water column. Fig. 7 shows the content of Cu and Zn, determined polarographically, drop to beteween a tenth and a fifth of values just beyond the 20‰ isohaline.

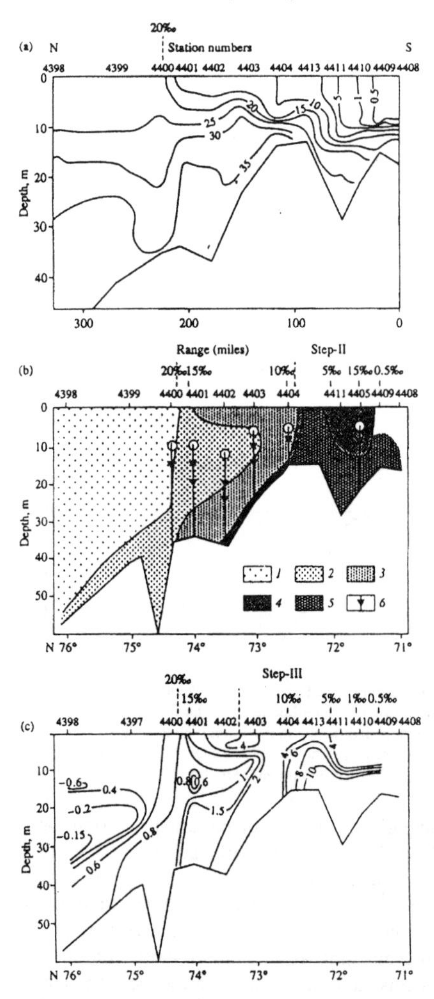

Fig.6 Transects across marginal filter of the Yenisey River (see location in Fig.3)
a) - salinity distribution [8]; B) - distribution of suspended particles (mg/l) and sedimentary matter fluxes, according to data of analyzing samples from sediment traps (mg $m^{-2} d^{-1}$). Suspended particles : 1- <0.5; 2- 0.5-1;3-1-2.5; 4-2.5-5, 5->5. c) - distribution of dissolved organic carbon (DOC) ("yellow substance"); one can explicitly see the area of removal from solutions to floccules (step IV of the marginal filter). The data hydrooptical observations [18,20].

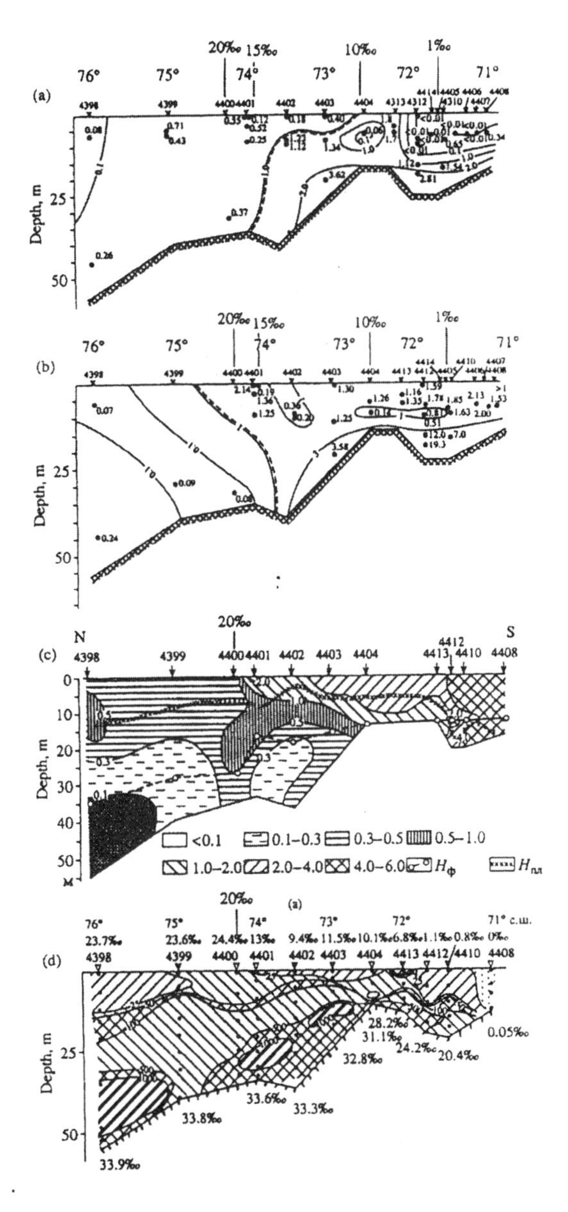

Fig7. Transect across marginal filter of the Yenisey River (method of inverse voltamperometry).a) - distribution of ions of dissolved copper forms (mg l^{-1}) [21]; b) - distribution of ions of dissolved zinc forms. The data of inverse voltammetering (µg l^{-1}) [29]. On A and B parts of Fig.7, sharp decrease in content with salinity rise up to 15-20‰ is seen. c) - distribution of "a" chlorophyll as an indicator of biotic part of the filter (step IV, the phytoplankton pump (mg l^{-1}) [22]. Chlorophyll content is shown in the Figure. Lower border of a photosynthesis layer; level of maximum density gradient. d) - distribution of zooplankton biomass (mesozooplankton, mg/m^3) - an indicator of zooplankton biological pump work. Dots indicate horizons for taking samples of 150 l with a bathometer; values of surface and near-bottom salinity

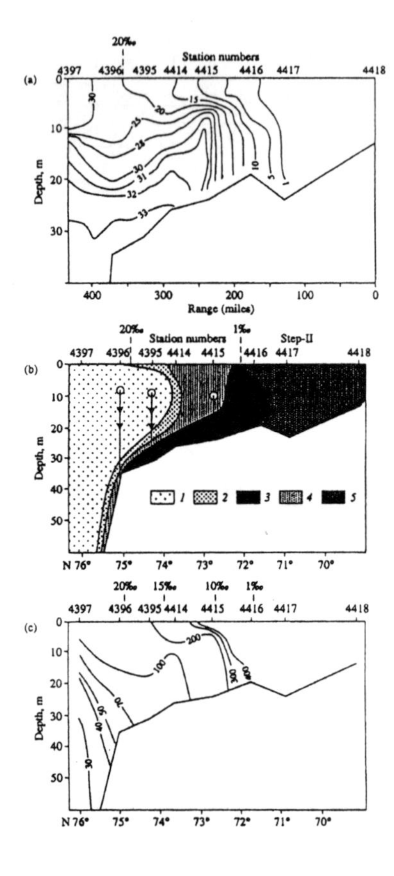

Fig.8. Transects across marginal filter of the Ob River (see location in Fig.3) a) - salinity distribution [23] b) - distribution of suspended particles (mg l^{-1}) and sedimentary matter fluxes according to data from sediment traps (mg $^{m-2}$ d^{-1}); : 1- <0.5; 2- 0.5-1;3-1-2.5; 4-2.5-5, 5->5 1) for total suspended matter; 2) for Corg [17]. Horizons are shown used for sampling with traps (triangles); "Silt plug" is seen at step I+II of the filter, settling out of the suspended matter bulk within the 2-20‰ salinity range (90-93% of riverine suspended particulate matter). One can see settling out of the main DOC part, with transition into floccules at step III of the filter. c) distribution of dissolved organic carbon. The date of hydroioptical observations. Transfer into floculi at step II (10-15 ‰).

The first stages of the filter are dependent on biological conditions. Phytoplankton biomass (here chlorophyll content) on transencts across the estuaries of the Yenisey and Ob rivers (Fig.7 d, 9b) declines with nutrient concentrations. The distribution of zooplankton is mostly governed by the distribution of their food, (i.e. phytoplankton and their distribution patterns are spatially concordant though varying seasonally fig.7 d, 9 c).

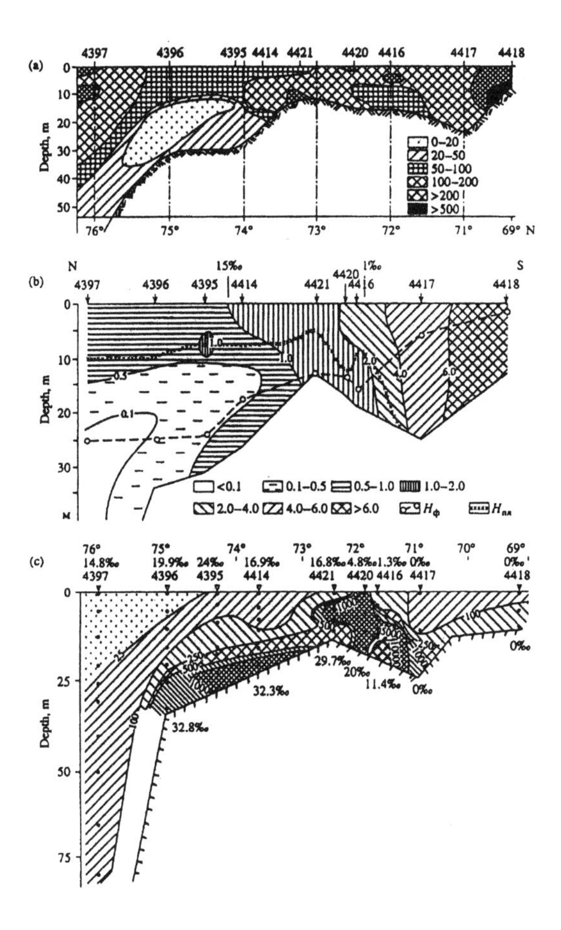

Fig.9. Transects across marginal filter of the Ob River (work of biological parts of the filter). a) - distribution of bacteria in water thick (10^3 cells in 1 ml) (data by Mitskevich, oral report); b) - distribution of "a" chlorophyll (mg/m^3). Unlike the Amazon River, a "biological plug" goes up to the Ob River mouth: water turbidity impedes phytoplankton development only at station 4418. Pertinent symbols are the same as in Fig.7 C [22] ; c) - zooplankton biomass (mesoplankton, mg/m^3). At 73° N latitude, replacement of marine zooplankton by saltish-water one occurs [23].

Fig.9 shows that bacterioplankton (number of bacteria ml^{-1} water) are abundant in all parts of the filter. Microscopic examination shows, however, that at the colloidal stage bacteria are rare in the water, but are concentrated in floccules, forming dense patches around accumulations of organic matter newly generated from solution (colloid) (Fig.10).

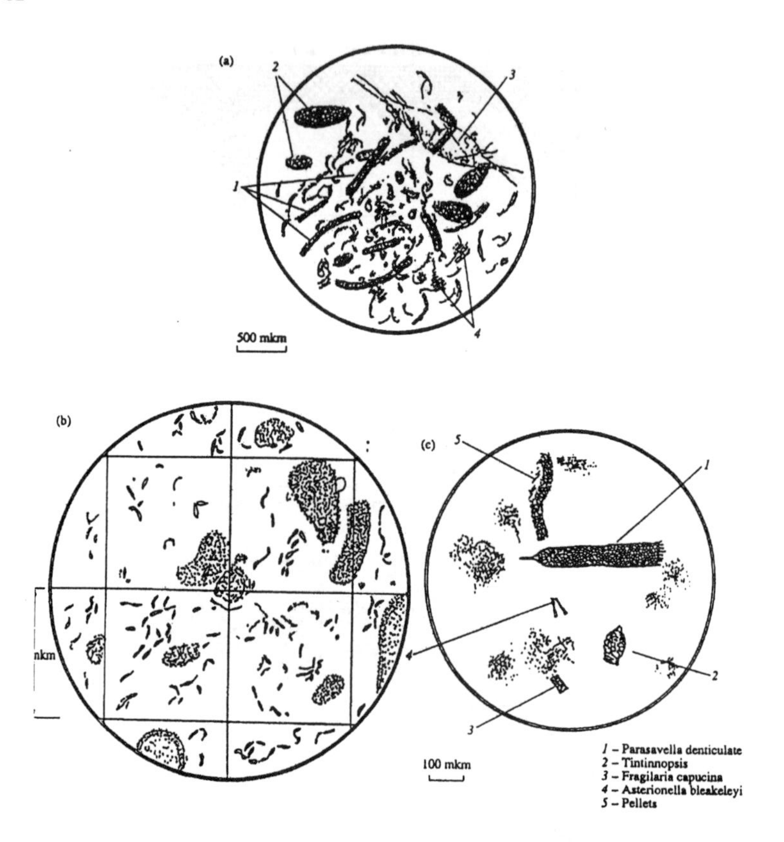

Fig.10. Sketches of particulate matter from sediment traps of the Kara Sea and the Yenisey River mouth. a) - composition of sedimentary matter taken from traps at station 4382, horizon of 60 m: 1) pellets of *Calanus* ; 2) pellets of *Euphausiacea*; 3) larva skin of 2Euphosiacea 0; 4) floccules and flakes. Magnification is indicated in the Figure. b) - the same station, horizon of 100 m. Bacterial cells and floccules. Magnification is in the Figure. c) - the Yenisey River mouth, station 4402, horizon of 20 m. Symbols are in the Figure [17].

Both the biological and abiotic portions of marginal filters of the Siberian rivers vary substantially within a year. The maximum influx of all components occurs during spring-summer floods (freshets), soon after the ice melts in the river and adjacent sea. A peak of freshets is followed by a phytoplankton bloom, (Fig.11) and later, a zooplankton bloom whose grazing activity together with decreasing influx of nutrients, leads to a 10- to 20-fold reduction in phytoplankton biomass.

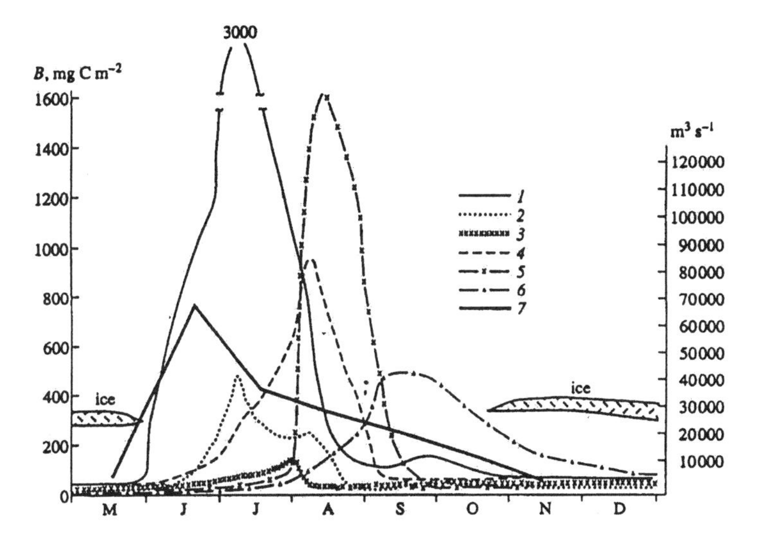

Fig. 11. Seasonal work dynamics of the Kara Sea ecosystem and marginal filter calculated by the point model of the plankton community main elements[24, 25]. 1) phytoplankton; 2) bacteria; 3) protozoans; 4) fine (small) mesoplankton; 5) coarse (large) euriphages; 6) big predators; 7) liquid flow, km^3/month. Intensive operation of the biological part of the Kara Sea marginal filter lasts 4 months, and during the remainder 8 months in a year the biological part of the filter does not virtually work. 1 - ice, 2 - km^3/month.

Deprived of food, zooplankton dies and its biomass sharply decreases. Part of this decrease is due to predation from larger organisms. Thus the phytoplanktonic part of the marginal filter continues here on large scales from early June until late August, with a maximum in July. Zooplankton grazing lasts as little as two months, with maximum in August. From late October when freeze-up starts to the beginning of the freshets and ice break-up late in May, the biological part of the filter is severely reduced, although does not disappear entirely. Filtering organisms such as molluscs and barnacles operate throughout this period.

In rivers of the tropical and equatorial zones where seasonal variations are minimal, the filter operates throughout the year. The processes can be described from the Amazon River.

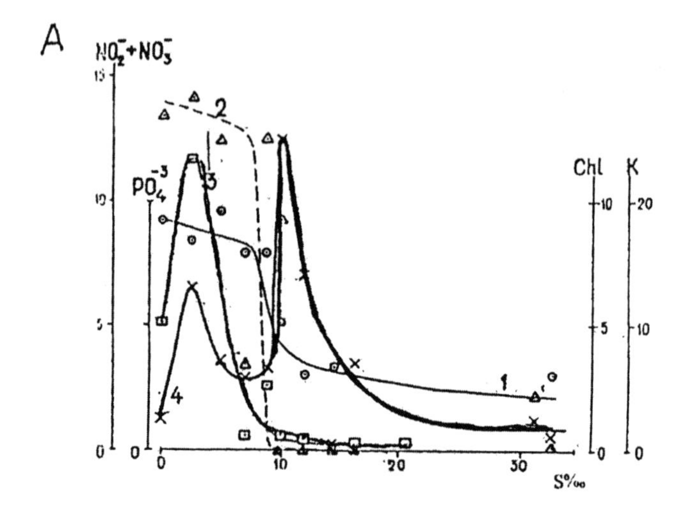

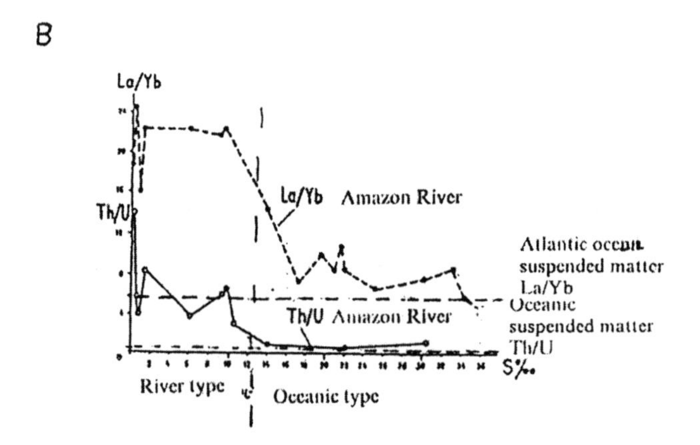

Fig.12. Marginal filter of the Amazon River. a) - main stages of the filter operation as related to salinity [15]. Biotic part of the filter: 1) decrease in phosphates contents, and 2) sums of nitrites and nitrates in connection with phytoplankton development; 3) chlorophyll content (maximum conforms to the "phytoplankton plug"). Abiotic part of the filter: 4) particulate matter content (maximum conforms to the "silt plug"), by optical characteristics. Surface layer, the high-water season (May 1983). b) - two types of particulate matter (fluvial and marine) in the Amazon River mouth, according to La/Yb and Th/U ratios.

Fig.12 shows low nutrient levels in the water (phosphates, nitrates and nitrites) since they are consumed by phytoplankton, together with fixation of a part of dissolved phosphates by colloidal iron at the colloidal (III) step. This is confirmed by the P/Al ratio and also by a drastic drop in the content of dissolved forms of nickel, copper and aluminium at a salinity of 10‰ [15]. When the salinity is about 10‰ in the Amazon mixing zone, losses of dissolved organic matter are maximal [1] caused by flocculation. Based on data related from these and other rivers (of the Black and Baltic seas, Far East seas, Asian rivers), a general model has been constructed for the marginal filter at the continent-ocean interface.

4. The Marginal Filter Model

The marginal filter consists of two main parts principally different in their functioning: 1) abiotic situated closer to a river mouth; and 2) biotic , located offshore, seaward. (Fig.13).

4. 1. ABIOTIC PARTS OF THE MARGINAL FILTER.

Here the riverine terrigenic material is of major importance while the role of biota in rivers (plankton and benthos) is insignificant. The abiotic (terrigenic) parts of a filter include three sequential steps:

coarse-grained
fine-dispersive and
physico-chemical (of newly formed sorbents).

a) STEP I. Coarse-grained part of a filter

This occurs at the mouth and adjacent part of the estuary. The sediment consists of particles larger than 0.01 mm (sands and aleurites) consisting of mineral grain fragments as well as terrestrial plant remains (peat, wood, ets.). The mineral grains generally have a coat of organic matter and Fe oxyhydrates which have originated from soils, mires, fluvial water and partly arise due to coagulation of organic substances and iron. The organic matter coat imparts a negative charge and sorption properties to the coare-dispersion material. Thus the first step of the filter has inert grains with sorption coats that retain some dissolved forms of elements and transfers them from solutions into suspended matter and thereafter to bottom sediments [26] [27] [28].

At the mouth of the river the coarse material sediments to the sea bed. The amount is governed by the extent of the river mouth and in the Arctic seas it reaches its maximal development, sometimes more than 50% of the total particulate fluvial discharge is deposited, whereas in seas of the temperate zone and tropical regions only 10-20% is deposited.

b) STEP II. The fine-grained part of a filter (step II)

The high particulate matter concentration in river mouths induces at least three processes essential to geochemistry, biology and sedimentation processes:

1.With high concentrations of particulate matter, as was noted above, water transparency decreases restricting phytoplankton development (evenwhere nutrient levels are high). Phytoplankton development begins when luminous flux of some 0.1% of solar radiation. As a first approximation corresponds to transparency of about 1 m as determined with the Secchi disk (H) or to particulate matter content of water of 4-5 mg/l ($E = 5\text{-}6\ m^{-1}$). Phytoplankton development is also reduced in turbid waters ($H = 2\text{-}4^{-1}$). Turbid river estuaries have low phytoplankton development. In river mouths maximal phytoplankton development (and biotic particulate matter formation) occurs wherever high transparency is coupled with adequate nutrients supplies, leading to a "biological plug". Clearly, the "biological plug" usually occurs offshore, seaward of the fine-grained abiotic part of the filter (Fig. 1,4,6).

2. In areas with high particulate matter content the amount of dissolved elements extracted from the water by sorbents (Table 2), never depletes the whole reservoir. Therefore, the maximum of sorbent saturation often does not occur together with the maximum of suspended material concentration and sedimentation but occurs further seaward.

TABLE 2. Average rates of sediment accumulation in mouths of the rivers (mm 1000 y^{-1})

Amazon	more than 100
Mississippi	10 000-1 000 000 (1 m/yr)
Parana (S.America)	10 000
Menam (Chao Phaya, Gulf of Siam)	30 000
Rhone	5 000-6 000
Potomac	16 000- 18 000
Nile	160-320 and more than 320
Niger	200
Lena	2 000

3. The third consequence is that with a higher particulate matter content water takes on properties of a heavy liquid, and a dense suspension spreads over the sea-floor rather than at the surface. Depending on the density of the underlying saline waters and of the material in suspension, these currents can flow either over the seafloor (dense flow) or along a density layer and isohalines (light flow). These flows begin with a particulate matter content of 25-2500 mg l^{-1} (low density turbidity currents) and 2-250 .10^{3} mg l^{-1}, (high density turbidity currents) [30].

As is seen in Table 3, particulate matter content for a number of rivers substantially surpasses this limit. It also follows from Table 3 that the greatest concentrations of particulate matter in river waters are found in rivers of both steppe and tropical zones (the Amazon and others), as well as in mountain rivers. This suggests that climatic and weathering processes are the key controlling factors. Rivers of the taiga and tundra (in Siberia) have low suspended particulate matter content.

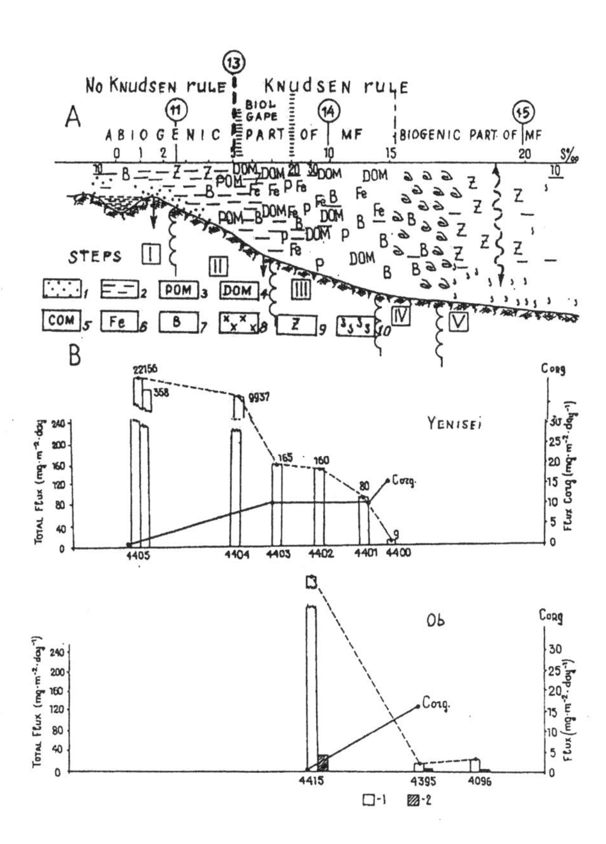

Fig.13. Model of marginal filter and quantitative parameters of its work (fluxes). a)- filter model. Abiotic parts of the filter (steps I and II): 1) settling out of coarse fractions in suspended particulate matter (sands and silt aleurites), step I; 2) settling out of the fine (pelitic) fractions of suspended particles, step II. Settling out of the colloidal part. Physico-chemical part of the filter, step III (colloidal pump): 3) suspended organic matter (SOM); 4) dissolved organic matter (DOM); 5) colloidal organic matter (COM); 6) ferric oxyhydrates. Biotic parts of the filter: 7) spread of bacterioplankton; 8) area of the maximum phytoplankton development (phytoplankton biopump), step IV; 9) area of development of zooplankton filter organisms (zooplankton biopump), step V - final filteration; 10) flux of pellets, i.e. transfer of fine suspended particles by zooplankton (selectiveless filtration) into bottom sediments. A zigzagged curve indicates daily vertical migrations of zooplankton, displacement of the filtration device in a vertical direction. Critical point of the marginal filter (in circles): 11) turbidity by the Secchi disk is less than 1 m, i.e. work start of biotic part of the filter; 12) critical salinity: drastic decrease in taxonomic diversity of organisms (salinity of 5-8‰); 13) beginning of the Knudsen law action (>5‰) see Fig 14; 14) start of massive coagulation and formation of floccules of colloidal (dissolved) organic matter and ferric oxyhydrates (some 10‰); 15) outer boundary of the marginal filter (some 20‰). b) - sedimentary matter fluxes in marginal filters as exemplified by the Ob and Yenisey rivers (mg/ m^{-2} d^{-1}): 1) total flux of sedimentary matter according to data from sediment traps; 2) Corg flux. 1

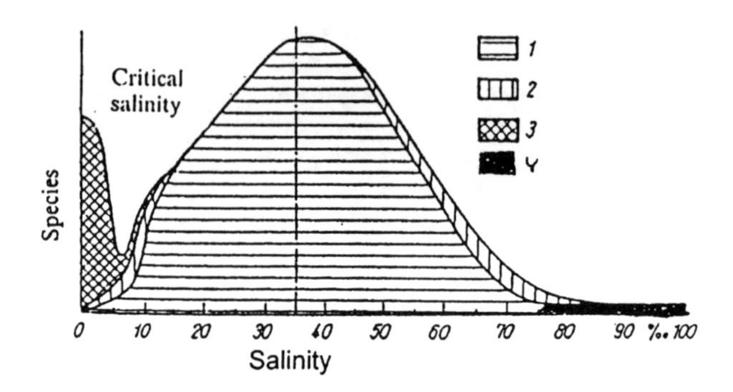

Fig.14. Numbers of organism species with salinity variation. 1) marine; 2) saltish-water and marine eurihaline; 3) fresh-water [29]. 1 - Critical salinity, 2 - Number of species, 3 - Salinity.

TABLE 3. Average particulate matter content (mg l^{-1}) in the largest rivers. Rivers of the tropical zone and a worm part of the temperate zone in comparison with rivers of Siberia.

Tropical Rivers	
Ganges-Brahmaputra	1 200
Indus	2 448
Demerara	3 000-10 000
Hwang Ho	14 975
Nile	833
Mississippi	156
Amazon	1 842
Rivers of Siberia (the zone of taiga and tundra)	
Lena	30
Ob	39.5
Yenisey	21.6

Another key feature of many estuaries is that in some areas they contain more particulate matter than that originating from the river or present in marine waters (Figs. 1, 4, 6) since biotic and abiotic particulate matter arise here. Therefore, areas of high turbidity begin in estuaries. Aleurite (silt) and pelite particles and suspended POM of appropriate "hydraulic" coarseness (peat particles, wood, bark, pollen, etc.) with their adsorbed elements also sediments in the estuary. The POM content of fluvial water ranges from 0.5 to 12% organic C.

Since these particles are fine, the adsorbtion capacity of their surface is high. On arrival at the seabed the organic carbon operates as a diagenetic pump. Elements of transitive valency are reduced, many becoming soluble and thereby are released from sediment to the above-bottom water. This is especially true for Mn.

The quantitatively important fine-dispersive fraction of the suspended particulate matter (<0.01 mm) in step II, gradually replaces the coarse-dispersive fraction as the river water flows seaward. The fine material is comprised largely of clay minerals with extremely high total surface area and high sorption properties which are usually increased due to a coating of organic substances and Fe hydroxides. That this mechanism holds is confirmed by comparing the absorbing capacity of pure minerals and of the same minerals from soils, suspended matter and bottom sediments, (Table 3). An analysis of elements on suspended particles sampled from the Amazon River showed that the bulk of Fe and Mn in the suspended matter was indeed contributed by particles finer than 10 μm and resided in hydroxide coats on the surface of the particles. Similar data are available from other rivers [21]. About 19 Gt y^{-1} are currently transported to this part of the filter where clay minerals and organic matter undergo coagulation on meeting marine water and some 17 Gt y^{-1} precipitate yearly over the a salsinity range from 1 to 15‰, with maximum of sedimentation between 3-5‰.

At this stage, the pelite fraction, like all particles of natural suspensions in aqueous medium have negative charge [31]. Coagulation of fine particles in riverine suspended material under the effect of an electrolyte (sea water), is of major significance. In this event the finest particles (which even pass through membrane filters) coagulate. Under conditions of massive coagulation, (which involves other coloidal and dissolved substances from river water such as organic matter and oxyhydrates of iron and aluminium), the "silt plug" arises. This is the area with maximal suspended mineral matter content which are even higher than those of the original river water.

At these first two stages of the pump biotic processes are negligeable and confined to bacterioplankton development and also (especially at the first stage) development of riverine plankton, which usually dies off with salinity increase (except for euryhaline forms).

We have derived data on similar fluxes in estuaries of the Yenisey and Ob rivers by means of sediment traps [32]. In the outer parts of estuaries of these rivers and adjoining areas of the Kara Sea average values of sediment matter fluxes at summertime were between 1-20 mg m^{-2} d^{-1} while at step II of the Ob and Yenisey marginal filters, 1321 and 22156 mg m^{-2} d^{-1}, respectively. These values are several orders of magnitude higher than on the adjacent shelf (Fig.6,8).

The high efficency of these two first steps of marginal filters of world rivers is confirmed by the fact that the average particulate matter content of the world rivers 460-500 mg l^{-1}. Beyond the estuaries on the shelf usually values do not exceed 1 mg l^{-1} and in the open ocean, some 0.1 mg l^{-1} [11].

STEP III: THE PHYSICO-CHEMICAL PART OF MARGINAL FILTERS: THE COLLOIDAL PUMP (C-PUMP)

The study of the colloidal part of river- and sea water (1-100 nm) is only just beginning. Up to now this part of the system was presumed to involve only dissoleved matter. The colloidal fraction differs from coarser material (>1 μm) primarily by the fact that particles of such size do not precipitate by themselves and are in Brownian motion. These particles have very large surface areas and can absorb many elements, and thus govern the distribution of the latter between a truly dissolved part and that adsorbed onto the particles.

The colloidal system consists of biotic and abiotic processes. The biotic part is in turn divided into living picoplankton and excretory products (bacteria, coccolithophors, viruses, etc.) and their dead residues, while the abiotic part consists of fine-particle clay minerals and a large group of minerals. Photochemical transformations are of major importance in this system and involve both biotic and abiotic parts of the system and colloids.

The mean content of colloid s in surface layers of sea water varies between 0.03 to 0.09 mg l^{-1}, while the average particulate matter content totals some 0.1 mg l^{-1}. The number of particles 1 ml^{-1}, determined using a Coulter Counter attains about 10 x 10^8 corresponding to some 8 m^2 of total mean surface of particles contained in 1 m^3 of sea water.

The first granulometric measurements of this fraction using an electron microscope showed that the number of particles rapidly increases with decrease in diameter. The highest number of particles is in the fraction < 120 nm, and many of them merge into coarser aggregates. In the colloidal phase the finest clay minerals, (most however, have precipitated at step II of the filter), colloids of organic matter and Fe oxyhydrates, and to a lesser extent, Mn and Al oxyhydrates. It co-occurs with fine-particulate (partly colloidal) argilleous matter. Here, within a salinity range from 3 to 10-20 ‰ dissolved organic matter (DOM) flocculates. Humic substances form floccules, and are powerful sorbents which selectively capture elements from solution, forming a series: Fe^{3+} > Al^{3+} > Cu^{2+} > Zn^{2+} > Ni^{2+} > CO^{2+} > Mn^{2+} and to fulvic acids: Fe^{3+} > Al^{+3}+ > Cu^{2+} > Ni^{2+} > Co^{2+} > Ca^{2+} > Zn^{2+} > Mg^{2+}.

Massive sedimentation of newly generated Fe oxyhydrate, also a strong sorbent, and co-precipitation of elements in the Fe group: Cu^{2+} > Zn^{2+} > Ni^{2+} > Cd^{2+} > Co^{2+} > Ca^{2+} > Mn^{2+} occurs. The following elements are bound to ferric hydrooxides: anions and cations of P, S, As, B, J, Br, F, U, Pb and Hg as well as oxihydrated complexes of Ti, Zn, and Cr. Univalent and bivalent cations of Mg, Ca, Ba, Co, Ni, Cu, Zn, Cd and oxianions of Mo, W, and Sn are bound to hydroxides of Mn. These bonds, naturally, are most clearly seen in suspensions and bottom sediments with small amounts of the other essential sorbent, organic matter [33]. In bottom sediments and suspensions most of the elements are bound to organic matter (Cu, Zn, Cr, Pb, Ni, Cd) and to ferric oxyhydrates, [34].

A complex of minor elements in suspended matter and bottom sediments is thus to a great extent controlled by, most importantly, clay minirals (colloidal part) organic matter, iron and manganese.

When river and sea water mix (as in steps I and II) ion-exchange reactions occur. Thus, one observes desorption from the surface of suspended particles and transition to solutions of Mn, Cu, Cd, Cs, Ba etc. At the next step (III), these elements are again captured from solutions and absorbed on humic substances and oxyhydrates of Fe. At stages I to III, in estuaries this re-packing of minor elements is confirmed by experiments with tagged atoms [35]. The main process for macroelements is replacement of Ca on clayey particles by Na and to a lesser extent, by K [36]. In arid zones this process of Ca release to water sometimes causes formation of $CaCO_3$ acoompanied by precipitation of some elements (within a salinity range 1-5‰ or, rarely to 10‰) [37].

It is important to note that at this step sorbents are not yet released from terrigenous material, as it is the case at steps I-II. They arise from solution in the marginal filter itself and this step is thus of great geochemical significance. Newly formed sorbents have bare surfaces and are added in substantial amounts. Thus there are key exchange reactions with replacement of some ions of particle surfaces. In this respect step III differs basically from step II where sorbents were supplied from soils and weathering crusts. At later steps of the filter, the fresh (plankton) autochthonous (estuarine-marine) organic matter, both particulate and dissolved, gains in importance, while riverine organic matter has reduced importance.

On membrane filters and in samples derived from sediment traps from this part of a filter one can see numerous floccules and aggregates of complex composition (Fig.10). These are largely organo-clays ferrous agglutinates and floccules which actively adsorb biotic elements and therefore are newly formed accumulations of food for bacteria. The floccule surface is usually covered by a layer, or several layers, of bacteria that can be observed using a microscope. These are like natural granules of fertilizers with a complex of biotic elements needed for bacteria development and removed from water by sorption. The heaviest particles contain clay mineral particles, whereas the lightest ones, floccules of organic matter and iron oxyhydrates.

Moving seaward, the heaviest floccules sediment, and are captured in near-bottom traps. At this third step of the marginal filter, physico-chemical factors becomes clear dominate over gravitational factors. Here flocculation of remains of clay minerals, dissolved and colloidal organic matter and Fe occurs. This leads to further transformation of dissolved (colloidal) forms to suspended particulate matter, formation of new suspended matter in addition to that contained in the original river water. Organic matter flakes are generally colourless or white but with Fe present they become yellow-brown.

Since newly formed sorbents have bare surfaces they extract from dissolved forms whole groups of elements (including those desorbed ąt step II) and concentrate them by several orders of magnitude. It is well-known that due to precipitation with Fe oxyhydrates, whole groups of elements (Fe- group) are extracted, and this is extensively used in analytical chemistry and technology. For example Al and Fe salts are utilized for tap water purification [38].

Thus the third stage is mostly a colloidal one since at the two earlier stages the major part of coarser particulate matter has sedimented. The water is usually still too turbid for good phytoplankton growth. Investigations undertaken in recent years have shown that the "dissolved" share of elements isolated with membrane-based filtration (filters of 0.45 mm) largely consisted of colloidal material (Table 5).

TABLE 4. Contents in metals in colloidal form (% of "dissolved" form). [39] [40] [41] [42]

River	Corg	Fe	Mn	Cu	Ni	Cd	Pb	Zn
Lena	57	-	-	-	-	-	-	69
Yenisey	-	97	-	-	60	76	22	70
Ob	-	89-92	-	-	50	50-57	42-52	70
Average for Russian rivers	80	78	68	-	-	-	-	47
Phone	-	8-30	-	20-40	0-18	0-38	-	70
Venetain Lagoon	-	87	54	46	18	34	58	88
Tamba (Australia)	-	99.6	-	62	-	-	-	80

The processes occuring at this stage are still inadequately studied, but some generalisations can be made. Organic matter, both dissolved and in colloidal form plays a highly important role (particularly for the Siberian rivers). It is not only present in river water in the large amounts but serves as the most powerful sorbent (humics), has protective properties, and prevents or inhibits coagulation. The data available are, however, contradictory for many estuaries since both particulate organic matter content varies according to the varying environmental conditions and with the varying physico-chemical properties of the elements being sorbed. As noted above, here physico-chemical rather than gravitational laws are dominant, and it is the "colloidal pump" which removes the elements from river (and partly marine) waters.

Tables 4 & 5 show that in the dissolved fraction of river water (passing through a filter with pore diameter of 0.45 μm) colloidal DOM plays the principal quantitative role. The average DOM content of world river waters is some 10 mg l^{-1}, i.e. about an order of magnitude as much as that of the most important elements! Oxyhydrates and

TABLE 5. Dissolved organic carbon content (DOC), water flow and dissolved carbon flux from various climatic zones

Morphoclimatic zones	Dissolved organic carbon content mg l^{-1}	Water flow, $m^3 y^{-1}$	Water % of total	Dissolved Corg flux, $10^6 t\ y^{-1}$	Dissolved Corg flux, % of total
Tundra	2	1222	3	2.2	1
Taiga	7	4376	11.7	30.6	13
Temperate	4	10285	27.5	44.1	17.6
Humid tropics	8	19186	51.3	153.5	65.5
Dry tropics	3	2169	5.8	6.5	2.8
Semiarid	1	262	0.7	0.3	0.1
Total		37400	100	234.2	

other Fe compounds (40 $\mu g\ l^{-1}$)rank second, aluminium 50 $\mu g\ l^{-1}$ and manganese 10 $\mu g\ l^{-1}$. Out of almost 50 trace elements studied in fluvial water only eight have concentrations higher than 10 $\mu g\ l^{-1}$, (in addition to the above also Ba, F, Sr and Zn, [43] [44].

Of most significance among river dissolved organic matter are the natural high-molecular hydrophilous substances fulvic acids (60-80% of DOM) and humic acids. The highest abundance of DOM is in river waters flowing through swamps and woodlands. In mountain rivers the role of DO; is lower and is a peculiar feature of filters of this river type. In swamp rivers the DOM content reaches 50-100 mg l^{-1} . Thus, organic matter influx to estuaries is goverened by climatic and altitude in mountain catchment areas, [1] [45].

As was noted, dissolved organic matter plays a very important part in the history of trace elements since it is a protective colloid and imparts negative charge to particles of fine (colloidal) suspension of any composition.

A new method for determination of DOM *in situ* is laser spectroscopy. Fig. 6 shows such data for the distribution of "yellow substance" (DOM) on transects across the Ob and Yenisey estuaries. In the Yenisey River water "yellow substance" content sharply decreases at 13-15‰ (from more than 200 to 50-70 units). Further decrease is gradual, up to less than 30 units in the open Kara Sea. A similar pattern is observed in the Ob River, but the DOM content of the river water is even higher (more than 400) because of the peat bogs and swamps in the hinterland. The largest decrease (the DOM coagulation front of the Ob) takes place within a salinity range from 12 to 16‰ [46]. The maximum sedimentation of colloidal Fe also occurs here, i.e. this is the area of maximal power of the colloidal pump operation and the start of the "geochemical plug".

The formation of organic matter floccules in the Kara Sea is associated with nutrition of not only bacteria but also zooplankton which utilizes for nutrition dissolved terrigenic organic matter transformed into floccules. This was shown by microscopic studies of the material from sediment traps (many fresh floccules overgrown by bacteria) and by examination of the stomach contents of small crustacean filterers. Within the area of step III of the Ob and Yenisey river filters, in copepods stomachs there were almost only floccules (not phytoplankton, as is generally the case in a biological part of a filter). The trophic significance of floccules for nutrition becomes particularly important in winter when phytoplankton is sparse. So, dissolved organic matter enters the zooplankton food chains via floccules and bacteria.

Dissolved organic matter is very essential at this and the following (IV) steps because it serves as a basis for organo-metallic compounds (including chelates) in which metals are bound inside a molecule and lose their chemical propertiesuntil they are released. Metals behave therefore, as if within a package, and thus are chemically inactive.

Thus organic matter is not only the most significant sorbent at this stage of a filter but it also determines many other mechanisms of the marginal filter operation. In particular organic matter causes precipitation of some 70-80% of iron, the other very important natural sorbent in the mixing zone. The sorption capacity of ferric oxyhydrate (Table 3) varies abruptly from 150 g^{-2} at pH 7 to 300-320 g^{-2} at pH 8, i.e. passing from acid river water (ph 5-7 in swampy rivers) to neutral sea water.

Experimentally, in the course of investigations into processes of water preparation [38], sorption of the cations most widespread in river water has been found to begin on Fe oxyhydrates at different pH values. As a result, the elements form the following series as regards threshold values of sorption (magnitudes of pH are indicated): Cu^{2+}, Zn^{2+}, Ni^{2+} , Mn^{2+} at pH 4, 5, 6, 8 respectively, [38]. It follows from examining this series that in acid waters of swamps and rivers (at pH lower than 5-7), Mn sorption on Fe oxyhydrates does not ocuur. Therefore Mn soluble forms enter estuaries in larger amounts than those elements whose sorption theshold begins at the pH of swamp and river waters such as copper, zinc,and nickel, etc.

In a number of publications it has been shown that iron is closely linked in river waters with fulvic acids and eventually up to 70-80% of elements such as Fe, Mn, Ni, Cu and Zn are transferred along with organic matter [47, 48,49].

The surface area of suspended particulate matter in the Gironde and the Loire rivers was 10 $m^2 g^{-1}$ whereas in the estuary up to 30 $m^2 g^{-1}$ [50] and in the Rhone, surface area increased from 4 $m^2 g^{-1}$ in the river to 13 $m^2 g^{-1}$ in the estuary. In the Tamar River, a relationship has been found between surface area of suspended particles and salinity. The largest surface area was 19.8 $m^2 g^{-1}$ occurred at a salinity of 2‰ and decreased to 10 $m^2 g^{-1}$ at salinities exceeding 2‰ [51]. This conforms to stage II and especially stage III marginal filters. Along with surface area, cation exchange capacity and wettability varied. The increased surface area of suspended matter suggests new formation of colloids.

Of great interest are the data on direct isolation of colloids from fluvial water and water from estuaries by means of filtration through cellophane filters under pressure of some 15 atm [52]. These data show up to 80% dissolved Fe, 47% Zn, 68% Cu and 73% Mn changed to the colloidal form. Data on transport in the colloidal form have been obtained for Norwegian rivers [53] where figrues were 81% Fe, 46% Zn and 91% Al (Tables 7 and 8).

The Fe content of the suspended form in estuaries is two orders of magnitude greater than that of the dissolved form. Detailed investigation of Fe forms in suspension has shown that iron was far from being entirely chemicaly inert, i.e. it is not fully bound (fixed) into molecules of detrital minerals. A substantial part of Fe in suspension occurs as surface-adsorbed films (up to 40-50%) and of amorphous hydroxides (both "fresh" and “old”) linked to organic matter (5-15%). Chemically inert forms of silicate generally contribute 30-70% to Fe in suspension. However, precipitation of suspended Fe, in contrast to dissolved Fe, generally occurs at steps I and II in the filter whereas the dissolved forms are transformed into sorbents.

In marginal filters of tropical rivers, dissolved aluminium is important . The role of dissolved forms of Mn in estuaries is minor with only 20% being retained here (Table 6).

TABLE 6. Removal of suspended and dissolved (colloidal) forms of elements (% of initial amount in river water) in marginal filters, average for world rivers, [49] [12]

Elements	Removal at the river-sea border (% of input by rivers)	
	Suspended forms	Dissolved (colloidal) forms
Na	90	10
K	90	10
Ca	90	10
Mg	90	10
Si	90	20
Al	90	30
Fe	90	80
Mn	40	20
Zn	50	10
Cu	60	40
Ni	-	20
Co	-	10
Cd	-	5
C org	80-90	25

4. 2. BIOTIC PARTS OF MARGINAL FILTER.

Plankton and benthic organisms inhabiting coastal areas of the ocean and in particular estuaries, have been selected for resistance to large environmental changes. Among these are resistance to toxic metals, for example from volcanic eruptions that occur in catchment areas (e.g. the Kamchatka Peninsula, Kurile Islands, etc.). Other important environmental factors are drastic increase in both turbidity and the level of water during freshets, stirring-up of bottom sediments during storms, impact of tidal and wind-induced surge-and-recess phenomena, emergence and disapearance of fast ice. Rather often, as was found when great amounts of ash were emitted on the Kamchatka Peninsula, the concentration of heavy metals in river and coastal waters increased as much as a hundred-fold. Volcanic ash contains many toxic compounds, especially during the first years after eruptions.

The biological communities of marginal oceanic zones are the first to encounter pollutants and other anthropogenic impacts. This group of organisms because of their resistance to environmental changes and their abiloity to detoxify the water, the organisms provide oportunity for development of other less susceptible pelagic communities [54]. These groups of organisms have been suggested to be called "contour-bionts" [55], since they delineate the boundary etween the continent and the ocean.

TABLE 7. Transfer of various elements into suspended particle matter (% of total) [43]

% of Total	Element
99-99.9%	Ga, Tm, Lu, Gd, Ti, Er, Nd, Ho, La, Sm, Tb, Yb, Fe, Eu, Ce, Al
90-99	P, Ni, Si, U, Co, Mn, Cr, Th, Pb, U, Cs
50-90	Li, N, Sb, As, Mg, B, Mo, F, Cu, Zn, Ba, K
10-50	Br, I, Cl, Ca, Na, Sr

4.2.1 STEPS IV and V : The final Biotic Part of the Filter

In these steps the remaining matter, basically the dissolved elements, is subjected to "purification". Here elements are involved in the biological cycle and are incorporated into cellular material, tests and metabolites losing their chemical properties (e.g. in chelated compounds). Finally they are dispersed when the oprganisms die. Only after decomposition of these organic residues, do the elements recover their chemical properties. This is the key distinguishing feature of.work of the biological part of a filter.

The second feature specific to this part of the filter is biological concentration of elements dissolved in water, up to 10^3- to 10^8- fold and also changes in ratios between elements present in water, according to the physiological needs of organisms (so-called biodifferentiation of elements).

The third feature is that organisms take from water the necessary amounts of elements to meet their physiological requirements and as a result of which the composition of organisms differs radically from that of the water in which they live. The fourth feature of the biological parts of marginal filters is that phytoplankton transfer dissolved forms of elements into suspended matter thus performing important geochemical changes to river and sea water. The fifth feature is unique, that the zooplancton and benthic organisms filter the whole estuarine water volume transferring the remainding suspended matter, including newly formed material into large clumps whixh sediments to bottom sediments.

There are two key steps within a biological part of marginal filters:

STEP IV: concentration, or assimilation (IV), with transfer by phytoplankton of dissolved forms of elements and contaminants into suspended particles, i.e. bodies and carapaces/flustules of bacterio- and phytoplankton (the phytoplankton pump) and

STEP V: biofiltration by zooplankton and benthos, which by means of special devices separate suspended particulate matter formed in steps II-IV, and also remaining particulate matter not yet captured, bind it and transfer it into bottom sediments (V).

4.2.1.1. STEP IV: The Phytoplankton part of a marginal filter pump.

The fourth step is mediated by phytoplankton it's primary feature being a decreasng role of mineral riverine particulate matter (removed mostly at steps I,II and III). At this stage, residues of dissolved elements are released, primarily biotic (besides the Liebig triad; P,N and K, some 50 other elements necessary for life). Phytoplankton cannot utilize the elements in suspension and this constitutes the basic difference between the processes occuring here and the processes in the first three steps, when both suspended and colloidal particles were removed. This biological pump captures material from the water, transforms it and desposits it to sediments. Only dissolved (and remains of colloidal) forms of elements that have passed through the first three steps are utilised. (In some cases, however, elements from marine water also occur, the quantity of which increases during the course of mixing).

When the biotic step in the filter starts, photosynthetic processes, (Fig.7). The deeper the illuminated layer (H), the higher the nutrient concentrations and the longer the growth season, the larger is the role of this step. The key regulating factors are rates of primary production and biomass. Here salinity is fairly high, 3-7‰, so that riverine plankton numbers are low and marine phytoplankton begins to dominate, Fig. 12 [29]. There is a dramatic desrease in number of species.

Phytoplankton development is regulated not only by mineral particulate matter content (the turbidity of surface waters) butis also self-limited. When nambers of plankton organisms are high water transparency decreases and despite high nutrient levels no furtherincrease in biomass occurs. The "biological plug" is formed by biotic particulate matter and is found near to the "silt plug".

In this step, as opposed to the first three filter steps, phytoplankton biomass is limited by zooplankton grazing, and can sharply reduce phytoplankton biomass, sometimes almost to zero values.

Here there is a large-scale transfer of dissolved forms of elements into cells and frustules of phytoplankton. Siliceous and carbonaceous frustules of diatoms and other algae, along with elements included in them, remain in suspension for a long time and for millions of years in bottom deposits.

Large amounts of organic matter are newly formed phytoplankton (autochthonous) and have a composition that differs from terrigenous material. The dominant planktonic organisms are marine, see Fig.12. The C:N ratio, isotopic composition and absolute age of organic matter are typically marine. Dead and dying cells and organic matter serve as additional powerful sorbents for several elements, mostly uranium and vanadium (Table 8). This means that the additional, newly formed biotic plankton sorbent continues to extract the remaining dissolved forms of elements from river water. Elements are bound primarily in carbonaceous and siliceous carapaces and frustules. The influx of fresh organic matter gives rise to high biomasses of zooplankton and benthic and nektonic organisms. The "biological plug" thus comprises maximal biomasses of phytoplankton, zooplankton and benthos. The typical food for benthos is by filtration and organisms collecting silt from the uppermost layer of sediments. This benthonic part of the filter is generally located in a "saline wedge" of an estuary although it obtains food from the upper less saline layer. Benthic organisms constitute the final step of biofiltration processing the material already filtered by zooplankton.

4.2.1.2. STEP V: The zooplankton part of a filter. Zooplankton pump (Z-pump)

The external boundary of the fourth step is predetermined by biological factors. These are development of filtering zooplanktonic organisms that feed on phytoplankton and develop only when plant biomass is high. The filtration is not selective, that is, both biotic and terrigenic particulate matter are collected from water and transferred to bottom sediment. Zooplankton filter immense volumes of water. Our estimates are that the whole volume of the World Ocean is filtered every six to twelve months and the water volume of the ocean's active layer where the main part of plankton dwells, every 20 days, [56]. The whole water column of the mixing zone in estuaries is usually filtered in two to five days. Fine particles of steps II-IV of the marginal filter, which have not had enough time to sediment in the estuary, and the bulk of phytoplankton are filtered out and deposited by this biological pump. Most sediments in rather small areas where the pump is most active, (deposition centres), (Figs .5, 11, 13).

TABLE 8. Specific surface (m^2 g^{-1}) and exchange capacity of clayey minerals, organic matter, soils, fluvial and marine suspended particles [57][58][59][60][51] [61]

	Specific surface (m^2 g^{-1})	Exchange capacity (mg-eq 100^{-1})
Minerals		
Kaolinite	10-50	3-15
Illite	30-80	10-40
Chlorite	-	10-40
Montmorillonite	50-150	80-150
Fresh ferric oxihydrates	157(1)-300	80-120
Mn oxihydrates	70-350(2)	-
Al oxihydrates	230-680	-
Opal (diatom)	90	10-34
Silica gel	580-610	-
Activated charcoal	890-1060	-
Organic matter		
Humic acids	1900	170-590
Fulvic acids	-	700
Sawdust	-	20-25
Peat	-	50-100
Soils		
Of taiga and tundra	-	6-10
Soddy and podzolic	-	35-56
Chestnut	-	17
Lateritic	-	11-13
Chernozemic	-	50-60
Suspended particulate matter of the rivers Gironde and Loire	10	25
Rhone	4-10	5-33
Delaware	-	9-86
Hudson	-	5-25

Suspended particulate matter of the estuaries		
Gironde and Loire	30	55
Rhone	13	33
Tamar	19.8 (S=2.2‰) 10 (S=28‰)	
Marine suspended particulate matter		
Average for marine suspended particulate matter	10-30	-
Average for suspended matter of the Pacific Ocean	-	10-30
Nearshore bottom sediments		
Noncarbonaceous sediments	-	25-89 (41.5) (3)
Organic matter of bottom sediments	-	10-71 (22.8)
Humic acids of bottom sediments	-	25-375 (301)
Organic matter from bottom sediments	-	23
Pelagic floor sediments		
Pacific Ocean, red clays (average)	-	27-49
Panama Basin, layer 0.5-3 cm	-	75
layer 15 -19 cm	-	98
Pacific Ocean, MANOP region	-	30
Sediments of Pacific Ocean,Bering Sea and Sea of Okhotsk (average)	-	13-39

(1) Goethite, ferrihydrite - 500; (2) Bernessite; (3) average

The importance of the zooplankton filtration mechanism can be judged by its biomass. The daily demand of zooplankton for food is 50-70% of the own weight of an organism. On the average, suspended particulate matter from surface layers contains some 20% of organic C. This means that to obtain an adequate ration, zooplankton organisms have to filter two to five times their body weight per day. Figs.7 and 9 show that mesoplankton biomasses in the mouths of the Ob and Yenisey rivers exceed 1000 mg m^{-3}. Thus to support the filterers contained in 1 m^3 of water there must be at least 2000 to 5000 mg (2-5 g) of particulate matter every 24 hours. As is seen in Figs. 6 and 8, maximum suspended matter content of these estuaries at the period of our research was 5 mg l^{-1}, or 5 g m^{-3}. It follows from these rough estimates that in order to satisly its food requirements, zooplankton in estuaries of the rivers in question have to filter the whole water volume removing phytoplankton and suspended matter within only one to two days. A dynamic model of the Kara Sea ecosystem has been made [25].

The suggestion of very high rates of a zooplankton filtration is consistent with data for the World Ocean as a whole. Oceanic zooplankton biomass reaches some 50 Gt wet weight and around 40 Gt dry weight. To provide the daily ration of organic matter ca. 20 Gt per 24 hours must be available, whereas rivers supply to the World Ocean 15-20 Gt y^{-1}! So, zooplankton processes yearly hundreds and thousands of times more suspended matter than is delivered by rivers.

The filtration system of estuaries thus has an enormous capacity. It is in estuaries that the major portion of suspended matter is filtered of off from water. At the time of investigation into estuaries of the Ob and Yenisey rivers in September throug early October 1993, the system was in pre-winter state when its capacity was one order of madnitude lower than at the bloom season in August (Fig.13).

The seasonal dynamics in the Kara Sea filter is governed by the phytoplankton bloom in July when food stores for zooplankton are created, and by gradual accretion of zooplankton biomasses in July to August. The following organisms play the major role among plankton biofilterers: copepods (50 to 80% of total biomass), ostracods, appendicularians, rotifers, as well as larger crustaceans and euphausids. These organisms are able to filter practically the entire range of suspended particles, of river and sea water from the coarsest (usually biotic ones) to the finest particles (less than 1 μm).

The minimum dimensions of particles captured by copepods depend on arrangement of their filtering apparatus. Many species capture particles down to 1 mm in size. Each individual filters on an average 100 to 200 ml of water per 24 hours but, as was noted earlier, this amount depends on the concentration and composition of suspended particulate matter [62].

At the phytoplankton bloom season when the quantity of food is much greater than required, gluttonous feeding is typical. Here crustaceans continue to filter material and produce pellets, with almost no change in the initial composition of suspended material. The organic C content of such pellets amounts to 20% or more, with some 1.7% of P. This material can serve as a food source for filter-feeders in the lower layer of the estuary, benthic organisms and bacteria.

Pellets made by small and large copepods are 3-5 μm and 1000 mm in size, respectively. Their sedimentation rate is 100 -500 m per day, and persistence is governed by a packaging envelope, the pellicule, which protects their contents from oxidation. Usually pellets exist about one week. After that, the pellets disintegrate. So, in shallow-water estuaries a pellet flux without substantial alteration reaches the bottom in less than 24 hours. Unlike the pelagic ocean, all their subsequent changes take place in bottom sediments and not in the water column.

Of great interest are organisms with especially fine filtration capacities, <1μm. In order to catch particulate matter they use thin nets and surfaces covered with mucus to which water flows are directed. Examples are appendicularians that build a tiny mucous "house" with a reticular orifice (a preliminary filter). After clogging of the preliminary filter, the organism leaves its "house" and builds a new one. Rotifers have a remarkable mechanism of particulate matter removal. With the help of a special apparatus they create two whirls from which suspended perticles are caught with thin cilia. Pelagic tunicates (salps) construct a mucous web-like net. They "trawl" water with this net thereby separating particles larger than 0.7 μm. Except for a small group of predators, all zooplankton filterers separate particular matter non-slectively. Benthic organisms form a biofilter belt in coastal areas of the ocean and often the highest biomasses are found in estuaries. Benthic organisms can also be divided into several groups as can be illustrated by data from the Black Sea. Here each square metre of mussel-bed filters from the bottom water 100 to 1000 l d^{-1}, and removes particles of up to 1 μm in size from it. Mussels do not endure particulate matter concentrations of >50 mg l^{-1}.

Many molluscs, (e.g. *Dreissena*)can remove bacteria from water. A colony of cirripedes of 1 m^2 surface area produces 20 to 50 g of pellet material per day and filters many hundred litres of water a day.

There are two key biofiltration steps, which ensure transfer of matter from dissolved and gaseous forms. Firstly, carbon fixation by phytoplankton is a major factor, which was not involved in the first three steps. Secondly, transfer of material from phytoplankton to zooplankton and benthos.

The benthic part of a marginal filter does not solely involve filter-feeders, an important role also is played by phytobenthos. These processes occur in the upper freshwater wedge of waters in stratified estuaries, where there are substantial transformations of riverine matter and the emergence of autochthonic material (floccules, agglutinates, etc). During storms the whole water column is mixed and bottom sediments are stirred up. There are also seasonal and between year changes associated with naural cycles and environmental changes. This model does not hold for estuaries which are homogeneous in their vertical profile and/or subject to strong tidal effects [63] [64] [65].

The composition of water, suspended particulate matter and organisms of the lower wedge of an estuary (the end member mixture step, EMS) substantsally differs from that of the upper wedge. Unlike the upper nonsaline wedge, one litre of water contains higher concentrations of many elements (Table 9),.although others have lower values than in river water. River water contains much more dissolved and particulate organic matter.

TABLE 9. Concentration factors in relation to fluvial water

Element	Concentration factor
Cl	2479
Na	1710
Mg	315
SO_4	243
K	173
Ca	27
Br	3350
Sr	168
Li	72
Rb	60
F	13
Mo	10
U	6
Fe	0.01
Mn	0.04
Cu	0.2
Zn	0.25
POM	10-10000
DOM	10-50

The particulate matter content of water in the lower wedge are three to four orders of magnitude less than the average for river water. In the lower layer, plankton is typically marine, and zooplankton organisms occur within a wide range of salinity and migrate diurnally from the upper layer estuary to the lower one and back.

The end member step (EMS) thus acts not only as a strong electrolyte; it brings to estuaries a variety of macro- and micro-elements, mainly in dissolved form. Therefore, here the bulk of the colloidal and dissolved material is transferred into suspended particulate matter and bottom sediments, and major fraction of riverine suspended matter sediments. Accordingly, there are seemingly two fluxes: one from the river to the sea (particulate matter, including organic substances, Fe and Mn) and the second a weaker flux from the sea to the river mouth (marine water salts and individual trace elements).

5. Conclusion

In conclusion marginal filters are composed of five steps, with a set sequence and with each step having its own characteristics and localization associated with variation in salinity, nutrients and other factors (Fig.14). This model can change with environmental conditions, however, and some steps overlap and vary seasonally some parts of a filter become more significant while others grow weaker. The importance of each step varies in space (according to natural climatic zones) and with time (depending on a seasonal and longer cycles). Thus, in rivers of the Arctic in the winter (nine months) biological parts of a filter are almost non-existent (Fig.11). The kay factor regulating sedimentary material in the Arctic is permafrost is which reduces the active layer of soil to 40-70 cm. In these environments organic matter in soils persists in the frozen ground over millennia, (e.g. remains of mammoths). Our measurements, by means of means of ^{14}C, have given ages in the soil layer below 10 cm of an average of 10000 to 18000 years. Terrigenic organic matter of soil, river and estuarine suspended matter is apparently age-tagged, and therefore, the age of the upper sediment layer of the Laptev and Kara seas "seems" to be very ancient, (some 6000-9000 years). The contemporary deposition area of the Lena River with sedimentation rates between17 000 to 20 000 B is situated 200-250 km north of the delta edge. Some 10 000 years ago it was further North (by 100-160 km) and the Lena delta moved to its current location some 5000-6000 years ago [66] which corresponds to age of the Mississippi delta and deltas of 36 other rivers around the world. These dates also coincide with the decrease in the rate of sea-level rise.

The unique combination of high suspended particle content, high dissolved organic matter and high organism abundance in estuaries is of great importance to sedimentation process. The finest fractions of suspended particulate matter (colloidal fraction) which are dominant at the third step of the filter include residual mineral particles (fine particles of detrital and the main share of clayey minerals). Organic matter that exists in colloidal form acts like glue to form large aggregates. Special investigations have been undertaken to compare sizes of aggregates with constituent finer particles before their inclusion into clots joined by organic binder. Deflocculation of particles shows that the mineral material included in them was substantially finer (and less in quantity) than before separation of floccules. It appears that there are two groups of floccules: 1) large flakes of "estuarine snow" and 2) small floccules comprising several smaller mineral grains.

As far as one may judge, domination by any type of floccules depends on the abundance of organic matter in water and concentration of mineral particles and their size. Coarse mineral grains (>0.05 mm) do not form floccules and are generally attached to surfaces. Dynamic conditions are also essential with wave-induced mixing floccules disintegrate, first the coarse and thereafter the smaller.

Data available, derived from both field and laboratory observations, suggest that sedimentation of individual mineral particles in environmental estuaries, the shelf and the upper part of the slope are rare processes. Their precipitation and sedimentation usually result from integration of such particles into floccules of different dimensions. The sedimentation rate of these pellets and floccules is much higher than that of their constituent separate mineral particles.

Thus, in estuaries there are two main types of "packages" 1) faecal pellets produced by live zooplankton and 2) floccules of dead colloidal organic matter on which bacteria development occurs. These conclusions have been supported by direct study of hundreds and thousands of suspension samples with a microscope as well as by examination of particulate matter from sediment traps.

Specific features of the quantitative and qualitative composition of sedimentary matter (suspended matter, solutions gages and biota) arriving from terrestrial catchment areas conform to both climatic and altitude. Estuaries and deltas can be classified into several types by composition of both live and dead parts of a filter and by their scales.

In essence of the continent-ocean marginal filter removes from water more than 90% of particulate matter and 20 to 40% of dissolved substances, transforms the most hazardous dissolved forms of substances into particulate matter and transfers diverse contaminants (suspended particles and solutions) into bottom sediments. Within the filter system, various water purification ways are used(precipitation, sorption, coagulation and flocculation, bioassimilation, filtration) and at the final stages, biological treatment, along with conversion of the most hazardous dissolved forms of elements into bodies, carapaces and frustules of organisms - into suspended particulate matter with final filtration by a system of filter organisms of zooplankton and benthos and transfer into bottom sediments with pellet influx.

The general direction of the process is massive transfer of sedimentary matter (suspended particulate, colloidal, truly dissolved), miscellaneous forms of elements into bottom deposits where contaminants are diluted by the immense amount of inert aluminosilicate substances present. The filter varies seasonally. In the high latitudes with 8-9 winter months the biological parts of filter are almost entirely non-existent. In flood periods the abiotic steps increase in importance and only after water becomes clear again do the biological steps start acting. In tropical regions such changes do not occur. Thorough studies of each step of the marginal filter are needed. Such studies are especially important for developing control mechanisms for anthropogenic pollution and for determining the most sensitive seasons and the states of filter system.

Another limitation of the suggested model is that it considers mixing processes without impacts produced by tides, offshore and onshore winds and waves. The influence of these processes on mechanisms of river and sea water mixing and on removal from water of different forms of elements shoud also be studied for specific filter.

If the most talented engineer tried to create such a perfect water purification system, all parts of which are interacting both in space and time, he would fail! It will not be an exaggeration to say that the system of marginal filters, which we are just beginning to understand, is one of nature wonders!

6. REFERENCES

1. Artemiev Ye.V. (1993) Geochemistry of Organic Matter in the River- Sea System (in Russian). Nauka, Moscow, pp. 204.
2. Lisitsin A.P. (1982) Avalanche sedimentation . Avalanche Sedimentation in Ocean (in Russian). Izd-vo Rostovsk. Un-ta, Rostov-on-Don, 3-59.
3. Lisitsin A.P. (1989) Sedimentary systems of ocean: the new approach to studying global and regional pollutions. Vestnik AN SSSR. 4, 58-67.
4. Lisitsin A.P. (1991) Processes of Terrigenic Sedimentation in Seas and Oceans (in Russian). Nauka, Moscow, pp. 270.
5. 5 Morozov N.P., Baturin G.N., Gordeev V.V., Gurvich Ye.G. (1974) On composition of suspended particulate matter and sediments in mouth areas of the North Dvina, Mezen, Pechora and Ob rivers. Gidrokhimicheskie Materialy, 60, 60-73.
6. Demina L.L., Artemiev V.Ye. (1984) Forms of migration of trace elements and organic matter in the Daugava River estuary. In Geological History and Geochemistry of the Baltic Sea (in Russia). Nauka, Moscow, 32-42.
7. Gordeev V.V., Miklichansky A.Z., Migdisov A.A., Artemiev V.E. Rate Element Distribution in the Surface Suspended Material of the Amazon River. Some of Its Tributaries and Estuary. Mitt. Geol. Palaeont. Inst. Univ. Hamburg, SCOPE/UNEP Sonderband, 58, 335-343.
8. Chudaeva V.A., Gordeev V.V., Fomina L.S. (1982) Phase state of elements of suspended particulate matter in some rivers of the Sea of Japan basin. Geokhimya. 4, 585-596.
9. Lisitsin A.P. (1964) Distribution and chemical composition of suspended particulate matter in the Indian Ocean waters. In Oceanologic Investigations during the International Geophysical Year (in Russian). Nauka, Moscow, 10, pp.135.
10. Lisitsin A.P. (1974) Sediment Formation in Ocean (in Russian). Nauka, Moscow, pp. 435.
11. Lisitsin A.P. (1978) Processes of Oceanic Sedimentation (in Russian). Nauka, Moscow, pp. 390.
12. Lisitsin A.P., Demina L.L., Gordeev V.V. (1983) The river-sea geochemical barrier and its role in a sedimentary process. in Biogeochemistry of Ocean (in Russian). Nauka, Moscow, 32-59.
13. Lisitsin A.P. (1986) Biodifferentiation of sedimentary matter in ocean and a sedimentation process. in Biodifferentiation of Sedimentary Matter in Seas and Oceans (in Russian). Izd-vo Rostovsk. Unta, Rostov-on-Don, 3-66.
14. Lisitsin A.P. (1977) Terrigenic sedimentation, climatic zonality and interaction of terrigenic and biotic material in oceans. Litologya i Polezn. Iskop. 6, 3-24.
15. Monin A.S., Gordeev V.V. (1988) The Amazon Region (in Russian), Nauka, Moscow, pp. 214.
16. Burenkov V.I., Gol'div Yu.A., Gureev B.A., Sud'bin A.I. (1995) Basic concepts of optical properties of the Kara Sea waters. Okeanology. 35, 376-388.
17. Lisitsin A.P. and Vinogradov M.Ye. (1994) The international high-altitude expedition to the Kara Sea (the 49th voyage of the research vessel "Dmitrii Mendeleev"). Okeanology. 34, 643-652.
18. Kuptsov V.M., Lisitsin A.P. (in press) Radiocarbon of the Quaternary deposits in shoreline area, bottom sediments of the Lena River and sediments of the Laptev Sea. Marine Chemistry.
19. Jouanneau J.M. (1987) The contribution of 14C dating to a better understanding of the POM behavior in estuaries. Marine Chem.21, 189-197.
20. Burenkov V.I. (1993) Optical properties of the Laptev Sea near the Lena River delta. Underwater Light Measurements. Proc. SPIE. 2047.
21. Kravtsov V.A., Gordeev V.V., Pashkina V.I. (1994) Dissolved forms of heavy metals in waters of the Kara Sea. Okeanology. 34,673-681.
22. Vedernikov V.I., Demidov A.B., Sud'bin A.I, (1994) Primary production and chlorophill in the Kara Sea in September, 1993.Okeanology. 34, 693-704.
23. Burenkov V.I., Vasil'kov A.P.(1987) Sud'bin A.I. Optical phenomena in riverine flow- origin lenses. Okeanology. 27, pp. 584.
24. Vinogradov M.Ye., Shushkina E.A., Lebedeva L.P., Gagarin V.I. (1994) Mesoplankton in the eastern part of the Kara Sea and estuaries of the Ob and Yenisey rivers. Okeanology. 34, 716-724.
25. Lebedeva L.P., Shushkina E.A. (1994) Assessment of autochtone detritus flux through plankton communities of the Kara Sea, Okeanology. 34, 730-735.
26. Lebedeva L.P., Shushnina E.A., Vinogradov M.Ye. (1994) A dynamic model of the Kara Sea pelagic ecosystem. Okeanologya. 34, 724-730.
27. Neihof R.A. and Loeb G.I. (1972) The surface charge of particulate matter in seawater. Limnol.

Oceanogr. 17, 7-16.
28. Loeb G.I. and Neihof R.A. (1977) Absorption of an organic film at the platinum-sea water interface. Journ. Mar. Res. 35, 283-291.
29. Martin J.M., Mouchel J.M., Nirel P. (1986) Some recent developments in the characterization of estuarine particulates. Water Sci. Technol. 18, 8-92.
30. Khlebovich V.V. (1974) Critical Salinity of Biological Processes (in Russian), Nauka, Leningrad, pp. 8.
31. Stow D.A. (1986) Deep clastic seas. in H.G.Reading (ed), Sedimentary Environments and Facies. 399-444. Blackwell Sci. Publ. Ltd., Oxford, pp. 615.
32. Hunter K.A. and Liss P.S. (1982) Organic matter and surface charge of suspended particles in estuarine waters. Limnol. Oceanogr.27, 322-335.
33. Lisitsin A.P., Shevchenko V.P., Vinogradov M.Ye., Severina O.V., Vavilova V.V., Mitskevich I.N. (1994) Sedimentary matter flux in the Kara Sea and the Ob and Yenisei estuaries. Okeanology. 34, 748-758.
34. Nolting R.F., Eisma D. (1988) Elementary composition of suspended matter in the North Sea. Neth. J. Sea Res. 22, 219-236.
35. Eisma D. (1993) Suspended Matter in the Aquatic Environment Springer Verlag. Berlin, Heidelberg, pp.315.
36. Li Yuan-Hui, Hisayuki Teraoka, Tsuo-Sheng Yang, Jing-Sheng Chen. The elemental composition of suspended particles from the Yellow and Yangtze Rivers. Geochim. Cosmochim. Acta .48, 1561-1564.
37. Sayles F.L. and Mangelsdorf P.C. (1979) Cation-exchange characteristics of Amazon River suspended sediment and its reaction with seawater. Geochim. Cosmochim. Acta. 43, 767-779.
38. Khrustalev Yu.P. (1982) Sedimentation features in the area of fluvial flow influence. in Avalanche Sedimentation in Ocean (in Russian). Izdvo Rostovsk. Un-ta, Rostov-on-Don, 59-71.
39. Babenkov Ye.D. (1977) Water Purification with Coagulants (in Russian). Nauka, Moscow, pp. 54.
40. Dai Min-Han and Martin J.M. (in press, a) First data on the trace metal level and behavior in two major Arctic river/estuarine systems (Ob and Yenisey) and in the adjacent Kara Sea (Russia). Earth and Planet Sci.Lett.
41. Dai Min-Han, Martin J.M., Cauwet G. (in press, b) The significant role of colloids in the transport and transformation of organic carbon and associated trace metals (Cd, Cu and Ni) in the Rhone delta (France). Marine Chem.
42. Gordeev V.V. (1983) Fluvial Runoff to Ocean and Features of its Geochemistry (in Russian). Nauka, Moscow, pp. 160.
43. Martin J.M., Guan D.M., Elbaz-Pouliohet F., Thomas A.J., Gordeev V.V. (1993) Preliminary assessment of the distributions of some trace elements (As, Cd, Cu, Fe, Ni, Pb and Zn) in a pristine aquatic environment: the Lena River estuary (Russia). Marine Chem. 43, 185-199.
44. Hart B.T., Sdrauling S., Jones M.J. (1992) Behavior of copper and zinc added to the Tamba River, Australia, by a metal-enriched spring. Aust. J. Freshwater Res. 43, 457-489.
45. Martin J.M. and Meybeck M. (1979) Elemental mass balance of material carried by major world rivers. Marine Chem. 7, 173-206.
46. Meybeck M. (1988) How to establish and use world budgets of riverine materials. in A.Lerman and M.Meybeck (eds), Physical and Chemical Weathering in Geochemical Cycles. Kuwler, 247-272.
47. Burenkov V.I., Kuptsov V.M., Sivkov V.V., Shevchenko V.P. (1994) Spatial distribution and dispersive of composition particulate matter in the Laptev Sea September. Okeanology.
48. Glagoleva M.A. (1959) Forms of migration of elements in fluvial waters. in On Cognition of Diagenesis of Sediments (in Russian). Izd. AN SSSR, Moscow, 2-28..
49. Yeremenko V.Ya. (1966) The methods to determine forms of heavy metals existence in natural waters. Gidrokhim. Mater. 41, 153-158.
50. Gordeev V.V., Chudaeva V.A., Shul'pin V.M. (1983) Behaviour of metals in mouth zones of two small rivers in the eastern Sikhote- Alin. Litologya and Mineral Deposits. 2.
51. Garnier J.M., Lipiatou E., Martin J.M., Mouchel J.M., Thomas A.J. Surface properties of particulates and distribution of selected pollutants in the Rhone delta and the Gulf of Lion. in: Martin J.M., Bartl H. (eds). Proc. EROS-2000. Worksh., Blancs (Spain). Febr. 6-9, 1990; Water Pollution Res. Rep. 2a. 20, 501-552.
52. Titley J.G., Glegg G.A., Glasson D.R., Millward C.E. (1987) Surface areas and porosities of particulate matter in turbid estuaries. Cont. Shelf Res. 7, 1363-1366.
53. Garanzha A.P., Konovalov G.S. (1997) The method to isolate a colloidal form of migration of microelements. 2nd All-Union Conf. on Methods of Analyzing Natural and Waste Waters. GEOKhI, Moscow.
54. Benes P. and Steinnes E. (1974) In situ dialysis, the determination of state of trace elements in natural waters. Water Res. 7, 947-53.
55. Khristoforova N.K. (1989) Bioindication and Monitoring of Marine Water Contamination by Heavy Metals (in Russian), Nauka, Leningrad, pp. 192.
56. Zaitsev Yu.P. (1979) Contour communities of seas and oceans. In Fauna and Hydrobiology of the Pacific Ocean Shelf Zones. Mater. 14th Pacific Ocean Sci. Cong. (in Russian), Khabarovsk, (in Russian) Id-vo DUNTs AN SSSR, Vladivostok, 51-54.
57. Vinogradov M.Ye., Lisitsin A.P. (1981) Global regularities of life distribution in ocean and their embodiment in bottom sediment composition. Regular trends both plankton and benthos distribution in ocean. Izv. An SSSR. Ser. Geol. 3, 5-25.
58. Balistriery and Murray. (1983) Metal-solid iteractions in the marine environment estimating apparent

equilibrium binding constants. Geochim. Cosmochim. Acta. 47, 1091-1098.
59. Forstner U. and Wittman G.T.W. (1979) Metal pollution in the aquatic environment. Springer Verlag, Berlin, Heidelberg, 4, pp.486.
60. Garnier J.M., Martin J.M., Mouchel J.M. (1995) Surface properties characterization of suspended matter in the Ebro delta (Spain), with application to trace metal sorption, (In press).
61. Gorbunov N.I.(1974) Mineralogy and Colloidal Chemistry of Soils (in Russian). Nauka, Moscow, pp. 312.
62. Rashid M.A. (1985) Geochemistry of Marine Humic Compounds. Springer, Berlin, Heidelberg, N.Y., pp. 300.
63. Petipa T.S. (1981) Trophodynamics of Copepods in Marine Plankton Communities (in Russian). Naukova Dumka, Kiev, pp.242.
64. Pritchard D.W. (1967) Observations of circulation in coastal plain estuaries. in Lauff G.H. (ed), Estuaries. Amer. Ass. Sci. Publ. 83, 37-44.
65. Dag J.H. (Ed) (1981) Estuarine ecology with particular reference to Southern Africa. Rotterdam.
66. Postma H. (1966) Sediment transport and sedimentation in the marine environment. in UNESCO (ed), Morning Review Lectures, 2nd Int. Oceanogr. Congr. 213-219.
67. Kuptsov V.M., Lisitsin A.P., Shevchenko V.P. (1994) 230Th as an indicator of particulate matter in the Kara Sea. Okeanology. 34, 759-766.

UNJUSTIFIABLY IGNORED: REFLECTIONS ON THE ROLE OF BENTHOS IN MARINE ECOSYSTEMS

W.E. ARNTZ, J.M. GILI & K. REISE
Alfred-Wegener-Institut für Polar- und Meeresforschung, Bremerhaven and Sylt, Germany and Instituto de Ciencias del Mar (CSIC), Barcelona, Spain

Abstract

The benthos is not highly regarded by most marine biologists. As a consequence, the in the past most funds have gone into pelagic research, despite the fact that the marine benthos has several properties which render its investigation particularly rewarding. It is far richer in species than is the water column. In addition due to the comparatively low mobility and longevity of many organisms which gives relative persistent assemblages, the behtos is far more suited for monitoring environmental conditions than the ephemeral associations that characterize a certain water body. However, on the rare occasions that benthic research was included in major international programmes, this was usually justified by the role as a temporary or final store for material sedimenting from the water column.

While the authors recognise that the connection with the pelagial ("pelago-benthic coupling") may indeed be one of the key facets of benthic research, they would like to stress that this is not simply a one-way interaction and the benthos influences pelagic processes in many ways. Furthermore, benthic processes have intrinsic values that go much beyond the value as a resource for human use or as a monitoring target for pollution. The study of marine benthos has contributed greatly to the development of ecological theory, e.g. to the definition of "stability" and the factors that may cause or prevent it.

The authors make use of their experiences in various marine ecosystems to illustrate their belief that benthic research is essential for understanding marine ecology, and should be given more attention - and improved funding!

1. Introduction

Already at first glance, the marine benthos reveals a number of properties which render its study not only exciting, but also particularly rewarding for the solution of ecological questions. The species richness of benthos is far greater then that of pelagic communities and many taxa are found exclusively at the seafloor. Benthic habitats are extremely diverse including, among others, soft and hard bottoms from the intertidal to the deep sea, mangrove forests, coral reefs, seagrass meadows, estuarine habitats and hydrothermal vents. This variety offers opportunities to study the effect of a wide range of environmental factors and processes from the polar seas to the tropics. Due to the large number of sessile or semi-sessile organisms with high site fidelity, benthic communities are easier to follow in space and time than the ephemeral assemblages in the water column, thus providing ideal subjects for the study of eutrophication and many other kinds of disturbance. Benthic living resources contribute substantially to human food and other articles of use to mankind, and our understanding of the population dynamics for managing these resources are ahead of those in terrestrial systems. Many hard-shelled organisms have growth rings that facilitate the study of population dynamics and thus, resource management. Furthermore, the culture of benthic organisms is rapidly growing industry in coastal areas with great prospects for the future. A large number of benthic suspension feeders all over the world accumulate

J.S. Gray et al. (eds.), Biogeochemical Cycling and Sediment Ecology, 105–124.

chemical substances and offer excellent opportunities for the monitoring of marine contamination, and many benthic species produce secondary metabolites which are of great interest to chemical and pharmaceutical industries.

Despite all these obvious advantages (and most likely, many more), benthic research has failed to receive levels of attention and support that have been given to water column research by the international science community and by funding agencies. Almost all large international programmes, such as BIOMASS, JGOFS, or GLOBEC, have either been restricted to research in the pelagial or, as in the case of JGOFS, justified a benthic component largely from the fact that the benthos receives, stores temporarily, transforms and buries material sedimenting out of the water column. International programmes that emphasize the benthos, such as SCAR's EASIZ, scrape a difficult living on single institutional or, at most, national funding.

We agree that pelago-benthic coupling illustrating the fate of pelagic production is an important facet of benthic research. However, biogeochemical fluxes and sedimentation are just two components out of a variety of interesting themes that should all deserve attention and appropriate funding. For this reason we attempt to revisit examples of other benthos studies carried out in the past or under way which support our view that this kind of research should be promoted in the framework of larger international programmes. Since we cannot cover the whole range referred to above, we will restrict these considerations largely to macrozoobenthos and to examples of basic research. We will not, or only marginally, include the equally important role of "applied research" including studies on benthic resources, resource management, culture of benthic organisms and use of benthic organisms for the study of pollution. Much of what we summarize and suggest may appear trivial to other experts in their respective fields, but this is inevitable in this kind of compilation. Moreover, among the targets are financing agencies and decision-makers who may be not so familiar with these themes.

2. Pelago-benthic coupling

The term "pelago-benthic coupling" is used in this paper for processes where the pelagial provides a major input to the benthos. To achieve an overall understanding of marine ecosystems, attempts have been made in recent decades to assess the individual components of food webs within or between systems [1] [2]. Attention has focused particularly on the energy transfer between the pelagic and benthic ecosystems [3]. In the open ocean energy flows vertically, gradually descending through the water column, with considerable loss of surface production by the time it reaches the bottom [4]. In contrast, in shallow ecosystems benthic organisms have more immediate access to fresh plankton production because of the proximity of the photic layer to the bottom, and because of mixing by tidal or wind generated currents.

Benthic communities occur in the interface zone between sea floor and water column. From the viewpoint of water column structure and dynamics, the benthos has always been regarded as a sink, where what remained of the production of the water column was deposited. However, in littoral systems, taking suspension feeder communities as an example, the benthos is an extremely active part of the system, receiving food particles settling out of the water column as well as actively exploiting production in the water column. Consequently, in terms of the boundary concept proposed by Margalef [5], communities of suspension feeders make up a highly active boundary system relatively distant from the concept of discontinuities or ecotones between communities [6] [7].

Theoretical notions of feeding by suspension feeders were previously very limited, differentiating between those that captured fine particles and those that captured zooplankton. More recent work has shown that the former capture a variety of particulate matter, e.g. plant detritus and small organisms such as bacteria [8] [9], and that zooplanktivores are also able to capture and assimilate a wide range of prey types, from particulate matter to phytoplankton [10] [11] [12]. Although a single prey type, e.g., particulate organic matter, may cover the demand in certain species [13] [14], nonselective diets appear to be the most suitable strategy for littoral suspension feeders [15].

Actually, the theory of optimal foraging developed for plankton organisms by Lehmann as far back as 1976 [16] has experienced a comeback and has been applied to benthic organisms, in particular to filter and suspension feeders [17] [18]. These organisms invest very little effort in food capture. The cost of capture is virtually nil in passive benthic suspension feeders, while in active suspension feeders pumping may account for up to 4 % of energy demand [19]. Another aspect of the optimal foraging strategy is to live on highly varied diets, the type of which is restricted only by morphological constraints such as size and shape. In the framework of optimal foraging theory [20], species that expend low levels of energy in foraging are considered ecologically highly successful [17].

Field and flume experiments in the past decade have contributed substantially to the study of pelago-benthic coupling in soft-bottom areas, and have revealed a close relation between sedimentation events and distribution patterns, on the one hand, and benthic biomass and metabolism on the other. Benthic organisms modify microtopography by changing grain sizes (pellet production, track formation, all kinds of constructions), but also by mucus production and other activities, thus increasing or decreasing bottom roughness and strongly modifying physically created fluxes [21]. Sedimenting plankton cells are rapidly invorporated into the sea floor sediments [22]. The penetration depth of particles and O_2 into the sediment, as well as the porewater transport to the sediment surface, are greatly influenced by structures (ripple marks, mounds etc.) at the surface [23] [24]. Large areas of tubes of polychaetes, for example, may reduce current speed and cause passive deposition of laterally drifting particles [25]. On the other hand, most investigations on the effects of tubes are still lacking a systematic separation of pure hydrodynamic effects from secondary biological effects [21].

The importance of benthic suspension feeders in regulating primary and secondary plankton production has been recognized, and sometimes quantified, in littoral and estuarine systems, [26] [27] [28] [29]. Suspension feeders have been observed to consume up to 90% of daily pelagic production during specific time periods. Experiments with artificial mussel beds, where the siphons of the bivalves and their activity were replaced by small tubes, and pumps [30], confirmed the enormous filtering capacity of bivalve aggregations. Further examples of large amounts of phytoplankton and seston biodeposited by dense clam populations are presented by Graf and Rosenberg [21]. Few quantitative estimates are available in the literature for other groups, because they require considerable effort to quantify food capture and ingestion as well as species production and biomass. They are lacking for most communities in which molluscs are not dominant, although density and diversity of suspension feeders may attain very high levels there. Consumption by certain species of tunicates has been calculated by Klumpp [13] and Kühne [31], and that of certain hydroid species by Gili and Hughes [32]. Furthermore, the results from many suspension-feeding sessile organisms have highlighted the fact that in coastal ecosystems these organisms need to be taken into account when assessing or modelling energy flow [33].

Another recently explored aspect of pelago-benthic coupling is related to the grazing by benthic invertebrates on small plankton in the water column. Planktonic cells, less than 5 µm in size, known as nano- and picoplankton, are the main contributors to marine productivity and biomass [34], and their interactions constitute the marine microbial food web. This trophic web has been extensively studied in the water column and has received much attention from the planktologists in the past decade. Recently, large experimental measurements have been carried out in tropical seas and in the Mediterranean that quantified the diet and the metabolic contribution of all types of plankton to several species of benthic suspension feeders. Many benthic invertebrates from a variety of phyla have the capacity to feed on the pico- and nanoplankton in the water column [35] [36]. However, this aspect has been quantified only recently by the use of flow cytometry [37] [38], demonstrating the high grazing efficiency of benthic invertebrates on all water column communities, including the microbial food web. The role of benthic suspension feeders in marine food chains has been demonstrated in several empirical papers [39]. If the total amount of water that is processed through the filtering devices of suspension feeders is considered, the impact of this activity should be recognizable in most of the water column, especially in littoral ecosystems. If these effects of the benthos are not considered in ecosystem approaches [40], the authors are forced to seek complicated explanations that could be avoided if all components of the system were considered.

An interesting subject under investigation in the framework of SCAR's EASIZ (Ecology of the Antarctic Sea Ice Zone) programme is the feeding behaviour of Antarctic suspension feeders on the Weddell Sea shelf. They appear not to utilise the rich rain of (fairly large) particles and aggregates sedimenting from the water column during the short Antarctic summer [41]. Moreover, thick layers of fluff consisting of diatom frustules do not seem to be used directly by other benthos, either [42]. Similar observations have been made in other benthic ecosystems, (Graf, pers. comm.). It can be concluded that multicellular organisms have problems in responding directly to algal blooms. There are two possible explanations, 1) the material is inadequate for use because it is too large or due to spines, skeletons or chemical composition; 2) suspension feeders have lost their capability of feeding on fresh material during glaciations when there were no algal summer blooms due to extended ice cover, and they had to live entirely on laterally advected (and broken down) material.

Moreover, the recent discovery that several littoral species in Antarctic waters continue feeding throughout most of the year [43] [44] suggest that depending on a less rich, but continuously available food source may be just as attractive as depending on rich food pulses during a short period of the year. At the present time, the feeding behaviour of high Antarctic suspension feeders is being studied to find out what alternative trophic sources ther are compared to related species in temperate and tropical areas. This approach has been one of the major tasks during the recent "Polarstern" EASIZ II cruise.

The input from the pelagial is not always favourable for the benthic fauna. Massive surface production, as in the large upwelling systems, causes oxygen deficiency at the seafloor [45] and toxic algae may cause considerable damage to continental shelf and shallow-water communities [46] [47].

One enigma of pelago-benthic coupling remains. Despite all efforts to quantify the input from the pelagial and its effects in the benthal [48], for most benthic systems, conservative estimates conclude that all material sedimenting from the plankton is needed just to allow for macrofaunal secondary production. The biomass of smaller benthic organisms may well be equivalent to that of the larger ones. Taking into account the much higher conversion efficiencies of the smaller organisms, it is by no means clear how they derive their productive energy. Lateral advection (see below) may account for some of it. Thus modelling encounters serious problems when it deals with this very basic aspect of pelago-benthic coupling.

3. Bentho-pelagic coupling

A benthic view of bentho-pelagic coupling (processes where the benthos provides a major input to the pelagial) has been presented by Graf [48] (who does, however, not use the term exactly as we are using it in this paper). As this paper concentrates on larger organisms, we only refer to the role of benthic bacteria in setting free nutrients and making them available again to life in the water column. A well-known example of bentho-pelagic coupling are benthic organisms that spend part of their life cycles in the water column, such as meroplanktonic larvae. The degree at which the benthos thus contributes to pelagic life differs latitudinally and among the benthic size fractions (see below) and is most important in tropical and temperate seas. Furthermore medusae, as the product of the vegetative benthic generation of hydrozoans, contribute significantly to pelagic predation.

Referring to some lesser known examples, canyons provide specific hydrodynamic conditions with fronts, up- and downwelling and fluxes that create a special habitat characterized by great local density and diversity of benthic and pelagic fauna, exceeding that of other habitats along the continental shelf and slope [49] [50]. Although most available information is only anecdotal, it suggests that canyons are areas of enhancement of biological productivity and species richness. Canyon hydrography seems to indicate that they are "collectors" from the upper water layers, (through downwelling) and distributors of such water coastwards, (by upwelling) [51]. This phenomenon is important in the context of the deposition of resting stages of macro- and nannoplankton in coastal canyons of the Catalan coast. Large amounts of cysts belonging to superficial and littoral species have been detected down to1200m depth in these canyons. They can be transported back to surface waters by upwelling, and may well be responsible for algal blooms observed at certain times at the surface, thus representing a good example of the importance of knowledge of life cycles for understanding ecosystem functioning [52] [53]. Benthic meiofauna, by predation, appears to an important impact on "recruitment" of the cysts in the water column [53]. The hypothesis of canyon-driven accumulation of resting stages of plankters at the seafloor and their return to the water column is a good example of bentho-pelagic coupling, which might turn out to be of major importance also for coastal zone management.

Coral reefs are another interesting case since despite living in nutrient-poor waters have a very high primary production due to the symbiosis between corals and dinoflagellates (zooxanthellae). This is a sensitive system that may break down ("bleaching") due to unusual warming or increased solar radiation, e.g. by El Niño. Sometimes the internal primary producers are expelled and, together with coral mucus, provide a major food source for zooplankton [54]. The mutual interactions between zooplankton and symbiotic algae in the corals seem to be closer than was suspected a few years ago.

In the Humboldt Current ecosystem, an oxygen minimum zone stretches from moderate depths to about 700 m due to the very high production in the euphotic zone, which provides organic matter in excess of what can be remineralized in the water column and at the seafloor [55]. This zone is colonized by thick mats of sulphur, "spaghetti", bacteria *(Thioploca* sp.), which have attracted much attention recently [56]. During El Niño events, flushing occurs at the seafloor by intrusion of O_2-rich waters from the equator and, possibly, by reduction of surface production, and a much richer macrobenthos community develops whereas the spaghetti bacteria decline [57]. Bentho-pelagic coupling strongly increases during El Niño as hake *(Merluccius gayi)*, which is normally pelagic in this system, and many invertebrate predators, start feeding on the benthos [58] [45].

In addition to the well-known provision of food to demersal fish and some mammals and birds by the secondary production of the benthos, biogenic reefs, kelp forests and seagrass beds serve as a shelter or refuge to shoals of fish and pelagic crustaceans. Furthermore, a number of benthic invertebrates perform regular, primarily nocturnal, migrations into the near-bottom layer of the water column. They thus directly interact with pelagic organisms, and may constitute an important food for demersal or even pelagic fish. This supra- or hyperbenthic fauna is dominated by harpacticoid copepods or peracarid crustaceans, and aggregates within 1 to 2 m above the bottom, [59] [60] [61] [62] [63]. Yet another example of benthos providing food for animals in the pelagial is Antarctic krill feeding at the seafloor [64].

In summary, pelago-benthic and bentho-pelagic coupling, including recirculating processes, are two-way systems that provide excellent opportunities for cooperation between benthologists and planktologists wherever they are willing to cooperate.

4. Life cycles, population dynamics, reproduction

The study of population dynamics and life strategies of benthic species is a traditional field of benthic research. Yet, it is the basis for understanding the dynamics and function of benthic communities [65], and it provides part of the information that is needed for stock management and aquaculture.

Many benthologists working in temperate areas tend to think that life-history strategies, population dynamic characteristics and reproduction are similar worldwide, since they usually only work with one local system. It is important, however, to see the local patterns in a global context in which the Baltic, the North Sea or Chesapeake Bay have their place on a continuum of high variability. Of course, there are also within-system variations due to the fact that organisms always exploit a variety of life-history strategies, but notwithstanding this fact there are latitudinal patterns which apply to a majority of species in a given area.

A large proportion of benthic macrofauna has pelagic (meroplanktonic) larvae, and there is a cline towards a much lower proportion in polar regions [66] [67]. Since Thorson started the discussion there has been debate about to what extent ecosystem seasonality, (polar or temperate and tropical), is responsible for the kind of pelagic larvae occurring, (lecithotrophic, planktotrophic), and for the longevity and the proportion of benthic species producing such larvae, [66] [67] [68] [69] [70]. Obviously there is a gradient in reproductive patterns between the tropics and polar regions, resulting in a higher share of species with delayed embryonic development, late first maturity, low number of well-equipped offspring and possibly a lower proportion of species with pelagic, in particular planktotrophic, larvae in polar areas. The validity of this latter paradigm, often referred to as "Thorson's rule", (which should, however, include also the other facets mentioned above), has been doubted recently [70] [71]. Many pelagic larvae have recently been found in polar areas, mainly from shallow water. While it is clear that in the future further meroplanktonic larvae will be found as the frequency of observation increases and improved gear becomes available, it is not clear whether this will invalidate "Thorson's rule". In fact, food availability may not be an insuperable problem for planktotrophic larvae as the microbial food web may persist throughout the winter. In any case life must be precarious in a water body almost devoid of larger particles over a long winter season [72].

There are, however, important trophic implications from these considerations. The mere fact that light conditions are continuous throughout the year in the tropics and extremely seasonal at polar latitudes, with intermediate conditions at temperate latitudes, implies very distinct patterns of primary production in the water column and food availability in these climatic zones. In tropical latitudes a pelagic larva or juvenile finds fairly constant, although not always rich, food supply throughout the year, whereas at high latitudes the larvae either adapt to a short summer input or uncouple themselves from this input altogether. In the latter case this means a benthic life, subsisting on the detritus food chain, or even remaining in the care of the adults. Also a large part of the adult fauna uncouples from fresh primary production, feeding on the detritus chain, as predators or scavengers. Necrophagy becomes a very important strategy under these circumstances [73]. For most polar species this means that on an annual average, they grow slower than their temperate or tropical relatives, (which is not contradicted by the fact that summer growth may be rapid). There are strong parallels here to the deep sea [74] [75] [76], and in both ecosystems there are exceptions to the rule of slow growth [77] such as boring bivalves [78], and the sponge *Mycale acerata*; [79]. Other facets of reproductive biology or population dynamics are slowed down as well, as has been confirmed for many Antarctic species, in some cases also on a gradient from high Antarctic to Subantarctic waters [80] [81]. First maturity occurs very late, e.g. in a number of Antarctic benthic crustaceans, spawning frequency is lower, and so is the number of eggs that are, however, large and yolky [82]. All these patterns together indicate that while a temperate shrimp such as *Crangon crangon* is not particularly sensitive to exploitation, Antarctic shrimps (or king crabs, and scallops) would be severely affected by a fishery. This is an important point for fishery managers and conservationists, because a large share of sensitive species also makes a fragile ecosystem (see also paragraph on community dynamics).

5. Benthic interactions

Benthic communities have the advantage of being fairly stable in space, at least in comparison with pelagic communities, and thus make a good framework for experimental manipulation of interactions. The artifacts that may be caused by the experimental setup (e.g., hard-bottom cages on soft bottoms) have been discussed at length [83] and are mostly well-known. There is no doubt that experimental manipulation at the seafloor can contribute substantially to the knowledge of ecosystem functioning if carried out in the right way and not interpreted too naively.

Biological interactions have been studied quite intensely in shallow-water systems from temperate regions to the tropics, in part by the use of experiments, but such studies have been scarce in polar regions, (but see [84]) and, for obvious reasons, in the deep sea (but see [85]. The contribution of such studies to ecological theory has been significant. Key processes that have been investigated include various forms of predation, parasitism, competition, mutualism, inhibition and facilitation in the context of succession theory, [2] [86] [87] [88] [89] [90] [91] [92] [93].

"Chemical ecology" has recently made important contributions to clarifying questions of species interactions. The capacity of many sessile benthic species to synthesize toxic substances may explain part of the observations on competition for space [94]. Many of these species are able to displace other species or to prevent being overgrown, by producing allelochemical substances, which inhibit growth or reproduction of competitors [95]. Quite a number of species invest energy in the production of these substances instead of reproduction or growth [96]. Other repellent or attractive substances are produced throughout the benthic realm, serving a great number of different purposes including protection from predators, symbioses, and reproduction. The existence of these "secondary metabolites", which are of great importance to the pharmaceutical industry, has called the attention of the chemists to the study of marine benthos. Chemical ecology is now contributing significantly to the study of structure and function of benthic communities.

For a long time, the question whether competition or predation is the more important factor in structuring benthic communities dominated discussions on the subject. There is, of course, no general answer, and instead of polarized debates it might be more productive to quantify the relative importance of the two processes in different types of communities [97] [98]. Coexistence among species in restricted space has been studied in epifaunal communities on hard bottoms, and led to the development of models such as the "competitor dominant overgrowth" [99]. Other studies revealed the mechanisms of species coexistence and exclusion in space and time in the rocky intertidal [100] [101]. Dayton [102] summarized these interactions for hard and soft bottoms: Hard-bottom communities tend towards competitive monopolization of space, which is controlled by hierarchies of grazers and predators in kelp communities or interrupted by physical disturbance (see below). In contrast, most soft-bottom communities exhibit little evidence of resource monopolization, and predation and disturbance go along with other biological factors, in particular biologically induced heterogeneity by spatial structures, such as tubes or by large infauna, that provide niches for smaller species [103] [104].

Furthermore, interactions between biotic processes, or between biotic processes and a variable physical environment, and also a wealth of indirect effects have been revealed by experimentation. Together these imply complex webs of interactions, albeit usually dominated by one or a few key species. On rocky shores, however, it often is the case that all these post-settlement processes are of minor importance because of recruitment limitation [105] [106] [107] [108] [109]. On soft bottoms, on the other hand, Ólafsson *et al.* [110] argue that recruitment limitation is less important than are post-settlement processes in determining population density and community organization.

6. Larval settlement and recruitment

Key interactive processes that are common to pelagic and benthic systems are settlement and recruitment. The "recruitment problem" has occupied pelagic researchers from Hjort [111], through Cushing [112] and Lasker [113], up to the present day [114]. On the seafloor it is by no means understood despite great efforts undertaken from those of Boysen Jensen [115] and Blegvad [116] to Thorson [66] [117] and Connell *et al* [118]. The measurement of hydrodynamic processes affecting benthic recruitment has been refined considerably [119]. Refuges and larval nurseries have been found to be important particularly in hard-bottom communities [102] and already decades ago great progress was made in finding out what factors may attract or reject settlement in meroplanktonic larva (e.g. Wilson, [120]). Alternative hypotheses to active habitat selection such as passive deposition, can be reconciled and are likely complementary [121]. However, there is still no answer to the question what determines year class strength and most likely there will be many answers [122]. The various phases involved in the recruitment process, (larval advection, pre-settlement, larval settlement and recruitment, post-larval dispersal) and their time scales have been identified, but their relative importance for the success of a year class in different communities (such as hard and soft bottoms) continues to be under discussion [106].

Many soft-bottom researchers seem to agree that while both larval and post-larval dispersal are important, it is the latter that principally determines recruitment [123] [124] [125]. In temperate coastal regions, strong bivalve recruitment often follows severe winters [126] [127], while repeated recruitment failure occurs after mild winters [128]. This pattern is caused by a temporal match-mismatch between juveniles of bivalves as prey and juvenile crabs and shrimps as predators. When settling after a severe winter, young bivalves may achieve large size immune to predation before their predators arrive. During a mild winter predators arrive earlier and the young bivalves are just of the right size for their predators, [129] [128]. In addition, female bivalves surviving a cold winter produced more eggs than after a mild winter when they suffer a higher weight loss due to elevated maintenance costs [130]. These two processes, fecundity and predation, operate in the same direction, and this may cause the high predictability of the recruitment pattern.

The field of larval settlement and recruitment is not only important because it connects pelagic and benthic research, but is also of utmost importance for benthic resource management all over the world. Sometimes environmental conditions and biotic interactions at a given moment are not very helpful in finding the causes for the structures observed. Instead, present structure was caused by physico-chemical and biological factors that were governing in the past. This means that, in contrast to plankton communities where most processes occur in short periods and reduced evolutionary complexity, long time scales may have to be considered to understand the present-day structure and function of many littoral ecosystems, particularly on hard bottoms. In some cases the fossil record can be helpful where shapes and distributions of the fauna may provide hints as to environmental conditions and ecological constraints in the past.

7. "Communities"

The separation of ever more "communities" or subcommunities has occupied and, obviously, fascinated generations of benthologists. However, this fascination has seldom spread to scientists working in other fields (and even less to students)! To make our point quite clear, we are not principally opposed to clustering and the delimitation of communities, it can be a useful tool for determining spatial and temporal gradients and in defining the consequences of pollution. We are just against all those fruitless clustering exercises where people are not able to tell you why they did it!

In fact, the original questions put forward by Möbius [131], Petersen [132], Thorson [133] and others were interesting ones: Do certain environmental conditions produce a predictable association, or assemblage, of benthic species? Will we be able to distinguish that same kind of community when we return after a decade or two (provided there has not been a radical change in the environment)? Do similar environmental conditions produce similar communities all over the world? However, it seems as though the answers to these questions have been available for quite some time now. Certain environmental conditions do produce certain kinds of assemblages, but the prediction of limits of assemblages are rather weak and the separation is sometimes easier by statistical methods than by eye (or the reverse). Benthic communities tend to be quite persistent if environmental fluctuations are low and disturbance is not too strong, but they increasingly exhibit other patterns (or stages) if the oscillations of the environment and the frequency or severity of disturbances increase. However, if earlier conditions return at least some communities will return to former patterns [134].

Examples for quite persistent communities include coral reefs, Antarctic sublittoral benthos below the ice scouring zone, and the deep-sea fauna. At intermediate levels of persistence we would consider lower subtidal shelf communities, upwelling, and tropical and subtropical shallow water where disturbance (by, e.g. monsoon rains) is low. Low levels of persistence are found for example in the Baltic or Southern North Sea associations (and others under strongly oscillating estuarine conditions). However, even Baltic communities reveal a certain degree of persistence if time scales are chosen long enough to include fluctuations as part of the pattern.

Similar environmental conditions produce similar communities in different areas of the world ocean. However, these "parallel communities" [133] [135] are characterized only in a few cases by the same genera; in most cases the dominant genera differ, but they often have the same function and they are quite often of similar appearance. Some French [136] and Russian [137] approaches also concur with this observation.

From a structural point of view, there are striking similarities between some marine benthic (mostly hard-bottom) communities, revealing parallels even with terrestrial ecosystems. For example the stratification of coralline structures has a stratum of erect and ramified organisms or colonies of corals and gorgonians, a second stratum of massive organisms, and a third stratum of encrusting organisms. Hidden within the substra, are epibiotic fauna and flora living between colonies, crevices and holes in the reef. This resembles a tropical rain forest with its different strata of vegetation and associated fauna [138]. Similar structures are the sublittoral coralligenous hard bottoms in the Mediterranean Sea [139] and in Ireland, Scotland and Norway, or the three-dimensional Antarctic epifaunal communities [84] [140], the latter living, however, on soft bottoms. The advantage of these parallel structures is that existing models can be applied on similar communities for comparison.

Similar environmental conditions not only create structurally or dynamically similar communities, but also morphologically similar species which are, however, genetically distinct. These organisms include the sibling and sympatric species, which are common in many benthic groups [141]. The elevated capacity for phenotypic adaptation among benthic species has led to the problem that their heterogeneity can sometimes only be detected using genetic methods, although distinct physiological and functional properties and life history traits are discernible. Particularly the latter are different from pelagic or terrestrial communities [142].

Benthic communities form a mosaic of organisms with different life strategies, which coexist in a relatively reduced space (at the ocean scale). This space provides the substrate for a large number of epifaunal, infaunal and colonizing species generating a three-dimensional structure whose complexity depends on the spatial heterogeneity, available energy (food), the amount of disturbance (see below) and other factors. Elevated complexity is organized in the way that the diversity of ecological niches tends to be very high and the rules for coexistence tend to be very complex. Study of the structure of both littoral and sublittoral communities, similar to the studies on interaction (see above), has stimulated ecologists to develop new theories, or to advance existent hypotheses. For example, Horn and MacArthur [143] DeAngelis and Waterhouse [144] have developed new concepts regarding the niche specialization theory, arguing that the high spatial heterogeneity in benthic habitats as compared to the water column enables many species to coexist in a multitude of specific ecological niches.

8. Community dynamics

Studies of dynamics, both at the population (see above) and the community level are particularly important in benthic ecology. Population and commuity dynamics are measured at various scales. Diel and seasonal community patterns, as well as interannual variability, have long been measured in the benthos. Knowledge of such patterns are important to judge whether a sample taken at a certain place or moment in time is representative. The response of species in a community (and thus, of the community as a whole) to environmental stresses depends on whether the stress is common in this kind of environment, (so the species could adapt), or whether it is unusual. Typical stresses are extreme temperatures, low salinity, oxygen depletion, wind, ice, and the ENSO (El Niño- Southern Oscillation). This may be the reason why man-induced disturbances have such a disastrous effect in many communities. Equally, the resilience of a community, (its capacity of swinging back into the *status quo ante,* is based on the reproductive biology, recruitment and migratory capacities and trophic characteristics of the species involved, seems to be ecosystem specific. (See Sutherland's [101] multiple stable points and Gray's [145] neighbourhood stability). Resilience is usually high in strongly fluctuating ecosystems with frequent stress (examples are the Baltic Sea, southern North Sea, or upwelling areas). In highly persistent systems subject to comparatively little disturbance under natural conditions (examples are the deep sea, Antarctic benthos below the ice scouring zone, and coral reefs) resilience is low. When making comparitive studies it is important to consider disturbance and resilience at similar spatial scales [118] [146]. In most resilient soft-bottom systems, opportunistic pioneer species with high growth rates, short generation times and high dispersal ability follow a disturbance in a series of successional stages tending towards higher system complexity [147] [148] [149] [150] [151] [152]. In the rocky intertidal, environmental disturbance initially reduces the levels of competition and coexistence [153], but the assemblages gradually recover along highly complex successions which occur stepwise or in a progressive gradient of structural complexity and competition [154] [155].

The severeness of the disturbance also plays an important role. The original, rather intuitive feeling that any kind of stress or disturbance should be disastrous, see Remmert, [156] was replaced by a view indicating that the effects of a disturbant depend very much on its size [157] [158] [155]. A good example is El Niño which along the latitudinal gradient can produce all kinds of effects from harmful (e.g., mass mortalities), to beneficial (e.g., enhanced recruitment and growth) of the same species. Beyond a certain threshold (amount or frequency of disturbance), even very resilient communities may fail to recover (examples are the Baltic deep-water communities suffering the anoxia). Return to "normal" conditions does not always mean recovery, but sometimes the reverse, (e.g. El Niño effects on macrobenthic communities in the Humboldt Current upwelling [57]). Strong El Niños also have a serious impact on tropical coral reefs, destroying the sensitive balance in the reef and initiating complicated interactions and successions among corals, sea urchins, alpheid shrimps and green filamentous algae, or between corals and *Acanthaster planci* [159].

From a practical point of view, it is important to know that many soft-bottom communities in boreal regions show a predictable development either to the more mature or to more impoverished communities [160] [161]. Communities can recover to their original appearance after considerable time, provided the environmental conditions return to their starting point, e.g., pollution inputs cease and recovery is initiated from surrounding, non-affected areas. However, no recent large systems have recovered after development of persistent hypoxia or anoxia, dissolved oxygen being the environmental variable in coastal ecosystems that has changed most drastically in recent times [162]. Recovery is scale dependent; e.g., large-scale disturbances that kill organisms playing a role in habitat stability are likely to result in very slow recovery [146].

The knowledge of community resilience is also important for ecosystem conservation. It is obvious that some marine ecosystems withstand much anthropogenic impact whereas others are very sensitive or even fragile. We have to know these properties to take preventive measures in time and not to exceed certain thresholds. Government officials have to be told that there are good reasons not to treat a coral reef or Antarctic benthic communities in the same way as those on a hypoxic seafloor in the Baltic or in the Benguela upwelling.

9. Diversity

After a period of neglect of traditional taxonomy the study of diversity has received increased attention recently due to the spectacular development of molecular genetics. In addition there is increasing environmental awareness and a growing perception among ecologists that there must be a connection between the variety of species in a community and its functioning (and of course, their perception that most good ecology is based on sound taxonomy, cf. Dayton, 1990 [163]). "Diversity" must be clearly defined in each case (the tendency to use the misnomer "biodiversity" rather has the opposite effect), and it is highly dependent on the methodology used and on scales. Structual complexity supports higher diversity [164] [165].

There are obvious differences in benthic diversity all over the world and ecologists have been interested in assigning a certain order to this variety. There has been a discussion on latitudinal clines and depth gradients in the world ocean for decades, and with few noteworthy exceptions [166] [167] [168] certain paradigms have been passed from textbook to textbook without being considered critically. One of these paradigms is the decreasing species richness, (we would prefer to refer to diversity, but the data situation there is even worse), of continental shelf benthos with increasing latitude [169], i.e. from the tropics to the poles ("bell-shaped curve"). In fact the evidence presented by some authors referred to calcareous organisms, such as molluscs and foraminiferans, but many authors misunderstood that and extended the argument to benthic fauna in general (see Clarke, [170]. That, however, does not hold true, we now know quite a few groups where species richness increases again towards the Antarctic (see IBMANT, [171] [172] [173]. The two polar regions seem to differ considerably in this respect [174] [163] [175], but also from the Arctic high diversities have occasionally been reported [176].

Another paradigm which has received some attention recently is that of higher diversities in the (upper) deep sea compared with continental shelves and slopes [169] [177]. This paradigm has been questioned [166] [167] and rejected using material from the Norwegian continental shelf and upper slope and shallow Australian waters [179] [178].

A third paradigm says that the open sea tends to have greater species richness than inshore habitats [180]. The important question here may be where the limit is drawn between the two areas, and what type of inshore habitats we are talking about.

In most cases the data on which the paradigms were based cannot really be compared. Benthic investigators have been ingenious for over a century in inventing ever new devices for the ideal collection of their subjects, but they have paid very little attention to comparability. Comparisons have been made between samples from van Veen grabs with those from anchor dredges or box corers or even Agassiz trawls, and each investigator uses a different way of washing the samples on a different mesh. Proper standardisation of methods is needed before the ecological importance that such comparisons should merit can be achieved. We must also take scales into consideration. Taking three one-square metre samples from the three principal sediment types in the North Sea tidal flats may yield 90 % or more of the non-motile local macrofaunal species, but the same exercise in the deep sea or on a tropical subtidal bottom may yield a small percentage only. A good example of what might be considered is the recent comparison between an Antarctic and an Arctic site by Starmans [181].

The diversity-stability relationship has occupied benthic ecologists for a long time. Here, too, is an original paradigm that higher diversity should mean higher stability (without defining very clearly, at first, what "stability" meant). Some interesting approaches to define the term "stability" into the three components, resistance, resilience, and persistence, which are mostly used today, include the papers by Peterson [182] and Boesch and Rosenberg [183]. Sanders [169] [183], in his stability-time hypothesis, was the first to present a statement which could be tested by comparing various communities and which has stimulated the discussion for decades. Sanders [169] developed a "rarefaction" technique (which was later modified by Hurlbert, [184]) to compare not only species richness but also the relations between species numbers and numbers of individuals of each species. He noticed distinct diversity differences among the communities. Although the material he had at hand was somewhat limited, he concluded that ecosystems with a much greater age revealed a higher diversity than young ones. Comparisons of subtidal communities in the tropical warmwater belt and the deep Antarctic shelf, (both old systems although of quite different age), with very young systems such as the Baltic Sea or the southern North Sea) support the hypothesis. However, he also referred to greater diversities in those ("predominantly biologically accommodated") systems was due to lack of disturbance. Dayton [102], judging from hard-bottom communities, and Dayton and Hessler [158] (1972), in a deep-sea study, questioned this argument and suggested that a certain amount of disturbance may even be necessary to keep diversities high, as would also be expected from the intermediate disturbance hypothesis [138]. Several authors have supported this conclusion (e.g., Rhoads *et al.*, [149]), and recently it has been supported once more by the mosaic cycle hypothesis [186] [187] which indicates that moderate disturbance may lead to the co-existence of many successional stages and thus increase diversity at larger spatial scales. The reason is that the, often quite different, species in the various stages of succession accumulate. A nice example is the iceberg-ridden zone on the Antarctic continental shelf [188]. However, it is uncertain whether soft bottoms (which are not as much subject to monopolisation of space as are hard bottoms, see above), at very low disturbance levels might not also be a source of high diversity, as was originally suggested by Sanders.

International, across-latitude studies of benthic diversity with identical methods are necessary to get a clearer image of the patterns [168]. Diversity is an important community property to study because it may contribute to illustrate community function. And finally, it may be important for people engaged in ecosystem conservation to learn that the more diverse communities are rather on the more sensitive side and may require more protection than the highly resilient ones composed of only a small number of species.

10. Biogeography and evolution

Benthic communities have undergone important changes in the past and will continue to change in the future. The only way to predict in which way this change may occur is to study past evolutionary patterns. The Antarctic-South American (Magellan) connection is a particularly interesting case because the final separation occurred only about 20 million years ago, with periods of presumably more intense and less intense interchange up to the present day due to the proximity of the two continents and climate change [189]. Under present-day conditions, Antarctic isolation is rather marked because of the circumantarctic current system, separated by the deep sea in the Drake Passage and a steep temperature gradient.

The result of this 20 million years of separation is reflected in quite distinct ecosystems with different structure and dominance patterns, despite a number of similarities at higher taxonomic levels. Much work has to be done to elucidate patterns and exact times of extinction, emigration and immigration between the two systems, and both traditional taxonomy and molecular genetics will be helpful in this respect. For some groups these patterns are now fairly clear, e.g., for arcturid and serolid isopods [190].

Besides providing bases for the predictions of climate change an alteration in present species compositions and dominance patterns, this research also contributes to the knowledge of latitudinal gradients referred to above. Finally, it explains strange patterns such as the apparent eurybathy of some Antarctic invertebrate groups, which may have been caused by shifts in the ice shield which during glacial periods covered most of the shelf and forced those benthic species that were able to do so down the slope [191].

11. Conclusions

We believe that the examples given in this paper, despite their highly selective character, sufficiently support our point that marine ecology should include, as one of its pillars, research at the seafloor even where there is no direct relation with water-column research. Obviously, pelagic and benthic research, including the coupling of the two subsystems, should be combined wherever possible. But as there are pelagic themes with little relation to the benthos, benthic research by itself includes a great variety of themes that are by no means boring or outside present fields of interest in marine ecology. We do not claim that progress in marine ecology has come only from benthic research, just consider the long path, referred to above, from Hjort's second hypothesis via Cushing's match/mismatch hypothesis and Lasker's studies on fish larvae to Sinclair's attempt to summarize explanations of the bottleneck in fish recruitment. However, the benthos offers a wide range of opportunities to study ecological questions, and the contribution of benthologists to ecological theory (no matter whether their ideas finally proved right or wrong) has been, and continues to be, important. Why, then, are there no large international programmes focussing on these questions?

One reason may be that pelagic researchers can cooperate much more easily with physical oceanographers and meteorologists, in fact even with marine geologists, because processes in the pelagial have a direct bearing on ocean-atmosphere interactions, ocean currents and the fossil record in marine sediments. Unfortunately, some important species of macrofauna do not fossilize very well in marine sediments. An example is Antarctic krill which, despite its enormous abundance and presumed several million years of existence, has not left traces in the sedimentological record, (M. Thomson pers. comm.). The reason is that rich benthic assemblages normally thrive where oxygen conditions are favourable, resulting in rapid breakdown and difficult fossilization of organisms. Furthermore, physical oceanographers, planktologists and geologists have a common preference for the deep sea where the manifold disturbance effects are missing that characterize coastal zones and the continental shelf. Marine science is largely interdisciplinary; as a consequence, programmes have the advantage that several disciplines can cooperate. Furthermore, the present preference for subjects related to global climate change supports those programmes that have a direct relation to atmospheric research and carbon cycling, (biological C pump). Although a large amount of carbon is fixed in biological structures on the seafloor, most planktologists argue that this amount is negligible compared to the role of plankton organisms such as pelagic diatoms or copepods. Cautious attempts to ascribe a certain importance to other organisms, e.g., in the APIS seal programme [192] have been turned down [193]. It might be worthwhile assessing the global role of benthic structures and processes.

Pelagic research also profits from the fact that many of the large ocean fisheries concentrate on pelagic shoaling fish such as anchovies, sardines, (jack) mackerels, herrings, blue whiting, capelin and even hake, which only in a few cases have a direct relation to the benthic ecosystem. Hake has certain demersal traits during El Niño events (see above), and so does herring due to its demersal eggs, but they are exceptions. Fishery managers are interested in the sedimentological record of scales and otoliths of shoaling fish, but unfortunately, this subject is mostly studied by geologists. Similarly, those species that have a good record in the sediment, such as radiolarians, planktonic diatoms and foraminiferans, have been of interest to geologists rather than biologists.

However, we should also consider that benthologists have committed mistakes themselves, apart from the fact that the low priority given to interchange between the two fields is not only a characteristic of the people working in the water column. Perhaps there has been too much emphasis on describing the structure of benthic communities (or assemblages, or associations) and on separating different kinds of communities. Of course we have to acknowledge that for realizing the function of benthic assemblages under present-day conditions it is important to know which species co-exist, and this is also important to palaeontologists who can use this information to interpret conditions in the past ("actuopalaeontology"). But we may have exaggerated. Similarly, the present desperate situation of classical taxonomy may have been caused by the fact that too many benthos taxonomists were sitting in an ivory tower without making an effort to relate their research to actual ecological problems. Furthermore, the benthologists themselves have to improve their ability to make rapid and precise identifications of species.

Another reason may be that presently there is a strong tendency towards "applied" ecology to which classical benthic ecologists, those outside resource management, have responded only marginally. To avoid misunderstanding, by no means do we suggest that all benthic research should be applied, basic research is a prerequisite for any kind of applied research. However, more efforts could be made to make advances in basic research available to, and suitable for, conservationists and managers. The Pearson & Rosenberg environmental assessment model is a positive example as it may allow quick management decisions. Models of this kind are badly needed also for tropical and polar marine environments.

Finally it seems as though regionalism has been more common among benthic researchers than among planktologists, (which, of course, reflects the fact referred to initially that there is much more variability among benthic ecosystems than in the water column). Few people have the opportunity of comparing benthic systems on a worldwide basis and to assign the appropriate place to their Kiel Bay, Limfjord, Chesapeake Bay or wherever they work. So, with a few notable exceptions, the Baltic or European Marine Biologists Associations are still regional clubs. Sadly many North Americans do not read European or Latin American literature, despite the fact that everybody is now making a great effort to write in English. Young scientists re-invent the wheel over and over because they do not consider the literature of even such a short period as 10 years. Of course all these shortcomings are not restricted to benthic researchers, but in view of the strong "innate" regionalism they have certainly contributed to the low acceptance of our field internationally.

12. Acknowledgements

The authors would like to thank John Gray for the invitation to a stimulating meeting, Gerd Graf for providing material for the chapters on coupling between the pelagial and the benthal, and Eike Rachor, Brigitte Hilbig and Julian Gutt for useful comments on an earlier version of the manuscript. The senior author is grateful to the Spanish Ministry of Education and Culture, which provided the basis for assembling great part of the information for this review within the framework of a Humboldt-Mutis Award. This is AWI publication no. 1487.

13. References

1. Margalef, R. (1978) General concepts of population dynamics and food links, in O. Kinne (ed.), *Marine Ecology*, Vol. IV. John Wiley, pp. 617-704.
2. Paine, R.T. (1994) Marine rocky shores and community ecology: an experimentalist's perspective. *Excellence in Ecology* **4**, Ecology Institute Oldendorf/Luhe
3. Bonsdorff, E. and Blomqvist, E.M. (1993) Biotic coupling on shallow water soft bottoms - Examples from the Northern Baltic Sea. *Oceanogr. Mar. Biol. Ann. Rev.* **31**, 153-176.
4. Josefson, A.B., Jensen, J.N. and Aertebjerg, G. (1993) The benthos community structure anomaly in the late 1970s and early 1980s - a result of a major food pulse? *J. exp. mar. Biol. Ecol.* **172**, 31-45.
5. Margalef, R. (1997) Our biosphere. *Excellence in Ecology* **10**, Ecology Institute Oldendorf/Luhe.
6. Margalef, R. (1979) The organization of space. *Oikos* **33**, 152-159.
7. Shelford, V.E. (1963) *The ecology of North America*. University of Illinois Press.
8. Leonard, A.B. (1989) Functional response in Antedon mediterranea (Lamarck) (Echinodermata: Crinoidea): the interaction of prey concentration and current velocity on a passive suspension feeder. *J. exp. mar. Biol. Ecol.* **127**, 81-103.
9. Gaino, E., Bavestrello, E., Cattaneo-Vetti, R. and Sarà, M. (1994) Scanning electron microscope evidence for diatom uptake by two Antarctic sponges. *Polar Biol.* **14**, 55-58.
10. Riisgård, H.U. (1991) Suspension feeding in the polychaete *Nereis diversicolor*. *Mar. Ecol. Prog. Ser.* **70**, 29-37.
11. Coma, R., Gili, J.M., Zabala, M. and Riera, T. (1994) Heterotrophic feeding and prey capture in a Mediterranean gorgonian. *Mar. Ecol. Prog. Ser.* **115**, 257- 270 .
12. Fabricius, K.E., Benayahu, Y. and Genin, A. (1995) Herbivory in asymbiotic soft corals supports unusally high growth rates. *Science* **268**, 90 f.
13. Klumpp, D.W. (1984) Nutritional ecology of the ascidian Pyura stolonifera: influence of body size, food quantity and quality on filter-feeding, respiration, assimilation efficiency and energy balance. *Mar. Ecol. Prog. Ser.* **19**, 269-284.
14. Asmus, R.M. and Asmus, H. (1991) Mussel beds: limiting or promoting phyto-plankton? *J. exp. mar. Biol. Ecol.* **148**: 215-232.
15. Coma, R., Gili, J.M. and Zabala, M. (1995) Trophic ecology of a marine benthic hydroid. *Mar. Ecol. Prog. Ser.* **119**, 211-220.
16. Lehman, J.T. (1976) The filter-feeder as an optimal forager, and the predicted shapes of feeding curves. *Limnol. Oceanogr.* **21**, 501-516.
17. Hughes, R.N. (1980) Optimal foraging theory in the marine context. *Oceanogr. Mar. Biol. Ann. Rev.* **18**, 423-481.
18. Okamura, B. (1990) Behavioural plasticity in the suspension feeding of benthic animals, in: R.N. Hughes (ed), *Behavioural mechanisms of food selection*, NATO ASI ser., pp. 637-660.
19. Riisgård, H.U. and Larsen, P.S. (1995) Filter-feeding in marine macro-invertebrates: pump characteristics, modelling and energy cost. *Biol. Rev.* **70**, 67-106.
20. Stephens, D.W. and Krebbs, J.R. (1986) *Foraging theory*. Princeton University Press.
21. Graf, G. and Rosenberg, R. (1997) Bioresuspension and biodeposition: a review. *J. Mar. Systems* **11**, 269-278.
22. Graf, G. (1987) Benthic energy flow during a simulated autumn bloom sedimentation. *Mar. Ecol. Prog. Ser.* **39**, 23-29.
23. Huettel, M., Ziebis, W. and Forster, S. (1996) Flow-induced uptake of particulate matter in permeable sediments. *Limnol. Oceanogr.* **41**, 309-322.
24. Ziebis, W., Forster, S., Huettel, M. and Jørgensen, B.B. (1996) Complex burrows of the mud shrimp *Callianassa truncata* and their geochemical impact in the sea bed. *Nature* **382**, 619-622.
25. Friedrichs, M. (1996) Auswirkungen von Polychaeten--röhren auf die Wasser-Sediment-Grenzschicht. Dipl. Thesis Univ. Kiel
26. Dame, R., Zingmark, R., Stevenson, R. and Nelson, D. (1980) Filter feeder coupling between the estuarine water column and benthic subsystems, in V.S. Kennedy (ed), *Estuarine perspectives*. Academic Press, San Diego, pp. 521-526.
27. Cloern, J.E. (1982) Does the benthos control phytoplankton biomass in South San Francisco Bay? *Mar. Ecol. Prog. Ser.* **9**, 191-202.
28. Officer, C. B., Smayda, T.J . and Mann, R. (1982) Benthic filter feeding: A natural eutrophication control. *Mar. Ecol. Prog. Ser.* **9**, 203-210.
29. Hily, C. (1991) Is the activity of benthic suspension feeders a factor controlling water quality in the Bay of Brest? *Mar. Ecol. Prog. Ser.* **69**, 179 - 188.
30. O'Riordan, C.A., Monismith, S.G. and Koseff, J.R. (1995) The bed hydrodynamics on model bivalve filtration rates. *Arch. Hydrobiol. Spec. Issues Adv. Limnol.* **47**, 247-254.
31. Kühne, S. (1997) Solitäre Ascidien in der Potter Cove (King George Island, Antarktis). Ihre ökologische Bedeutung und Populationsdynamik. *Ber. Polarforsch.* **252**, 153 pp.
32. Gili, J-M. and Hughes, R. G. (1995) The ecology of marine benthic hydroids. *Oceanogr. Mar. Biol. Ann. Rev.* **33**, 353-426.
33. Gili, J.M., Alvà, V., Coma, R., Orejas, C., Pagès, F., Ribes, M., Zabala, M., Arntz, W., Bouillon, J., Boero, F. and Hughes, R.G. (1998). The impact of small benthic passive suspension feeders in shallow marine ecosystems: the hydroids as an example. *Zool. Verh.* Leiden **323**, 1-7.
34. Valiela, I. (1995) *Marine ecological processes*. Springer-Verlag, New York.
35. Jørgensen, C.B. (1990) *Bivalve filter feeding: Hydrodynamics, bioenergetics, physiology and*

ecology. Olsen & Olsen, Fredensborg.
36. Jørgensen, C.B., Kiørboe, T., Møhlenberg, F. and Riisgård, H.U. (1984) Ciliary and mucus net filter feeding with special reference to fluid mechanical characteristics. *Mar. Ecol. Prog. Ser.* **15**, 283-292.
37. Pile, A.J., Patterson, M.R. and Witman, J.D. (1996) In situ grazing on plankton <10 μm by the boreal sponge Mycale lingua. *Mar. Ecol. Prog. Ser.* **141**, 95-102.
38. Ribes, M., Coma, R. and Gili, J.M. (in press) Heterotrophic feeding in symbiotic gorgonian corals. *Limnol. Oceanogr.*
39. Gili, J-M. and Coma, R. (1998) Benthic suspension feeders: Their paramount role in littoral marine food webs. *Trends Ecol. Evol.* **13**, 316-321.
40. Gasol, J.M., del Giorgio, P.A. and Duarte, C. (1997) Biomass distribution in marine planktonic communities. *Limnol. Oceanogr.* **42**, 1353-1363.
41. Alvà, V., Orejas, C., and Zabala, M. (1997) Feeding ecology of Antarctic cnidarian suspension feeders (Hydrozoa, Gorgonacea, Pennatulacea), in W.E. Arntz and J. Gutt (eds), The expedition Antarktis XIII/3 (EASIZ I) of RV "Polarstern" to the eastern Weddell Sea in 1996. *Ber. Polarforsch.* **249**, 14-16.
42. Barthel, D. (1997) Presence of fluff in an Antarctic shelf trough, at 600 m depth, in W.E. Arntz and J. Gutt (eds), The expedition Antarktis XIII/3 (EASIZ I) of RV "Polarstern" to the eastern Weddell Sea in 1996. *Ber. Polarforsch.* **249**, 16 f.
43. Barnes, D.K.A. and Clarke, A. (1995) Seasonality of feeding activity in Antarctic suspension feeders. *Polar Biol.* **15**, 335-340.
44. Clarke, A. and Leakey, R.J.G. (1996) The seasonal cycle of phytoplankton, macronutrients, and the microbial community in a nearshore Antarctic marine ecosystem. *Limnol. Oceanogr.* **41**, 1281-1294.
45. Arntz, W.E. and Fahrbach, E. (1991) *El Niño-Klimaexperiment der Natur*. Birkhäuser, Basel, pp. 264.
46. Olsgard, F., (1993) Do toxic algal blooms affect subtidal soft-bottom communities? *Mar. Ecol. Prog. Ser.* **102**, 269-286.
47. Rosenberg, R., Linddahl, O. and Blanck, H. (1988) Silent spring in the sea. *Ambio* **17**, 289 f.
48. Graf, G. (1992) Benthic-pelagic coupling: A benthic view. Oceanogr. Mar. Biol. Ann. Rev. 30, 149-190.
49. Greene, C.H., Wiebe, P.H., Burczynski, J. and Youngbluth, M.J. (1992) Acoustical detection of high-density demersal krill layers in the submarine canyons off Georges Bank. *Science* **241**, 359-361.
50. Vetter, E.W. (1995) Detritus-based patches of high secondary production in the nearshore benthos. *Mar. Ecol. Prog. Ser.* **120**, 251-262.
51. Hickey, B.M. (1995) Coastal submarine canyons, in P. Müller and D. Henderson (eds), *Topographic effects in the Ocean*. SOEST Special Publication, University of Hawaii, Manoa , pp. 95-110.
52. Lewin, R.(1986) Supply-side ecology. *Science* **234**, 25-27.
53. Boero F., Belmonte, G., Fanelli, G., Piraino, S. and Rubino, F. (1996) The continuity of living matter and the discontinuities of its constituents: do plankton and benthos really NOT FULL REFERNCE
54. Sorokin, Y.I. (1995) *Coral reef ecology*. Springer, Berlin.
55. Rosenberg, R., Arntz, W.E., Chumán de Flores, E., Carbajal, G., Finger, I. and Tarazona, J. (1983) Benthic biomass and oxygen deficiency in the upwelling system off Peru. *J. Mar. Res.* **41**, 263-279.
56. Fossing, H., Gallardo, V.A., Jørgensen, B.B., Hüttel, M., Nielsen, L.P., Schulz, H., Canfield, D.E., Forster, S., Glud, R.N., Gundersen, J.K., Küver, J., Ramsing, N.B., Teske, A., Thamdrup, B. and Ulloa, O (1995) Concentration and transport of nitrate by the mat-forming sulphur bacterium *Thioploca*. *Nature* **374**, 713-715.
57. Tarazona, J., Arntz, W.E. and Canahuire, E. (1996) Impact of two "El Niño" events of different intensity on the hypoxic soft bottom macrobenthos off the central Peruvian coast. *Mar. Ecol.* **17**, 425-446.
58. Arntz, W.E., Valdivia, E. and Zeballos, J. (1988) Impact of El Niño 1982-83 on the commercially exploited invertebrates (mariscos) of the Peruvian shore. *Meeresforsch.* **32**, 3-22.
59. Sibert, J.R. (1981) Intertidal hyperbenthic populations in the Nanaimo estuary. *Mar. Biol.* **64**, 259-265.
60. Buhl-Jensen, L. and Fosså, J.H. (1991) Hyperbenthic crustacean fauna of the Gullmarfjord area (western Sweden): species richness, seasonal variation and long-term changes. *Mar. Biol.* **109**, 245-258.
61. Chevrier, A., Brunel, P. and Wildish, D.J. (1991) Structure of a suprabenthic shelf-community of gammaridean Amphipoda in the Bay of Fundy compared with similar sub-communities in the Gulf of St. Lawrence. *Hydrobiologia* **223**, 81-104.
62. Dauvin, J.-C. and Zouhiri, S. (1996) Suprabenthic crustacean fauna of a dense Ampelisca community from the English Channel. *J. mar. biol. Ass. U.K.* **76,** 909-929.
63. Brandt, A. (1997) Abundance, diversity and community patterns of epibenthic- and benthic-boundary layer peracarid crustaceans ar 75° N off East Greenland. *Polar Biol.* **17**, 159-174.
64. Gutt, J. and Siegel, V. (1994) Observations on benthopelagic aggregations of krill (*Euphausia*

superba) on the deeper shelf of the southeastern Wedell Sea. *Deep-Sea Res.* **41**, 169-178.
65. Giangrande, A., Geraci, S. and Belmonte, G. (1994) Life-cycle and life-history diversity in marine invertebrates and the implications in community dynamics. *Oceanogr. Mar. Biol. Ann. Rev.* **32**, 305-333.
66. Thorson, G. (1936) The larval development, growth, and metabolism of Arctic marine bottom invertebrates compared with those of other seas. *Medd. Grönland* **100** (6), 1-155.
67. Mileikowsky, S.A. (1971) Types of larval development in marine bottom invertebrates, their distribution and ecological significance: a re-evaluation. *Mar. Biol.* **10**, 193-213.
68. Pearse, J.S., McClintock, J.B. and Bosch, I. (1991) Reproduction of Antarctic benthic marine invertebrates: tempos, modes and timing. *Amer. Zoologist* **31**, 65-80.
69. Arntz, W.E., Brey, T. and Gallardo, V.A. (1994) Antarctic zoobenthos. *Oceanogr. Mar. Biol. Ann. Rev.* **32**, 241-304.
70. Pearse, J.S. (1994) Cold-water echinoderms break "Thorson's rule", in C.M. Young and K.J. Eckelbarger (eds), *Reproduction, larval biology, and recruitment of the deep-sea benthos.* Columbia Univ. Press, New York, pp.26-39.
71. Schlüter, M. (1998) Die räumliche und zeitliche Verteilung des Meroplanktons (Larven des Evertebraten-Benthos) in der zentralen Barentssee. Dipl. Thesis, Univ. Bremen.
72. Gieskes, W.W.C. and C. Veth (1987) Secchi disc world record shattered. *EOS* **68**, 123.
73. Arnaud, P.M. (1970) Frequency and ecological significance of necrophagy among the benthic species of Antarctic coastal waters, in M.W. Holdgate (ed), *Antarctic ecology, Vol.1.* Academic Press, London, pp. 259-267.
74. Grassle, J.F. (1977) Slow recolonization of deep-sea sediment. Nature 265, 618 f.
75. Grassle, J.F. (1980) In situ studies of deep-sea communities, in F.P. Diemer, F.J. Vernberg and D.Z. Mirkes (eds), Advanced concepts in ocean measurements for marine biology. The Belle Baruch Library in Marine *Science* **10**, 321-331.
76. Lipps, J.H. and Hickman, C.S. (1982) Origin, age, and evolution of Antarctic and deep-sea faunas, in W.G. Ernst and J.G. Morin (eds), *The environment of the deep sea.* Prentice-Hall, Eaglewood Cliffs, pp. 324-356.
77. Turner, R. (1973) Wood-boring bivalves, opportunistic species in the deep sea. *Science* **180**, 1377-1379.
78. Dayton, P.K. (1989) Interdecadal variation in an Antarctic sponge and its predators from oceanographic climate shifts. *Science* **245**, 1484-1486.
79. Rauschert, M. (1990) Ergebnisse der faunistischen Arbeiten im Benthal von King George Island (Südshetlandinseln, Antarktis). *Ber. Polarforsch.* **76**, 1-75.
80. Wägele, J.W. (1987) On the reproductive biology of Ceratoserolis trilobitoides (Crustacea: Isopoda): latitudinal variation of fecundity and embryonic development. Polar Biol. 7, 11-24.
81. Gorny, M., Arntz, W.E., Clarke, A. and Gore, D.J. (1992) Reproductive biology of caridean decapods from the Weddell Sea. *Polar Biol.* **12**, 111-120.
82. Arntz, W.E., Brey, T., Gerdes, D., Gorny, M., Gutt, J., Hain, S. and Klages, M. (1992) Patterns of life history and population dynamics of benthic invertebrates under the high Antarctic conditions of the Weddell Sea, in G. Colombo, I. Ferrari, V.U. Ceccherelli and R. Rossi (eds), *Marine eutrophication and population dynamics.* Olsen & Olsen, Fredensborg, pp. 221-230.
83. Hulberg, L.W. and Oliver, J.S. (1980) Caging manipulations in marine soft-bottom communities: importance of animal interactions or sedimentary habitat modifications. *Can. J. Fish. Aquatic Sci.* **37**, 1130-1139.
84. Dayton, P.K., Robilliard, G.A., Paine, R.T. and Dayton, L.B. (1974) Biological accommodation in the benthic community at McMurdo Sound. *Ecol. Monogr.* **44**, 105-128.
85. Grassle, J.F. and Morse-Porteous, L. (1987) Macrofaunal colonization of disturbed deep-sea environments and the structure of deep-sea benthic communities. *Deep-Sea Res.* **34**, 1911-1950.
86. Rhoads, D.C. and Young, D.K. (1970) The influence of deposit-feeding organisms on sediment stability and community trophic structure. *J. Mar. Res.* **28**, 150-178.
87. Connell, J.H. and Slatyer, R.O. (1977) Mechanisms of succession in natural com-munities and their role in community stability and organization. *Am. Nat.* **111**, 1119-1144.
88. Gray, J.S. (1981) The ecology of marine sediments. An introduction to the structure and function of benthic communities. Cambridge Univ. Press, Cambridge.
89. Levin, L.A. (1982) Interference interactions among tube-building polychaetes in a dense infaunal assemblage. *J. exp. mar. Biol. Ecol.* **65**, 107-119.
90. Reise, K. (1985) *Tidal flat ecology.* Ecological Studies 54, Springer-Verlag, Berlin.
91. Whitlatch, R.B. and Zajac, R.N. (1985) Biotic interactions among estuarine infaunal opportunistic species. *Mar. Ecol. Prog. Ser.* **21**, 299-311.
92. Chapman, A.R.O. (1986) Population and community ecology of seaweeds. *Advances in Marine Biology* **23**, 1-161.
93. Ambrose, W.G. (1991) Are infaunal predators important in structuring marine soft-bottom communities? *Amer. Zool.* **31**, 849-860.
94. Porter, J.W. and Targett, N.M. (1988) Allelochemical interactions between sponges and corals. *Biol. Bull.* **175**, 230-239.
95. Jackson, J.B.C. and Buss, L. (1975) Allelopathy and spatial competition among coral reef invertebrates. *Proc. Nat. Acad. Sci. USA* **72**, 5160-5163.
96. Uriz, M.J., Turon, X., Becerro, M.A., Galera, J. and Lozano, J. (1995) Patterns of resource

allocation to somatic, defensive, and reproductive functions in the Mediterranean encrusting sponge *Crambe crambe* (Demospongiae, Poecilosclerida). *Mar. Ecol. Prog. Ser.* **124,** 159-170.

97. Tilman, D. (1990) Constraints and tradeoffs: toward a predictive theory of competition and succession. *Oikos* **58,** 3-15.
98. Jumars, P.A. (1993) *Concepts in biological oceanography. An interdisciplinary primer.* Oxford Univ. Press, Oxford.
99. Sebens, K.P. (1988) Competition for space: effects of disturbance and indeterminate competitive success. *Theor. Pop. Biol.* **32,** 430-441.
100. Dayton, P.K. (1971) Competition, disturbance, and community organization: the provision and subsequent utilization of space in a rocky intertidal community. *Ecol. Monogr.* **41,** 351-389.
101. Sutherland, J.P. (1974) Multiple stable points in natural communities. *Am. Nat.* **108,** 859-873.
102. Dayton, P.K. (1984) Processes structuring some marine communities: are they general? In Stong Jr., D.R., Simberloff, D., Abele, L.G. and Thistle, A.B. (eds), *Ecological communities: conceptual issues and the evidence.* Princeton Univ. Press, Princeton, pp. 182-197.
103. Peterson, C.H. (1979) Predation, competitive exclusion, and diversity in the soft-sediment benthic communities of estuaries and lagoons, in R.J. Livingstone (ed*), Ecological processes in coastal and marine systems*. Plenum Press, New York.
104. Gaines, S. and Roughgarden, J. (1985) Larval settlement rate: a leading determinant of structure in an ecological community of the marine intertidal zone. *Proc. Nat. Acad. Sci., USA* **82,** 3707-3711.
105. Reise, K. (1985) *Tidal flat ecology.* Ecological Studies 54, Springer-Verlag, Berlin.
106. Connell, J.H. (1985) The consequences of variation in initial settlement vs. post-settlement mortality in rocky intertidal communities. *J. exp. mar. Biol. Ecol.* **93,** 11-45.
107. Menge, B.A. and Sutherland, J.P. (1987) Community regulation: variation in disturbance, competition, and predation in relation to environmental stress and recruitment. *Amer. Nat.* **130,** 730-757.
108. Underwood, A.J. and Fairweather, P.G. (1989) Supply-side ecology and benthic marine assemblages. *Trends Ecol. Evol.* **4,** 16-20.
109. Gaines, S.D. and Bertness, M.D. (1992) Dispersal of juveniles and variable recruitment in sessile marine species. *Nature* **360,** 579-580.
110. Olafsson, E.B., Peterson, C.H. and Ambrose, W.G. (1994) Does recruitment limitation structure populations and communities of macro-invertebrates in marine soft sediments: the relative significance of pre- and post-settlement processes. *Oceanogr. Mar. Biol. Ann. Rev.* **32,** 65-109.
111. Hjort, J. (1914) Fluctuations in the great fisheries of northern Europe viewed in the light of biological research. *Rapp. P.-v. Réun. Cons. Int. Explor. Mer* **20,** 1-228.
112. Cushing, D.H. (1975) *Marine ecology and fisheries.* Cambridge Univ. Press, Cambridge, pp. 278.
113. Lasker, R. (1978) The relation between oceanographic conditions and larval anchovy food in the California Current: identification of factors contributing to recruitment failure. *Rapp. P.-v. Réun. Cons. Int. Explor. Mer* **173,** 212-230.
114. Sinclair, M., Tremblay, M.J. and P. Bernal (1985) El Niño events and variability in a Pacific mackerel (Scomber japonicus) survival index: support for Hjort's second hypothesis. *Can. J. Fish. Aquat. Sci.* **42,** 602-608.
115. Boysen Jensen, P. (1919) Valuation of the Limfjord. I. Studies on the fish food in the Limfjord, 1909-1917, its quantity, variation and annual production. *Rep. Danish biol. Sta.* **26,** 1-44.
116. Blegvad, H. (1928) Quantitative investigations of bottom invertebrates in the Limfjord 1910-1927 with special reference to the plaice food. *Rep. Danish biol. Sta.* **34,**
117. Thorson, G. (1950) Reproductive and larval ecology of marine bottom invertebrates. *Biol. Rev.* **25,** 1-45.
118. Connell, J.H., Hughes, T.P. and Wallace, C.C. (1997) A 30-year study of coral abundance, recruitment, and disturbance at several scales in space and time. *Ecol. Monogr.* **67,** 461-488.
119. Eckman, J.E. (1983) Hydrodynamic processes affecting benthic recruitment. *Limnol. Oceanogr.* **28,** 241-257.
120. Wilson, D.P. (1952) The influence of the nature of the substratum on the metamorphosis of the larvae of marine animals, especially the larvae of Ophelia bicornis Savigny. *Ann. Inst. Oceanogr. Monaco* **27,** 49-156.
121. Butman, C.A. (1987) Larval settlement of soft-sediment invertebrates: The spatial scales of pattern explained by active habitat selection and the emerging role of hydrodynamical processes. *Oceanogr. Mar. Biol. Ann. Rev.* **25,** 113-165.
122. McEdwards, L. (1995) *Ecology of marine invertebrate larvae.* CRC Press, Boca Raton.
123. Günther, C-P. (1992) Dispersal of intertidal invertebrates: a strategy to react to disturbances of different scales? *Neth. J. Sea Res.* **30,** 45-56.
124. Armonies, W. (1994) Drifting meio- and macrobenthic invertebrates on tidal flats in Königshafen: a review. *Helgoländer Meeresunters.* **48,** 299-320.
125. Olivier, F., Vallet, C., Dauvin, J-C. and Retiére, C. (1996) Drifting in post-larvae and juveniles in an *Abra alba* (Wood) community of the eastern part of the Bay of Seine (English Channel). *J. exp. mar. Biol. Ecol.* **199,** 89-109.
126. Ziegelmeier, E. (1970) Über Massenvorkommen ver--schiedener makrobenthaler Wirbelloser während der Wiederbesiedlungsphase nach Schädigungen durch "katastrophale" Umwelteinflüsse. *Helgol. Wiss. Meeresunters.* **21,** 9-20.
127. Beukema, J.J. (1982) Annual variation in reproductive success and biomass of the major

macrozoobenthic species living in a tidal flat area of the Wadden Sea. *Neth. J. Sea Res.* **16,** 37-45.

128. Beukema, J.J. (1992) Expected changes in the Wadden Sea benthos in a warmer world: lessons from periods with mild winters. *Neth. J. Sea Res.* **30,** 73-79.
129. Reise, K. (1987) Experimenta! analysis of processes between species on marine tidal flats, in Schulze, E.-D. and Zwölfer, H. (eds), *Potentials and limitations of ecosystem analysis.* Springer-Verlag, Berlin, pp. 391-400.
130. Zwarts, L. (1991) Seasonal variation in body weight of the bivalves *Macoma balthica, Scrobicularia plana, Mya arenaria* and Cerastoderma edule in the Dutch Wadden Sea. *Neth. J. Sea Res.* **28,** 231-245.
131. Möbius, K. (1871) Das Thierleben am Boden der deutschen Nord- und Ostsee. *Sammlg gemeinverst. wiss. Vorträge,* Hamburg **6,** 3-32.
132. Petersen, C.G.J. (1914) Valuation of the sea. II. The animal communities of the sea bottom and their importance for marine zoogeography. *Rep. Danish biol. Sta.* **21,** 1-68.
133. Thorson, G. (1957) Bottom communities (sublittoral or shallow shelf), in J.W. Hedgpeth (ed), Treatise on marine ecology and palaeoecology, Vol.1. *Mem. Geol. Soc. America,* **67,** pp. 451-534.
134. Rosenberg, R. (1976) Benthic faunal dynamics during succession following pollution abatement in a Swedish estuary. *Oikos* **27,** 414-427.
135. Rhoads, D.C. and Young, D.K. (1970) The influence of deposit-feeding organisms on sediment stability and community trophic structure. *J. Mar. Res.* **28,** 150-178.
136. Pérès, J.M (1961) *Océanographie biologique et biologie marine* (I). Paris.
137. Golikov, A.N. and O.A. Scarlato (1973) Comparative characteristics of some ecosystems of the upper regions of the shelf in tropical, temperate and arctic waters. *Helgol. Wiss. Meeresunters.* **24,** 219-234.
138. Connell, J.H. (1978) Diversity in tropical rain forests and coral reefs. *Science* **1,** 1302-1309.
139. Gili, J.M. and Ros, J. (1985) Study and cartography of the benthic communities of Medes Islands. *P.S.Z.N. I: Mar. Ecol.* **6,** 219-238.
140. Schickan, T. (1996) Epibiontische Vergesellschaftungen im Weddell- und Lazarevmeer, Antarktis. Dipl. Thesis, Univ. Bremen.
141. Knowlton, N. (1993) Sibling species in the sea. *Ann. Rev. Ecol. Syst.* **24,** 189-216.
142. Strathmann, R.R. (1990) Why life histories evolve differently in the sea. *Am. Zool.* **30,** 197-207.
143. Horn, H.S. and MacArthur, R.H. (1972) Competition among fugitive species in a harlequin environment. *Ecology* **5,** 749-752.
144. DeAngelis, P.L. and Waterhouse, J.C. (1987) Equilibrium and nonequilibrium con-cepts in ecological models. *Ecol. Monogr.* **57,** 1-21.
145. Gray, J.S. (1977) The stability of benthic ecosystems. *Helgol. Wiss. Meeresunters.* **30,** 427-444.
146. Thrush, S.F., Whitlatch, R.B., Pridmore, R.D., Hewitt, J.E., Cummings, V.J. and Wilkinson, M.R. (1996) Scale-dependent recolonization: the role of sediment stability in a dynamic sandflat habitat. *Ecology* **77,** 2472-2487.
147. Thrush, S.F. et multiple authors (1997) Scaling-up from experiments to complex ecological systems: where to next? *J. exp. mar. Biol. Ecol.* **216,** 243-254.
148. Rachor, E. and Gerlach, S.A. (1978) Changes in a sublittoral sand area of the German Bight, 1967 to 1970. *Rapp. P.-v. Réun. Cons. Int. Explor. Mer* **172,** 418-431.
149. Rhoads, D.C., McCall, P. and Yingst, J.Y. (1978) Disturbance and production on the estuarine seafloor. *Amer. Sci.* **66,** 577-586.
150. Pearson, T.H. (1981) Stress and catastrophe in marine benthic ecosystems, in G.W. Barrett and R. Rosenberg (eds), *Stress effects on natural ecosystems.* John Wiley & Sons, Chichester, pp. 201-214.
151. Arntz, W.E. (1981) Biomass zonation and dynamics of macrobenthos in an area stressed by oxygen deficiency, in G. Barrett and R. Rosenberg (eds), *Stress effects on natural ecosystems.* John Wiley and Sons, Chichester, pp. 215-225.
152. Boesch, D.F. and Rosenberg, R. (1981) Response to stress in marine benthic com-munities, in G.W. Barrett and R. Rosenberg (eds), *Stress effects on natural eco-systems.* John Wiley & Sons, Chichester, pp. 179-200.
153. Sousa, W.P. (1984) The role of disturbance in natural communities. *Ann. Rev. Ecol. Syst.* **15,** 353-391.
154. Menge, B.A. and Lubchenco, J. (1981) Community organzation in temperate and tropical rocky intertidal habitats: prey refuges in relation to consumer pressure gradients. *Ecol. Monogr.* **51,** 429-450.
155. Paine, R.T. and Levin, S.A. (1981) Intertidal landscapes: disturbance and the dynamics of pattern. *Ecol. Monogr.* **51,** 37-63.
156. Remmert, H. (1992) *Ökologie.* Ein Lehrbuch. Springer-Verlag, Berlin.
157. Connell, J.H. (1975) Some mechanisms producing structure in natural communities: a model and evidence from field experiments, in M. Cody and J. Diamond (eds), *Ecology and evolution of communities.* Harvard Univ. Press, Cambridge, pp. 460-490.
158. Dayton, P.K. and Hessler, R.R. (1972) Role of biological disturbance in maintaining diversity in the deep sea. *Deep-Sea Res.* **19,** 199-208.
159. Glynn, P. (1985) El Niño-associated disturbance to coral reefs and post disturbance mortality by *Acanthaster planci. Mar. Ecol. Prog. Ser.* **26,** 295-300. .
160. Pearson, T.H. and Rosenberg, R. (1978) Macrobenthic succession in relation to organic enrichment and pollution of the marine environment. *Oceanogr. Mar. Biol. Ann. Rev.* **16,** 229-311.

161. Arntz, W.E. and Rumohr, H. (1982) An experimental study of macrobenthic colonization and succession, and the importance of seasonal variation in temperate latitudes. *J. exp. mar. Biol. Ecol.* **64**, 17-45.
162. Diaz, R.J. and Rosenberg, R. (1995) Marine benthic hypoxia: a review of its ecological effects and the behavioural responses of benthic macrofauna. *Oceanogr. Mar. Biol. Ann. Rev.* **33,** 245-303.
163. Dayton, P.K. (1990) Polar benthos, in W.O. Smith Jr. (ed*), Polar oceanography, part B: Chemistry, biology, and geology.* Academic Press, San Diego, pp. 631-685.
164. Dittmann, S. (1990) Mussel beds - amensalism or amelioration for intertidal fauna. *Helgol. Meeresunters.* **44,** 335-352.
165. Günther, C-P. (1996) Development of small Mytilus beds and its effect on resident intertidal macrofauna. *Mar. Ecol.* **17**, 117-130.
166. Gray, J.S. (1994) Is deep-sea diversity really so high? Species diversity of the Norwegian continental shelf. *Mar. Ecol. Prog. Ser* . **112**, 205-209.
167. Gray, J.S., Poore, G.C.B., Ugland, K.I. and Wilson, R.S. (1997) Coastal and deep-sea benthic diversities compared. *Mar. Ecol. Prog. Ser.* **159**, 97-103.
168. Warwick, R.M. (1997) Biodiversity and production on the sea floor, in G. Hempel (ed) *The ocean and the poles, grand challenges for European cooperation.* Fischer, Stuttgart, pp. 217-226.
169. Sanders, H.L. (1968) Marine benthic diversity: a comparative study. *Am. Nat.* **102**, 243-282.
170. Clarke, A. (1992) Is there a diversity cline in the sea? *Trends Ecol. Evol.* **9**, 286 f.
171. IBMANT (*Marine biological investigations in the Magellan region related to the Antarctic*) (in press), W.E. Arntz and C. Ríos (eds),Scientia Marina spec. issue.
172. Brey, T., Klages, M., Dahm, C., Gorny, M., Gutt, J., Hain, S., Stiller, M., Arntz, W.E., Wägele, J.W. and Zimmermann, A. (1994) Antarctic benthic diversity. *Nature* **368**, 297.
173. Arntz, W.E., Gutt, J., and Klages, M. (1997) Antarctic marine biodiversity: an over-view, in B. Battaglia, J. Valencia and D. Walton (eds), *Antarctic communities.* Cambridge Univ. Press, Cambridge, pp. 3-14.
174. Knox,G.A. and Lowry, J.K. (1977) A comparison between the benthos of the Southern Ocean and the North Polar Ocean with special reference to the Amphipoda and the Polychaeta, in: M.J. Dunbar (ed), *Polar Oceans.* Arctic Inst. of North America, Calgary, pp. 423-462.
175. Piepenburg, D., Voß, J. and Gutt, J. (1997) Assemblages of sea stars (Echinodermata: Asteroidea) and brittle stars (Echinodermata: Ophiuroidea) in the Weddell Sea (Antarctica) and off northeast Greenland (Arctic): a comparison of diversity and abundance. *Polar Biol.* **17**, 305-322.
176. Kendall, M.A. (1996) Are Arctic soft-sediment macrobenthic communities impoverished*? Polar Biol.* **16,** 393-399.
177. Grassle, J.S. and Maciolek, N.J. (1992) Deep-sea species richness: regional and local diversity estimates from quantitative bottom samples. *Am. Nat.* **139**, 313-341.
178. Gray, J.S. (1993) Is coastal biodiversity as high as that of the deep sea? In Eleftheriou, A., Ansell, A.D. and Smith, C.J. (eds), *Biology and ecology of shallow coastal waters.* Olsen and Olsen, Fredensborg 1995, pp. 181-184.
179. Coleman, N., Gason, A. and Poore, G.C.B. (1997) High species richness in the shallow marine waters of south-east Australia. *Mar. Ecol. Prog. Ser.* **154**, 17-26.
180. Levinton, J.S. (1995) *Marine biology. Function, biodiversity, ecology.* Oxford Univ. Press, New York.
181. Starmans, A. (1997) Vergleichende Untersuchungen zur Ökologie und Biodiversität des Mega-Epibenthos der Arktis und Antarktis. *Ber. Polarforsch.* **250**, 1-150.
182. Peterson, C.H. (1975) Stability of species and of community for the benthos of two lagoons. *Ecology* **56**, 958-965.
183. Sanders, H.L. (1969) Benthic marine diversity and the stability-time hypothesis. *Brookhaven Symp. Biol.* **22**, 71-80.
184. Hurlbert , H.S. (1971) The nonconcept of species diversity: a critique and alternative parameters. *Ecology* **52**, 577-586.
185. Huston, M. (1979) A general hypothesis of species diversity. *Am. Naturalist* **113**, 81-101.
186. Remmert, H. (1991) The mosaic-cycle concept of ecosystems - an overview, in Remmert, H. (ed), *The mosaic-cycle concept of ecosystems.* Springer-Verlag, Berlin, pp. 1-21.
187. Reise, K. (1991) Mosaic cycles in the marine benthos, in Remmert, H. (ed), *The mosaic-cycle concept of ecosystems.* Springer-Verlag, Berlin, pp. 61-82.
188. Gutt, J., Starmans, A. and Dieckmann, G. (1996) Impact of iceberg scouring on polar benthic habitats. *Mar. Ecol. Prog. Ser.* **137**, 311-316.
189. Crame, A. (1993) Latitudinal range fluctuations in the marine realm through geo-logical time. *Trends Ecol. Evol.* **8**, 162-166.
190. Brandt, A. (1991) Zur Besiedlungsgeschichte des antarktischen Schelfes am Bei-spiel der Isopoda (Crustacea, Malacostraca). *Ber. Polarforsch.* **98,** 240 pp.
191. Brey, T., Dahm, C., Gorny, M., Klages, M., Stiller, M. and W.E. Arntz (1996) Do Antarctic invertebrates show an extended level of eurybathy? *Antarctic Science* **8**, 3-6.
192. APIS (Antarctic Pack Ice Seals) (1995) *Report of the 1995 APIS program planning meeting,* SCAR Group of Specialists on Seals, pp. 26.
193. Franeker, J.A. van, Bathmann, U.V. and Mathot, S. (1997) Carbon fluxes to Antarctic top predators. *Deep-Sea Res.* **44**, 435-455.

UNDERSTANDING SMALL-SCALE PROCESSES CONTROLLING THE BIOAVAILABILITY OF ORGANIC CONTAMINANTS TO DEPOSIT-FEEDING BENTHOS

THOMAS L. FORBES
Department of Marine Ecology and Microbiology, National Environmental Research Institute, Pob 358, Frederiksborgvej 399, DK-4000, Roskilde, Denmark
tf@dmu.dk

Abstract
The bioavailable fraction of a sedimentary contaminant is defined as that portion present in the environment which is available for uptake by organisms. This simple definition obscures a number of difficulties which impede our present ability to predict uptake and bioaccumulation of organic contaminants by benthic deposit-feeding organisms. Mathematical modelling in combination with laboratory results demonstrates that physico-chemical equilibria are disrupted due to animal activity at small spatial (μm->mm) and temporal (sec->min) scales forcing consideration of short-term kinetic factors and casting doubt on simple equlibrium partitioning predictions. These analyses further suggest that uptake by ingestion of contaminanted sediments may be the most important pathway for many infaunal benthos. Laboratory absorption efficiency experiments indicate a trophic uptake rate at least 20-30 times that predicted for pore water alone.

1. Bioavailability of Organic Contaminants

One of the most significant disturbances of the coastal zone is the contamination of nearshore sediments by anthropogenic organic compounds [1]. These compounds, like the natural organic matter they closely associate with, tend to accumulate in hydrographically quiet depositional environments close to their source of input from land. Within a coastal environment experiencing relatively homogeneous contaminant loading, contaminant concentration tends to correlate strongly with sedimentary organic content. The temporal and spatial scale of organic contaminant behavior is tightly coupled to local oceanographic conditions and contaminant biogeochemistry [2]. The effects of benthic communities are often superimposed on this coupling to generate the pattern of contaminant distribution and fate that we observe [2] [3]. Moreover, the benthic inhabitants themselves may often be affected by sedimentary contamination creating the potential for feedback relationships between pollutant fate and effect

In order for a contaminant to become a pollutant it must have a deleterious effect on living organisms. In order to exert a direct effect on an organism the contaminant must first be bioavailable. The bioavailable pool is simply defined as that fraction of the total amount of contaminant present in the environment which is available for uptake by the biota. This simple definition masks a number of difficulties which impede our ability to accurately predict exposure and uptake of contaminants by sedimentary organisms. Below I highlight and discuss some recent research and modeling results that indicate the most important temporal and spatial scales at which the key processes affecting the uptake and accumulation of organic contaminants can be expected to operate.

J.S. Gray et al. (eds.), Biogeochemical Cycling and Sediment Ecology, 125–136.

2. Key Factors Controlling Bioavailability

2.1. ENVIRONMENTAL FACTORS

The three principle physico-chemical factors controlling the deposition of both natural and contaminant particle-bound organic matter are 1) particle size (and covariate surface area), 2) contaminant surface chemistry, and 3) the local hydrodynamic regime. Relatively small particles with high specific surface area, and thus contaminant-binding capacity, tend to accumulate in muddy environments. High levels of organic loading create environments which are characterized by hypoxic/anoxic pore waters and high rates of anaerobic metabolism, often resulting in a tendency to preserve or accumulate contaminants that degrade slowly under anoxic conditions. Given the current state of knowledge of the organic geochemistry of marine sediments it is usually quite difficult to predict the kinetics and distribution of contaminants at the small spatial and temporal scales relevant to bioavailability.

In terms of biomass and abundance as well as with regard to geochemical effects, deposit-feeding invertebrates are frequently the dominant form of metazoan life in marine sediments [3] [4] [5]. With the exception of environments which undergo very high rates of organic enrichment, the biomass of deposit-feeding animals is highly correlated with the organic content of sediments [6] [7]. This means that deposit-feeding benthic invertebrates, their food (i.e., sedimentary organic matter), and the organic contaminants associated with their food will all tend to concentrate in the same place.

Thus at relatively large scales we can make reasonable predictions about where organic contaminants will tend to accumulate and to a lesser degree what suites of benthic species will tend to co-occur in the contaminated area. At smaller scales we cannot accurately predict uptake rates and degree of bioaccumulation. I suggest that this is due to a lack of understanding of small scale spatial processes occurring at the individual organism or local population level at temporal scales of seconds to perhaps days.

2.2. ORGANISMAL FACTORS

Many sediment-dwelling animals are known to ingest sediments in a highly selective manner. Even though the exact mechanism is still unclear and may differ across taxa, most deposit-feeders probably selectively ingest the organic-rich fraction of the available sediment [4]. The precise degree of selectivity varies but many species are able to enrich the ingested fraction of organic material by a factor of two or more [5] [8] [9] [10]. Thus if animals do not actively reject contaminant in favor of natural organic matter, contaminant concentration in ingested sediment will often be two or more times that of the bulk sediment. Deposit feeders also exhibit relatively high weight-specific rates of ingestion when compared to other metazoans of similar size. Rates range from several to over 100 body weights of sediment ingested per day [4]. These values make even very low absorption efficiencies (i.e., a few percent) potentially significant with regard to total contaminant uptake and exposure.

Pore water is inherently toxic to most benthic infaunal organisms. High concentrations of the end products of anaerobic metabolism (e.g., sulfide and ammonium), combined with the need for molecular oxygen to support efficient metazoan metabolism, produce a tendency for infaunal benthos to actively isolate themselves from pore water. The precise nature and degree of isolation is probably species dependent and somewhat variable. This local environmental control exerted by infaunal benthos is largely achieved by construction of burrows with semi-permeable linings capable of acting as molecular sieves in addition to active ventilation of the burrow system with overlying water. Further effects can occur at the population level as burrow spacing, size, population density and composition interact to control pore water concentrations at scales larger than the individual. In combination these activities result in a burrow water composition that often differs substantially from that of the surrounding pore water and larger scale concentrations that differ from equilibrium expectations [11] [12] 13]. The picture is further complicated because animals also transport sediment particles, and thus particle-bound contaminants, through their burrowing and feeding activities. For example, periodic episodes of burrow irrigation, typically occurring at intervals of minutes to hours, will strongly influence the conditions immediately surrounding the animal and make the role of contaminant sorption kinetics important.

Finally, even when a contaminant has been taken up by an organism, cellular metabolism is often capable of transforming or excreting the substance. This biochemical-level processing further complicates accurate prediction of bioaccumulation rates. An example from our own work involves the ability of the polychaete *Capitella* species I to metabolize the PAH fluoranthene to a more soluble form which is excreted into the water phase [14]. Thus in addition to lowering the body burden to below detection limits even when worms continued to occupy highly contaminated sediments (ca. 300 μg fluoranthene (g dw sed)$^{-1}$), the potential for system feedback was created due to the direct effect of the worms on fluoranthene fate.

2.3. IMPLICATIONS FOR CONTAMINANT REGULATION

The above considerations have immediate and important implications for contaminant regulation and sediment quality assessment because current regulatory guidelines and practices for the assessment of contaminated sediments are typically based on the following two assumptions [15] [16] [17]. First, a thermodynamic equilibrium is assumed to exist among the organism, sediment, and pore water compartments with respect to the partitioning of the contaminant and second, the contaminant dissolved in the pore water is assumed to be is the most bioavailable. In contrast to most current regulatory practice, the biotic factors outlined briefly above strongly suggest that 1) true thermodynamic equilibrium may in fact be a rare occurrence and 2) it may be unwise to uncritically assume that pore water contaminant is the most bioavailable. If thermodynamic equilibrium at the scale of the individual organism is rare, then determining the pathway of uptake becomes very important in the prediction of uptake rates and body burdens. This line of reasoning provided the rationale for the modeling work discussed below.

2.4. ENVIRONMENT-ORGANISM INTERACTIONS AND FEEDBACKS

Organism-driven mass transport of particles and solutes, collectively termed bioturbation, can modify the local sedimentary environment and effect the fate of the contaminant. The detailed nature of this effect will be dependent on the structure of the benthic community, but some simplification may be possible through consideration of functional groups of bioturbators [18] [19]. An example of the effect of bioturbation on contaminant fate from our own laboratory is the effect of irrigation by a single individual *Arenicola marina* on the pore water profile of the fluoranthene in a small microcosm (Figure 1) [20]. Fluoranthene was uniformly distributed with depth at the start of the two week experimental period.

The pore water concentration of fluoranthene is decreased by roughly half due to the >flushing= effect of worm irrigation. The lower concentration of sediment-bound fluoranthene at the surface of the worm microcosm is caused by digestion and absorption of fluoranthene which has been ingested at depth (10-12 cm) and defecated at the sediment-water interface. This surficial depletion can be used as a crude estimate of the absorption efficiency of fluoranthene by *Arenicola* in this particular sediment (> 60%). Note that precise prediction of the bulk pore water and sediment fluoranthene concentrations will require knowledge of a rather large number of chemical and biological parameters including, worm density, irrigation rate and periodicity, fluoranthene particle-sorption behavior, etc. From the perspective of bioavailability, these factors will interact to control the local small scale dissolved and particulate contaminant concentration near the worms. Thus one would expect the periodicity and strength of irrigation behavior, combined with the chemical nature of the burrow wall to combine to exert the major control on contaminant exposure through pore water. Within a particular habitat, it is these local small-scale considerations which are most important in determining contaminant exposure. This tight coupling between contaminant fate and benthic macrofaunal activity suggests the formation of feedback relationships between contaminant fate and effect.

2.4.1. The Scale of Observation

Given the often complex nature of animal-sediment relationships, accurate prediction of exposure and uptake requires knowledge of concentration gradients and kinetics at both small spatial (μm-mm) and short temporal scales (sec-min). These scales are extremely difficult to handle experimentally, nevertheless model predictions can be made relatively easily and can help to focus the resources and time of future experimental work. The effects of organisms on their immediate environment will further modify the distribution of contaminant in the sediment at scales relevant to exposure. The scale difficulties can be readily appreciated by examining the following two figures.

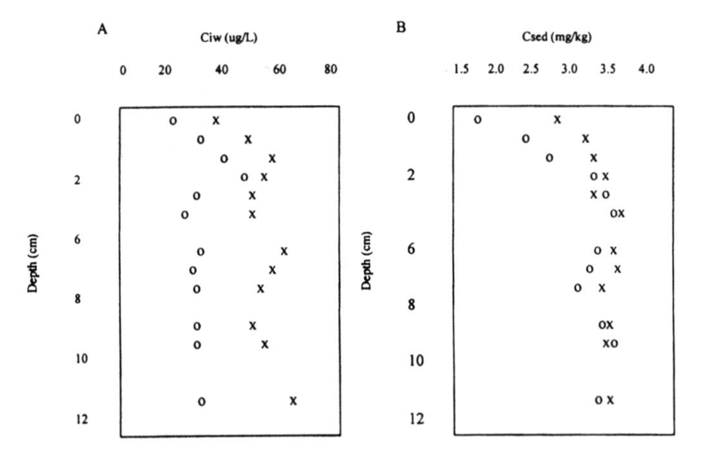

Figure 1. Pore water (C_{wi}) and particulate (C_{sed}) fluoranthene profiles for microcosms with (o) and without (x) *Arenicola marina.* Modified from Kure (1997).

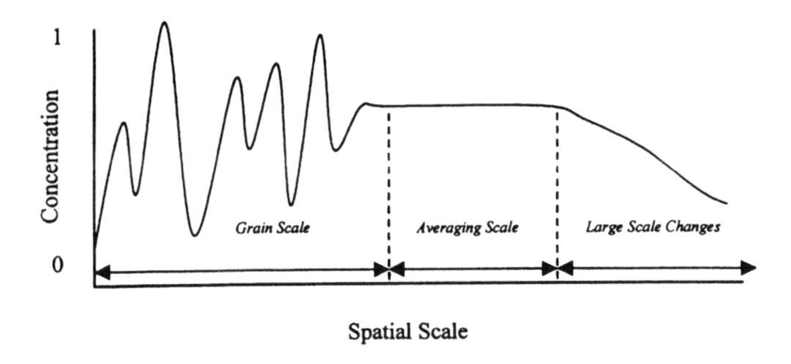

Figure 2. Schematic representation of the concentration of a particle-sorbed organic contaminant as the scale of observation is increased from a single sediment grain to scales associated with routinely measured diagenetic changes (centimeter to meter scales). Modified from the original conception of Boudreau [1997, Figure 3.2].

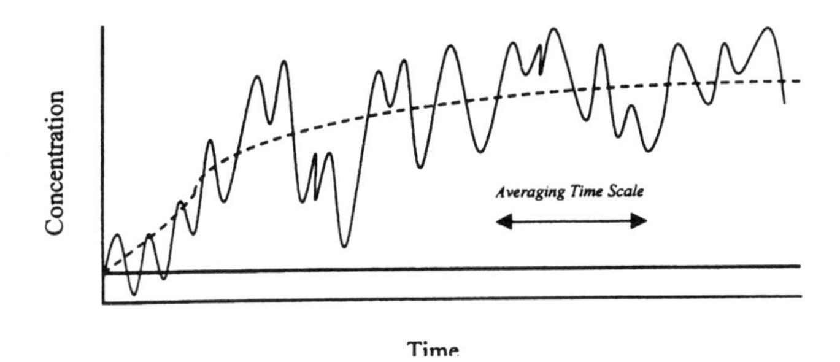

Figure 3. Schematic representation of the effect of the temporal scale of bioturbation on sedimentary organic contaminant concentration. See text for detailed explanation.

Figure 2 demonstrates how measured concentration depends on the spatial scale of observation [13]. This is a schematic representation of the concentration of an organic contaminant plotted as a function of sample spatial scale. Scale is defined as the size of the sediment parcel under observation. In practice this is usually a few milliliters of surface sediment or a similar-sized portion of a depth section within a core. The concentration units are arbitrary and range here from zero to a maximum of one. The concentration begins at zero because we have arbitrarily begun by measuring a single sediment grain which happens not to be contaminated. Increasing the scale tends to increase the odds of including more highly contaminated particles, but small-scale patchiness causes a relatively large variability in concentration values. As more and more grains are included in the sample, one reaches an 'averaging scale' over which the small-scale variability vanishes. At even larger scales of observation, new phenomena come into play which may cause additional deviations at the 'macroscopic scale'. The macroscopic or larger-scale changes in concentration can be due to spatial changes in geochemical properties resulting in a different contaminant degradation rate and can be caused by bioturbation, hydrodynamic differences or a number of other factors.

The solid horizontal line in Figure 3 depicts the expected contaminant concentration at a particular point in the absence of bioturbation. The concentration predicted by thermodynamic equilibrium for example. With an active macrobenthic community the concentration at any chosen point in the sediment column will typically differ from that predicted solely by consideration of physical and environmental factors. This organism-mediated change in expectation is due to both purely physical effects (i.e., biological particle mixing and water transport) and indirect faunally-induced biogeochemical changes affecting both reaction rates and their three dimensional spatial distribution. Changes in reaction dynamics can be brought about through macrofaunal effects on the microbial community [21] as well as through direct macrofaunal metabolism and processing of the contaminant itself [14]. The dotted, curved line shows the actual bulk sediment contaminant concentration integrated over 'averaging' spatial (Figure 2) and time (Figure 3) scales. The difference between the dotted curved and solid horizontal lines is thus the effect of bioturbation. The solid, wavy line illustrates the random deviations from the mean due to the small-scale effects of individual or small groups of organisms that occur as a function of time. These fluctuations are analogous to the grain-scale spatial deviations shown in Figure 2. These organism-caused deviations can occur over a range of scales - from less than a millimeter to centimeters and from seconds to months or years. Bioturbation causes these fluctuations by transporting matter and associated contaminants (solids or pore water) from adjacent points with higher or lower concentrations. Note that, as depicted in Figure 3, the mean concentration can deviate strongly and systematically from the expectation in the absence of organisms. This benthos-caused deviation can either be positive (as shown here), negative or fluctuating and is expected to be dependent on the structure of the benthic community and the resulting tempo and mode of bioturbation.

Chemical studies, even those explicitly incorporating the effects of macrobenthos, try to focus sampling at the 'averaging' time and spatial scales (Figures 2 & 3). This is typically the most meaningful scale to observe when constructing diagenetic models of biogeochemical processes [13]. The central theme of this chapter is to point out and demonstrate that the most meaningful scale for the investigation of bioavailability may often be quite different scale of observation typical of standard sedimentary chemical analyses. When considering contaminant bioavailability the appropriate scale of observation must be matched to the individual organism or local assemblage. The scale problems can be readily highlighted through a discussion of some of our recent work employing diagenetic reaction-diffusion models as a tool for the investigation of the relative role of pore water versus trophic uptake of organic contaminants by deposit feeders.

3. Modeling Small-Scale Processes Important in Determining Bioavailability

3.1 SEDIMENT QUALITY CRITERIA AND THERMODYNAMIC EQUILIBRIUM

At present most regulatory assessment of the quality of coastal marine sediments is carried out under the assumption that an organic contaminant is partitioned among the organismic, sedimentary, and pore water matrices as if it were at thermodynamic equilibrium [3] [15]. If equilibrium conditions exist, one can easily estimate organismal uptake based on lipid content in the biota and the sedimentary organic carbon content [15] [22] [23]. The presence of thermodynamic equilibrium also obviates the need to worry about the pathway of uptake of the contaminant. The same equilibrium concentration within the organism should be achieved regardless of the mode of uptake. However our current knowledge of the biogeochemisty of organism-sediment relationships suggests that the conditions conducive to thermodynamic equilibrium partioning may be rare. Field measurements of infaunal body burdens and sedimentary contaminant concentrations fail to support the existence of widespread equilibria in coastal environments [23].

I suggest that the cause of this failure is most probably due to bioturbation, trophic uptake and/or metabolism of contaminants by macrofauna [3]. Benthic animal behavior resulting in transport of solutes and particles in combination with the internal metabolic processing of contaminant that is taken up interact to disrupt or prevent the formation of equilibria.

3.2 MODEL RESULTS AND ASSUMPTIONS

Below I briefly summarize some recent results derived from models designed to estimate the the appropriate temporal and spatial scales relevent to the bioavailability of organic contaminants to infauna [2]. The model assumptions were framed in such a way as to generally favor the role of pore water in determining uptake. A primary goal was to provide an initial estimate of the environmental 'grain' of the processes determining uptake and thus generate first order predictions to help guide future experimental work. Such questions as - Does irrigation behavior act on time scales that would be expected to significantly disrupt establishment of steady state pore water concentrations as suggested by the *Arenicola* microcosm data? and - What is the expected thickness of the boundary layer around an unprotected worm inhabiting a contaminated sediment in the absence of bioturbation? - can be easily addressed. The basic model discussed below consisted of the standard transient-state diagenetic equation with diffusion and sorption terms for fluoranthene [2] [13] [24]. Model results were then evaluated in light of parallel experiments investigating the absorption efficiency, selectivity and feeding rate of *Capitella* on particle-bound fluoranthene. The worm absorption efficiency and sedimentary sorption kinetic experiments were performed on the same batch of fluoranthene-contaminated sediment to produce estimates of the relative role of pore water and deposit feeding in the uptake of fluoranthene.

3.2.1. Model Assumptions for Calculating Uptake from Pore Water.

The following six assumptions were used to construct the model for calculation of pore water uptake of fluoranthene. 1) Diffusion within the pore water is Fickian. That is diffusion into the worm was assumed to be proportional to the concentration gradient at the worm's surface. 2) The ability of the worm to take up fluoranthene is unlimited. 3) The worm cannot deplete the pore water of dissolved fluoranthene. Thus the pore water concentration of fluoranthene is determined solely by the sediment-bound concentration, worm uptake at the body surface, and the kinetics of the particle-sorption process. 5) Uptake occurs across the entire body surface which is assumed to be cylindrical and 6) that surface is smooth. The role of ingested pore water was neglected. Ingestion and absorption of dissolved contaminant will increase the relative contribution of the trophic pathway (i.e., deposit feeding) in determining total uptake.

These assumptions were chosen to bias the calculation in the direction of the overestimation of pore water uptake. Some of these assumptions are known to be violated to varying degrees [2] [23]. If worms reach thermodynamic equilibrium, deplete the local pore water concentration or simply alter burrow or pore water concentration through bioturbation then uptake via pore water will decrease. The possible exception to the general bias is assumption 6). To the extent that the epidermal body surface is not smooth the relative contribution of pore water uptake will be increased. Given the current lack of knowledge of the morphological nature of the external surface of *Capitella* and to simplify the calculations, the choice was made to model the body surface as a smooth cylinder – which will act to underestimate the importance of pore water uptake. Nevertheless, the net effect of the above six assumptions should be to strongly favor overestimation of pore water uptake rate.

3.2.2. Calculation of Uptake from Ingested Sediment.

In contrast to the model estimates of uptake from pore water, techniques presently exist for the accurate measurement of ingestive uptake of organic contaminants by deposit feeders. The methods are based on feeding animals sediment labeled with absorbed (^{14}C organic matter) and unabsorbed (^{51}Cr) tracers. The gamma emitting isotope of chromium is strongly particle reactive but is not absorbed to any significant degree by *Capitella* species I or most other deposit feeders [2]. The change in ratio of absorbed to unabsorbed tracer before and after passage of sediment through the gut can be used to accurately measure absorption efficiency of the ^{14}C tracer - in this case ^{14}C-fluoranthene.

In addition to absorption efficiency it is also possible to directly measure selectivity for and ingestion rate of the contaminant or type of organic matter under study (e.g., heterotrophic microbes, algae). Further details and additional references regarding this powerful technique can be found in Lopez [25] and will not be discussed further here. These methods were originally developed for the study of the nutrition of deposit feeders but in some ways are much more amenable for use in determining contaminant uptake. This is because the target compound (e.g., fluoranthene) is known and thus characterization of the radioactive label is unambiguous. Absorption efficiencies, ingestion and absorption rates were determined on the same batch of sediment used for sorption kinetic experiment. After radio-labeling with fluoranthene, this sediment was incubated for 8 months at 4°C to allow strong binding of the contaminant. Feeding rate data for medium-sized adult *Capitella* (2.5 mm^3 body volume, [26]) were taken from the literature and used to calculate dietary uptake rates for comparison with pore water uptake modeling results. Sediment porosity and temperature were accounted for [2].

Results indicated that *Capitella* species I is capable of absorbing 56% (∀3.4% s.d.) of the particle-bound fluoranthene on a single gut passage. For worms of this size gut residence time is approximately 1 to 1.5 hours. The desorption experiment using the same batch of sediment demonstrated that only 4% of the bound fluoranthene was released into the pore water over a 90 minute interval indicating that the gut environment greatly enhanced the bioavailability of ingested fluoranthene [2]. In addition, a forage ratio of 2.4 for fluoranthene was also determined . Thus worms were capable of absorbing over 50% of the ingested contaminant which had been concentrated by a factor of 2.4 relative to the ambient bulk sediment.

3.2.3. The Time and Spatial Scales of Pore Water Fluoranthene Uptake.

An example profile from the transient state pore water model is shown in Figure 4. A short-term steady-state concentration of fluoranthene of 0.404 $\mu g\ L^{-1}$ was determined from the desorption experiment. This concentration was then used as the initial pore water concentration at the start of the model run. Figure 4 thus predicts short-term changes in concentration about an unprotected worm exposed to a steady-state concentration of pore water fluoranthene. Such a situation might occur, for example, when animals are added to sediment microcosms at the start of a toxicity test and before worms have had a chance to construct irrigated burrows.

Steady state is achieved rapidly. Approximately 80% of the total change to the steady-state profile occurs within the first 10 minutes with steady state effectively present by 40 minutes. This has important implications for estimates of exposure because of worm behavior such as burrowing and irrigation. The boundary layer can be seen to extend out to approximately 1 mm. However, because of the strongly nonlinear nature of the boundary layer 87% percent of the total change (i.e., from 0 to 0.404 μg fluoranthene L^{-1}) occurs within 200 Φm of the surface of the worm. Because of the rapid progess to steady state conditions one would expect that disruption of the boundary layer by an unprotected worm crawling freely through sediment (effectively increasing the near-surface concentration gradient) may not have a large effect on total flux into the animal.

This can be seen by using the model data shown in Figure 4 to calculate the time dependence of flux into a worm [2]. Estimates of the near-surface gradient, pore water diffusivity, and worm surface area were used to calculate the time-dependence of uptake flux over the same 90 minute interval shown in Figure 4 and the values are plotted in Figure 5.

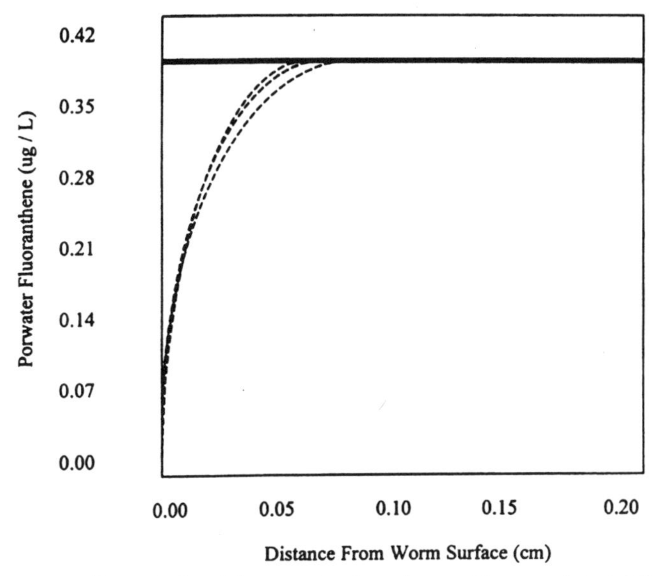

Figure 4. Time evolution of the pore water fluoranthene profile surrounding an individual worm calculated from the transient-state diffusion model incorporating sorption kinetics. Initial fluoranthene concentration was set at the asymptotic value of 0.404 Φg L^{-1}. Profiles were plotted for times 0, 10, 20, 30, 40, 50, 60, 70, 80, and 90 minutes. The solid horizontal line at [FLU] = 0.404 Φg $L^{-1}$is the asymptotic value measured in the desorption experiment.

These 'instananeous flux' calculations show an initial peak value after only 5 minutes followed by another, smaller transient peak at 20 minutes with a rapid decay to values indistinguishable from steady state by 40 minutes (Figure 5). Note that steady state fluxes or pore water profiles are relatively short term phenomena and do not indicate the presence of thermodynamic equilibrium. Though it at first appears large, the initial peak value of 0.1817 ng $worm^{-1}$ d^{-1} is less than a 4% increase in flux over the steady state value. Thus changes in boundary layer thickness caused by worm movement alone would not be expected to significantly increase total uptake flux. Alternatively, burrow construction could act to effectively displace the boundary layer away from the surface of the worm effectively extending it outward from the edge of an irrigated burrow wall. The spatial and temporal characteristics of the layer would then be controlled primarily by irrigation periodicity and the chemical compostion of the of the burrow lining.

Though little is known of the irrigation activity of Capitellid species it is probable that ventilatory activity is controlled by oxygen tension within the burrow itself. Given known burrow thicknesses (ca. 1 mm) and the oxygen demand of sediments of the type typically inhabited by *Capitella* species I, O_2 levels within recently irrigated worm burrows should decrease to zero within seconds to at most a few minutes (TF unpub.). These calculations combined with the levels of oxygen known to be required for positive worm growth rates suggest that worms irrigate on time scales that would likely disrupt the formation of a steady state pore water fluoranthene profile [27].

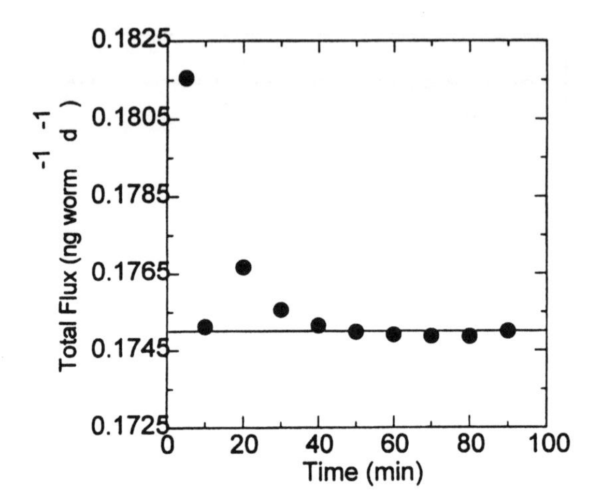

Figure 5. Plot of the time evolution of fluoranthene flux into an individual worm. Horizontal baseline is the calculated steady state flux of 0.175 ng worm^{-1} d^{-1}.

3.2.4. Comparison of uptake fluxes due to pore water and ingested sedment.

Measured data on fluoranthene absorption efficiency, selectivity, and feeding rate of *Capitella* feeding in contaminated sediment can be compared to model estimates of pore water uptake flux by constructing a ratio of ingestive to pore water flux rates. To picture how changes in selectivity and absorption efficiency might act to change uptake fluxes these ratios are plotted as a function of both in Figure 6.

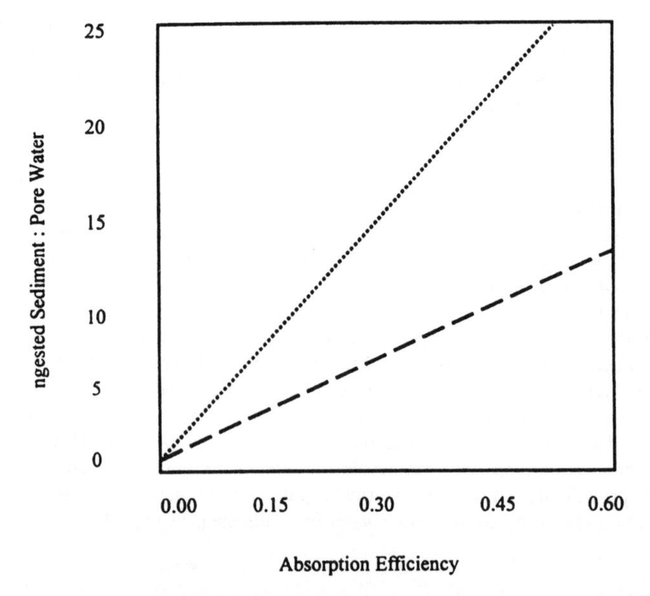

Figure 6. Ratio of fluoranthene uptake flux due to ingested sediment relative to that expected from pore water versus absorption efficiency. Dotted line: Uptake during selective feeding using the measured forage ratio of 2.4 for fluoranthene. Dashed line: non-selective feeding.

Compared to the non-selective situation (dashed line) one can see that absorption rate increases even more rapidly with selective deposit feeders as absorption efficiency increases (Figure 6). At the experimentally measured absorption efficiency of 56%, individual worms are calculated to take up fluoranthene via deposit feeding at rates that are 20 to 30 times greater than those predicted from pore water. Absorption efficiencies well below present detection limits (< 1%) are required for the contribution of pore water to equal that of ingestion.

The above analysis suggests that within a particular habitat or environment short time (seconds to minutes) and small spatial (microns to millimeters) scales will have the greatest role to play in determining contaminant uptake. These are the scales most strongly affected by animal activity and behavior. Ingestion selectivity, for example, is a process that occurs at a spatial scale on the order of sedimentary grains but is nevertheless influenced by behavioral and morphological characteristics which in turn can be a function of ontogenetic, ecological and evolutionary time scales (Figure 2). Boundary layer thickness and gradient steepness, which are expected to strongly affect pore water uptake are also under the behavioral control of organisms. Available data and model calculations both suggest that animal activity will occur at scales tending to disrupt physico-chemical equilibria immediately surrounding individuals thereby invalidating simple equilibrium partitioning calculations.

Consideration of animal-sediment interactions at small scales suggests that there is no chemically well-defined bioavailable fraction that can be unambiguously defined - even given perfect geochemical knowledge. Bioavailability is a function of organism-sediment interactions and these interactions occur at relatively small spatial and temporal scales. Because the chemical analysis of sediment typically occurs at >averaging= scales the data necessary for accurate prediction of contaminant uptake rates are currently unavailable. Until processes occuring at scales relevant to individual organisms can be measured or estimated, improved prediction will remain an elusive goal.

4. Acknowledgements

The ideas and thoughts presented here on bioavailability have benefitted greatly from discussions over time with Anders Giessing, Rikke Hansen, Liv Kure, Larry Mayer, Henriette Selck and Marianne Thomsen. I thank the Danish National Science Council for support of the present (11-1047-1 to TLF) and previous work on bioavailability.

5. References

1. Rosenberg, R., Cato, I., Förlin, L., Grip, K. and Rodhe, J. (1997) Marine environment quality assessment of the Skagerrak-Kattegat, *J. Sea Res.* **35**, 1-8.
2. Forbes, T.L., Forbes, V.E., Giessing, A., Hansen, R. and Kure, L. (in press) The relative role of pore water versus ingested sediment in the bioavailability of organic contaminants in marine sediments, *Env. Toxicol. Chem.*
3. Forbes, V.E. and Forbes, T.L. (1994) *Ecotoxicology in Theory and Practice*, Chapman and Hall, London.
4. Lopez, G.R. and Levinton, J.S. (1987) Ecology of deposit feeding animals in marine sediments, *Q. Rev. Biol.*, **62**, 235-260.
5. Lopez, G.R., Taghon, G., Levinton, J.S. (eds.), (1989), *Ecology of Marine Deposit Feeders*, Springer-Verlag, New York.
6. Sanders, H.L. (1958) Benthic studies in Buzzards Bay. I. Animal-sediment relationships, *Limnol. Oceanogr.*, **3**, 245-58.
7. Pearson, T.H. and Rosenberg, R. (1978) Macrobenthic succession in relation to organic enrichment and pollution of the marine environment, *Oceanogr. Mar. Biol. Ann. Rev.*, **16**, 229-311.
8. Lopez, G.R. and Cheng, I-J. (1982) Ingestion selectivity of sedimentary organic matter by the deposit-feeder *Nucula annulata* (Bivalvia: Nuculanidae), *Mar. Ecol. Prog. Ser.*, **8**, 279-282.
9. Lopez, G.R. and Cheng, I-J. (1983) Synoptic measurements of ingestion rate, ingestion selectivity, and absorption efficiency of natural foods in the deposit-feeding molluscs *Nucula annulata* (Bivalvia) and *Hydrobia totteni* (Gastropoda), *Mar. Ecol. Prog. Ser.*, **11**, 55-62.

10. Lee II, H., Boese, B.L., Randall, R.C. and Pelletier, J. (1990) A method for determining gut uptake efficiencies of hydrophobic pollutants in a deposit-feeding clam, *Env. Toxicol. Chem.*, **9**, 215-219.
11. Aller, R.C. (1983) The importance of the diffusive permeability of animal burrow linings in determining marine sediment chemistry, *J. Mar. Res.*, **41**, 299-322.
12. Winsor, M.H., Boese, B.L., Lee II, H., Randall, R.C. and Specht, D.T. (1990) Determination of the ventilation rates of interstitial and overlying water by the clam *Macoma nasuta, Env. Toxicol. Chem.*, **9**, 209-213.
13. Boudreau, B.P. (1997) *Diagenetic Models and Their Implementation: ModellingTransport and Reactions in Marine Sediments*, Springer-Verlag, Berlin, Germany.
14. Forbes, V.E., Forbes, T.L. and Holmer, M. (1996) Inducible metabolism of fluoranthene by the opportunistic polychaete *Capitella* species I, *Mar. Ecol. Prog. Ser.*, **132**, 63-70.
15. DiToro, D.M., Zarba, C.S., Hansen, D.J., Berry, W.J., Swartz, R.C., Cowan, C.E., Pavlou, S.P., Allen, H.E., Thomas, N.A. and Paquin, P.R. (1991) Technical basis for establishing sediment quality criteria for nonionic organic chemicals using equilibrium partitioning, *Env. Toxicol. Chem.*, **10**, 1541-1583.
16. Bierman Jr., R.N. (1990) Equilibrium partitioning and biomagnification of organic chemicals in benthic animals, *Environ. Sci. Technol.*, **24**, 1407-1412.
17. Webster, J. and Ridgeway, I. (1994) The application of the equilibrium partitioning approach for establishing Sediment Quality Criteria at two UK sea disposal and outfall sites, *Mar. Poll. Bull.* **28**, 653-661.
18. Forbes, T.L. and Kure, L. (1997) Linking structure and function in marine sedimentary and terrestrial soil ecosystems: Implications for extrapolation from the laboratory to the field, in N.M. van Straalen and H. Løkke (eds.), *Ecological Risk Assessment of Contaminants in Soil*, Chapman and Hall, London, pp. 127-156.
19. Kure, L. and Forbes, T.L. (1997) Impact of bioturbation by *Arenicola marina* on the fate of particle-bound fluoranthene, *Mar. Ecol. Prog. Ser.*, **156**, 157-166.
20. Kure, L. (1997) Interactions between particle-bound organic pollutants and bioturbating macrofauna, Ph.D. thesis, Institute of Biology, Odense University, Denmark.
21. Hansen, R., Forbes, T.L. and Westerman, P. (submitted to *Mar.Ecol. Prog. Ser.*) Importance of the polychaete *Capitella* species I and sediment-associated microorganisms in the degradation of di(2-ethlhexyl)phthalate.

22. Lake, J.L., Rubenstein, H. and Pavignano, S. (1987) Predicting bioaccumulation: Development of a simple partitioning model for use as a screening tool for regulating ocean disposal of wastes, in K.L. Dickson, A.W. Maki and W.A. Brungs (eds.), *Fate and Effects of Sediment-Bound Chemicals in Aquatic Systems*, Pergamon Press, Elmsford, NY, pp. 181-166.
23. Landrum, P.F., Harkey, G.A. and Kukkonen, J. (1996) Evaluation of organic contaminant exposure in aquatic organisms: The significance of bioconcentration and bioaccumulation, in M.C. Newman and C.H. Jagoe (eds.), *Ecotoxicology: A Hierarchical Treatment*, Lewis Publishers, Boca Raton, FL, pp. 85-132.
24. Berner, R.A. (1980) *Early Diagenesis: A Theoretical Approach*, Princeton University Press, Princeton.
25. Lopez, G.R. (1993) Absorption of microbes by benthic deposit feeders using the 14-C:51-Cr dual-labelling method, in P.F. Kemp *et al.* (eds.), *Handbook of Methods in Aquatic Microbial Ecology*, Lewis Publishers, Boca Raton, FL, pp. 739-744.
26. Forbes, T.L. and Lopez, G.R. (1987) The allometry of deposit feeding in *Capitella* species I (Polychaeta: Capitellidae): The role of temperature and pellet weight in the control of egestion, *Biol. Bull.*, **172**, 187-201.
27. Forbes, T.L., V.E. Forbes and Depledge, M. (1994) Individual physiological responses to environmental hypoxia and organic enrichment: Implications for early soft-bottom community succession, *J. Mar. Res.*, **52**, 1081-1000.

THE ROLE OF THE MARINE GASTROPOD *CERITHIUM VULGATUM* IN THE BIOGEOCHEMICAL CYCLING OF METALS

A. NICOLAIDOU* AND J. A. NOTT**
**Department of Zoology - Marine Biology, University of Athens, Panepistimiopolis, GR 15784, Athens, Hellas ** Plymouth Marine Laboratory, Citadel Hill, Plymouth, PL1 2PB, U.K.*

Abstract

The gastropod *Cerithium vulgatum* plays an important role in metal cycling in shallow water marine ecosystems. *C. vulgatum* detoxifies metals by incorporating them into insoluble granules formed in the digestive gland. When in granular form metals become unavailable to predators and thus are removed from food chains. Granules are incorporated in the durable faecal pellets of *C. vulgatum* and thus are retained in the sediment. The fate of metals in granules released from the faecal pellets depends on the properties of the metals and the characteristics of the sediment, especially the state of oxidation.

1. Introduction

Cerithium vulgatum is a very common littoral Mediterranean gastropod. It lives in all types of substrata including sandy mud [1] and is often associated with the sea grasses *Cymodocea nodosa* and *Zostera noltii,* where it feeds among their rhizomes. Examination of its gut contents by the authors showed that its diet consists of diatoms and epibenthic microalgae. In its turn it is an important prey of other gastropods such as *Pisania* and *Murex* [2]. Another species *C.litteratum,* is known as prey of larger animals such as lobsters [3] and fish [4] and there is no reason to believe that *C.vulgatum* is not consumed by larger animals. Therefore, it must be an important component of littoral food chains.

Despite its wide distribution, the use of *Cerithium* in studies related to heavy metals has been very limited. However, the tolerance of *Cerithium* to heavy metals, especially in relation to its genetic structure, has been studied [5] [6] [7]. In a preliminary survey of heavy metals in sediment and biota near a ferro-nickel smelting plant in Larymna, Hellas, *Cerithium vulgatum* was found to accumulate a variety of metals to higher concentrations than other gastropods. A series of papers which followed using material from this area and from a clean site in Vravrona, [9] [10] [11] 12] [13] [14], elucidated the role of *C. vulgatum* in the metal cycling in shallow water marine ecosystems. These findings, together with some unpublished data, are presented here.

2. Cerithium vulgatum as a metal accumulator

Atomic Absorption Spectroscopy of the digestive gland of *C.vulgatum* from Larymna showed higher concentrations of Mg, Mn, Zn, Co, Ni and Cr compared to those of a clean site (Table 1).

J.S. Gray et al. (eds.), Biogeochemical Cycling and Sediment Ecology, 137–146.

Table 1. Metal concentrations (ug/g dry weight) in pooled samples of digestive glands of *C.vulgatum* from the polluted and clean sites and of *Murex trunculus* from the polluted site, measured by atomic absorption spectroscopy [9].

	C. vulgatum	C. vulgatum	Murex trunculus
	Polluted site	Clean site	Polluted site
Ca	14543	11295	10269
Mg	25555	1230	2333
Fe	4940	3035	766
Mn	9569	5440	20
Zn	15728	3510	1899
Cu	183	358	7956
Co	510	60	7
Ni	553	25	41
Cr	165	49	14

Investigations of the digestive gland by x-ray microanalysis showed that the metals were localised in concentrically structured phosphate granules and amorphous residual lysosomes, both bound by a membrane [9]. Moreover, when the cells reached the end of a digestive cycle they disintegrated into the lumen of the gland, complete with granules, for excretion in the faeces. Using species available at Plymouth it was established that the phosphate granules are insoluble and when the digestive gland is eaten by a predator they pass through the gut, to be voided complete with metals, in the faeces [15]. Returning to *Cerithium vulgatum* this work was enlarged to include the passage of residual lysosome through the gut of a predator [11].

3. Transfer of metals from *Cerithium vulgatum* to higher trophic levels

One of the predators of *C. vulgatum* in Larymna is the gastropod *Murex trunculus* which was also analysed for heavy metals [9]. From the results in Table 1 it is obvious that some of the metals existing in excess in *C.vulgatum,* such as Mg, Fe, Mn, Zn, Co, Ni and Cr, were not transferred to its predator. This is also shown by the XRMA in Figure 1. *Murex* species store Cu in the digestive gland [16], which accounts for its high concentration. Experiments were designed to confirm these observations in the laboratory. However, although *Murex* were seen feeding on *C.vulgatum* in the field, they refused to do so when the two species were kept together in aquaria. Thus, the experiments were carried out using the hermit crab *Clibanarius erythropus* as a carnivore [11]. The hermit crabs were collected from a clean site and fed individually with *C.vulgatum* digestive gland from the polluted site for 10 days. The digestive gland of *C.vulgatum* was analysed by AAS and x-ray microanalysis (XRMA) before they were offered to the crabs. Digestive glands of hermit crabs were also analysed by AAS and XRMA. Finally, faecal pellets of *C.erythropus* were examined by scanning electron microscopy (SEM) and XRMA.

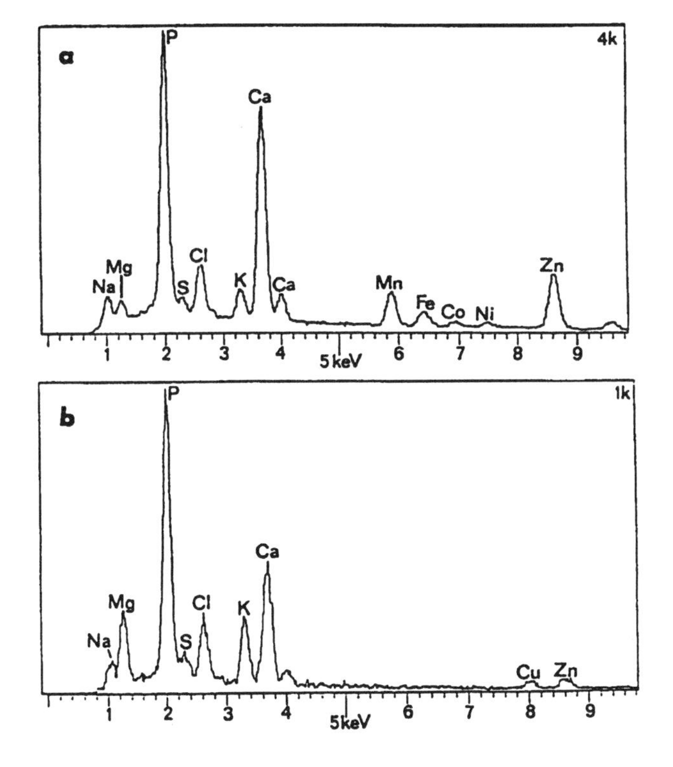

Figure 1. X-Ray microanalyses in the scanning electron microscope at 25 kV of phosphate granules in digestive glands dried onto stubs. Vertical scale=X-ray counts; horizontal scale=X-ray energy in keV. a. *Cerithium vulgatum* from polluted site. b. *Murex trunculus* from polluted site. This carnivore eats *C.vulgatum* but most of the metals are not transferred. [9]

Table 2. Mean concentrations (ug/g dry weight) and standard deviation (n=10) of metals in the digestive gland of the gastropod *C.vulgatum* and the hermit crab *C.erythropus* before and after feeding with *C.vulgatum*, measured by atomic absorption spectroscopy. Significance of difference between means in hermit crabs were tested by the Student's t-test. NS: P>0.05, * : 0.01<P<0.05 and *** P<0.001 [11].

Digestive	glands Cr	Ni	Content Cd	of Mn	Zn	Ag
C.vulgatum	93.3±32.3	499.6±105.3	68.5±19.2	4116±1400	6745±2792	5.2±1.6
Hermit crab	3.7±1.7	28.9±24.5	11±5.6	1596±1082	2351±1138	24.2±16.2
Hermit crab fed C.v	18.8±7.2 ***	40.4±33.9 NS	17±6.4 *	1166±336 NS	2553±828 NS	16.9±5.2 NS

The concentration of four metals, Ag, Mn Ni and Zn, measured by AAS in the digestive gland of the hermit crabs before and after feeding with *C.vulgatum* (Table 2) showed no statistically significant change. Cr had a statistically significant increase while Cd increased marginally (0.01<P<0.05). XRMA showed that the digestive gland of *C.erythropus* was packed with phosphate granules composed predominantly of calcium and phosphorus but also contained sulfur, chlorine, potassium and a trace of zinc (Fig.2). The phosphate granules of *C.vulgatum* showed peaks for Mn, Fe, Co, Ni and Zn. The faecal pellets of the hermit crabs before feeding with *C.vulgatum* consisted of mineral particles with a small proportion of organic material, which consistently gave an X-ray peak for iron. After feeding with *C.vulgatum* the faecal pellets contained masses of calcium phosphate granules of the type produced by *C.vulgatum* (contaminated with Mn, Fe Co, Ni Cu and Zn). It can be proposed that metals in *C.vulgatum* are detoxified by incorporation in granules as insoluble phosphates and that this reduces the bioavailability of the metals when the tissue is digested by the carnivore. Cadmium occurred in the digestive gland of *C.vulgatum* as shown by AAS but was not detected by XRMA. This indicates that it was dispersed in the tissues and did not attain the minimum localised level of concentration required for detection by XRMA. It is possible that most of the Cd is bound to soluble, high sulphur metalloproteins as in other molluscs[17,18] and as such it accumulates in the predators. Chromium occurs in association with phosphate and sulfide granules and yet it remains available to the crabs. If it is insoluble in the granules it can be proposed that there is an additional pool of Cr which is free and available within the cytosol.

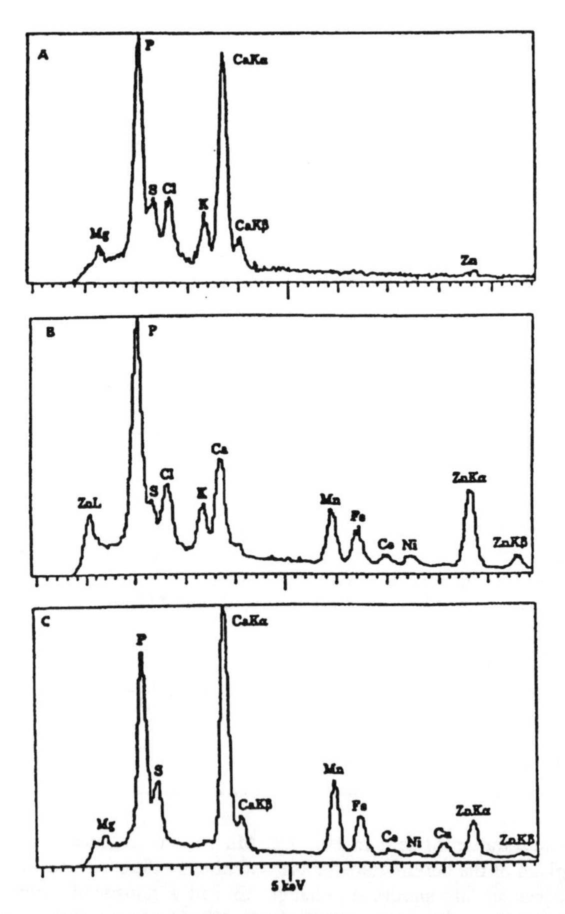

Figure 2. X-Ray microanalytical spectra of granules. a. *Clibanarius erythropus*. Granules in digestive gland. Major peaks for phosphorus and calcium and a range of minor peaks. b. *Cerithium vulgatum*. Granules in digestive gland with major peak for phosphorus; other peaks include range of pollutant metals. c. Granule derived from *C.vulgatum* but analysed in a *Clibanarius erythropus* faecal pellet; only chlorine and potassium have been removed. Full vertical scale=2000 X-ray counts; horizontal scale=X-ray energy. [11].

4. The fate of granules in the sediment

Upon lysis of senescent cells of the digestive gland the granules are released in the lumen and incorporated into the faeces [15]. To follow up the fate of the metals in the faeces discharged to the environment, fresh faecal pellets collected in aquaria and granules extracted from the digestive gland of *C.vulgatum* were incubated in sediments with contrasting characteristics [12]. Oxic ("white") and anoxic ("black") sediments from the polluted site of Larymna and the clean of Vravrona were used. The sediment in Larymna was silt with median diameter 40 um and organic carbon ~8% and from Vravrona it was sand with median diameter 750 um and organic carbon ~4.5%. XRMA analysis of the faecal pellets before the experiment and after incubation in the sediment for 5 days, 5 weeks and 15 weeks showed that the metal concentrations remained substantially unchanged. Individual granules, however, showed some changes after four days of incubation. Of the metals analysed by XRMA (Fig 3), Co and Ni were reduced in all four treatments. Mn and Fe were retained except in Larymna white sediment where they were reduced. Zn also was retained except in the black sediment of the clean site.

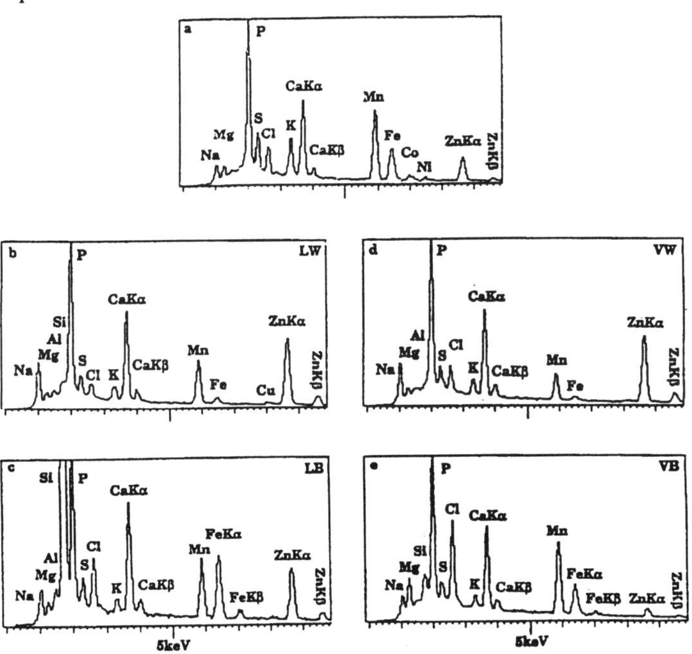

Figure 3. X-Ray microanalytical spectra of single, sphaerical, phosphate granules produced by *Cerithium vulgatum*: a. Control granule in digestive gland smeared on a graphite stub. b-e Granules incubated in the following sediments for 4 days: b.Larymna white; c.Larymna black; d.Vravrona white; e.Vravrona black. Individual granules probed with stationary, focused electron beam. Full vertical scale=2000 X-ray counts; horizontal scale=X-ray energy. [12].

5. The effect of *Cerithium vulgatum* on sediment metals

The effect of reworking by *C.vulgatum* on the metal content of clean and polluted sediments was examined experimentally. Polluted sediment was collected from Larymna and clean from Vravrona. They were placed to a depth of 1.5 cm in plastic aquaria measuring 40x20x20 cm and filled with artificial sea-water. There were two aquaria for each type of sediment. Twenty *C. vulgatum* were placed in one sediment of each type while the others served as controls. Animals used in the two sediment types were taken from the respective sites. The water was aerated with air stones and changed weekly by careful siphoning. After three weeks, five sediment samples and ten *C.vulgatum* (viscera only) were taken from each aquarium and analysed individually for Cd, Cr, Fe, Mn, Ni and Zn by AAS. In the animals, Cd showed a high level of variability and it was not taken into consideration. The same number of sediment samples and animals were also analysed before the experiment.

The results are shown in Table 3. The metals (except for Cd) in the polluted sediment decreased, probably lost through water changes. One way Analysis of Variance and Tukey's Least Significant Difference test showed that the differences were statistically significant between the original sediment and the control for all the metals except for Fe and Mn for which no significant change occurred. Metals in animals increased but changes were significant only for Fe and Mn. Although some of the metal was taken up by the gastropods the losses were smaller than in the controls. It appears that in the contaminated sediment animals assist the retention of some metals to the sediment. It is suggested that the gastropods, through the formation of granules, return metals accumulated from the water and food to the sediment as faeces.

In the clean sediment the metals in animals increase, however, this increase is significant only for Ni. In the sediment the loss is greater in aquaria with *C.vulgatum* although statistically significant difference occurs only for Mn. The latter could be associated with the observed consumption of detrital *Cymodocea nodosa* leaves known to concentrate metals selectively [14]. It is proposed that at low levels of contamination *C.vulgatum* accumulate metals but at higher concentrations some of the metals are returned to the environment.

Table 3. Mean concentrations (ug/g dry weight) of metals in the sediments and in *Cerithium vulgatum* viscera. Values in shaded boxes are significantly different with *0.01<P<0.05, ** 0.001<P<0.01 and *** P<0.001 . + Fe in sediments is expressed in mg/g dry weight. LA, VR: sediments and animals from Larymna (polluted site) and Vravrona (clean site).

Samples	Cd	Co	Cr	Fe+	Mn	Ni	Zn
LA Original sediment	*0.06±0.02* *	*49.90±2.01* *	*677±47* *	*39.2±1.8*	*340±12*	*889±40* **	*153±7* *
LA With Cerithium	*0.17±0.13*	*46.82±2.35*	*619±33*	*36.9±2.1*	*328±11*	*817±30*	*144±5*
LA Control sediment	*0.10±0.03*	*45.08±3.13*	*602±29*	*36.6±2.8*	*314±21*	*801±48*	*138±10*
VR Original sediment	*0.10±0.03* *	*6.74±0.32*	*105.0±13.4*	*13.2±1.6*	*271±13* ***	*60.5±6.2*	*44.3±3.8*
VR With Cerithium	*0.07±0.01*	*5.44±1.22*	*87.6±22.5*	*11.1±1.9*	*231±14*	*49.2±11.9*	*41.0±8.5*
VR Control sediment	*0.07±0.01*	*6.12±0.56*	*92.1±12.3*	*12.0±1.1*	*236±12*	*54.6±4.2*	*43.6±3.4*
LA Cerithium before		*64.20±48.5*	*36.8±12.0*	*1174±363* *	*1709±381* ***	*148±81*	*2316±992*
LA Cerithium after		*51.90±23.3*	*41.8±6.4*	*1579±470*	*2297±419*	*168±64*	*2390±970*
VR Cerithium before		*9.64±3.90*	*5.64±2.94*	*535±99*	*2070±857*	*13.8±4.0* **	*719±345*
VR Cerithium after		*16.63±10.15*	*5.91±4.57*	*531±186*	*2775±942*	*22.3±9.5*	*1083±428*

The overall pattern of the effect of *C.vulgatum* on the sediment may be appreciated by using multivariate analysis. The ordination of the three treatments in the polluted and clean sediments are shown in the MDS diagrams in Figures 4a and b respectively. In the polluted sediment, three distinct groups are formed with a gradient of metals from the higher concentrations, at the beginning of the experiment, occupying the left part of the diagram, to the lower concentrations of the controls on the right. Sediments with *C.vulgatum* occupy the space between them. In the clean sediment four of the five samples of sediment with *C.vulgatum* tend to be separated from the rest of the sediments. Although the original sediment tends to occupy the left part of the diagram and the control the right, they overlap in the middle to form a continuous group

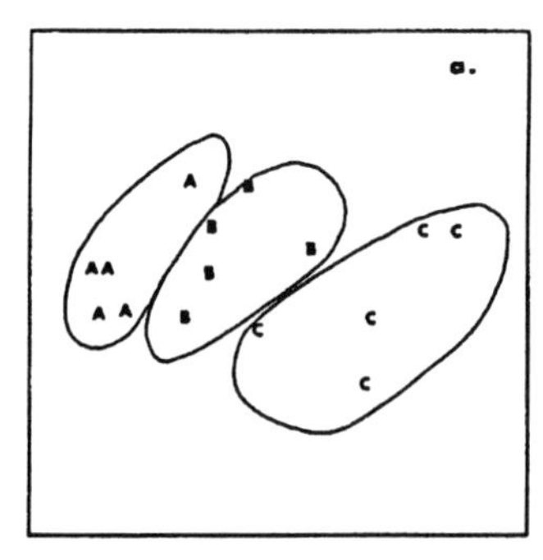

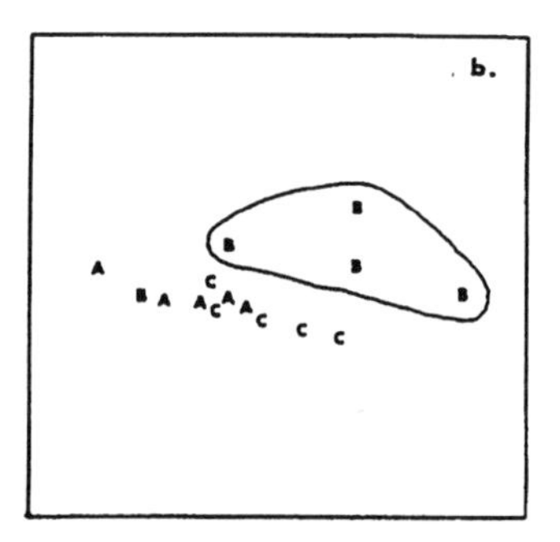

Figure 4. Multidimensional Scaling (MDS) plot of a. Larymna polluted and b. Vravrona clean sediment samples based on their metal concentrations. A. Original sediment at the beginning of the experiment, B: Sediment with *Cerithium vulgatum* and C: Control sediment without *C.vulgatum* both at the end of the experiment.

6. Discussion

The role of organisms in the cycling of metals has been recognised both in fresh waters (for example [19,20]) and in the marine environment. In the latter a number of papers have demonstrated the importance of bioturbating macrofauna in the exchange of ions between the sediment and the pore water in both shallow[21] [22] and pelagic sediments [23]. The relevant processes are reviewed [24] [25].

The importance of *Cerithium vulgatum* in the biogeochemical cycling of metals lies mainly in the reduction of their bioavailability. *C.vulgatum* receives metals from its food, water and sediment [14] and detoxifies them by incorporating them into insoluble phosphates within mineralised granules. Such granules are formed within basophil cells in the digestive gland as in other gastropods [26] [27] [28]. In bivalve molluscs they occur in the epithelial tissue of the kidney [29], while in decapod crustacea they are formed in the R-cells of the hepatopancreas [30] [31] [32] [33].

The experiments described above indicate that *C.vulgatum,* by incorporating metals into granules, renders them unavailable to the predators. The scheme of metal detoxification by the formation of insoluble granules has also been successfully tested [15] by feeding *Nassarius reticulatus* with kidney tissue from *Chlamys opercularis* and digestive gland of *Littorina littoralis* spiked with Zn [34], both tissues known to possess granules[29] [35]. It was also confirmed by *Nucella lapillus* feeding on the barnacles *Balanus balanoides* which possesses granules rich in Zn in the tissue surrounding the mid-gut [36]. Conversely, metals such as Cd which are bound to soluble low molecular weight, high sulphur proteins [17] [18] [37] are available to the predators. (For a review on metals in marine food chains see [38]).

The granules of *C.vulgatum,* as with other animals [20] [33], are released during cyclical cell breakdown and incorporated in the faeces. Faecal material plays an important role in the trophic relationships of coastal benthic communities [39] and concentrations of metals may exert a significant influence on coprophagus organisms and other members of the detrital food webs [40]. The pellets produced by *C.vulgatum* are durable structures remaining intact for days and weeks in sea water and sediments. They measure 1.6-2.5 mm in length by 0.25-0.35 mm diameter. They are two orders of magnitude larger than the bulk of the particles in the diet of *C.vulgatum*. Thus, the faecal pellets prevent the enclosed sedimentary particles from being re-ingested by *C.vulgatum* or taken up at least by microphagous sediment feeders.

Once the granules are released from the faecal pellets the fate of the incorporated metals depends on both metal properties and sediment characteristics, especially the state of oxidation. Under reduced conditions metals form insoluble compounds with sulphides which are retained in the sediment [41] [42]. It has been seen that Fe and Mn are retained in the granules of *C.vulgatum* in the reduced sediments while they are mobilised and decline under the oxidised conditions.

The same was observed in the sediment reworking experiments where smaller losses of Mn and Fe to the water occurred in the polluted more anoxic sediment. The other metals, Co, Cr, Ni and Zn showed significant losses from the sediment in the absence of *Cerithium vulgatum*. Although the above experiment was not very conclusive, in conjunction with the previous findings, it confirms the role of *C.vulgatum* in assisting the retention of metals to contaminated sediments.

In conclusion, *C.vulgatum* detoxifies metals by incorporating them in granules, removes them from the food chains and assists in retaining them to the sediment. Considering that *C.vulgatum* is a common and abundant component of shallow water Mediterranean communities, its importance in the biogeochemical cycling of metals becomes evident.

7. References

1. D'Angelo,G. and Gargiullo, S. (1978) Guida alle Conchiglie Mediterranee. Fabri Editori S.p.A., Milano. 224 pp.
2. Taylor, J.D. (1987) Feeding ecology of some common intertidal neogastropods at Djerba, Tunisia. *Vie Milieu*, **37**, 13-20.
3. Cox, C., Hunt, J.H, Lyons, W.G. and Davis, G.E. (1996) Nocturnal foraging in the Caribbean lobster *Panulirus argus* , in S.A. Woodin, D.M.Allen, S.E. Stancyk, J. Williams-Howze, R.J. Feller, D.S. Wethey, N.D. Pentcheff, G.T. Chandler, A.W. Decho and B.C. Coull, eds. *Proceedings 24 Annual Benthic Ecology Meeting, Columbia, South Carolina 7-10 March 1996*. Abstract only.
4. Wainwright, P.C. (1987) Biomechanical limits to ecological performance: Mollusc-crushing by the Caribbean hogfish, *Lachnolaimus maximus* (Labridae). *J.Zool.* **213**, 283-297.
5. Nevo, E., Noy, R., Lavie, B. Beiles, A. and Muchtar, S. (1986) Genetic diversity and resistance to marine pollution. *Biol.J. Linn. Soc.*, **29**, 139-144.
6. Lavie, B. and E. Nevo (1988). Multilocus genetic resistance and susceptibility to mercury and cadmium pollution in the marine gastropod, *Cerithium scabridum*. *Aquat. Toxicol.* **13**, 291-296.
7. Karnaukhov, V.N., Milovidova, N.Y. and Kargopolova, I.N. (1977) Carotenoids and resistance of marine molluscs to sea pollution. *Zh. Evolint. Biokhim. Fiziol.* **13**, 134-138. Transl. in English.
8. Nicolaidou, A. and Nott J.A., (1989) Heavy metal pollution induced by a ferro-nickel smelting plant in Greece. *Sci. Total Envir.* **84**, 113-117.
9. Nott, J.A. and Nicolaidou, A. (1989) The cytology of heavy metal accumulations in the digestive glands of three marine gastropods. *Proc. R. Soc. Lond.* B. **237**, 347-362.
10. Nott, J.A. and Nicolaidou, A. (1989) Metals in gastropods: metabolism and bioreduction. *Mar. environ. Res.*, **28**, 201-205.
11. Nott, J.A. and Nicolaidou, A. (1994) Variable transfer of detoxified metals from snails to hermit crabs in marine food chains. *Mar. Biol.* **120**, 369-377.
12. Nott, J.A. and Nicolaidou, A. (1996) Kinetics of metals in molluscan faecal pellets and mineralised granules, incubated in marine sediments. *J. exper. Mar. Biol. Ecol.* **197**, 203-218.
13. Nicolaidou, A. and J.A. Nott, (1990). Mediterranean pollution from a ferro-nickel smelter: Differential uptake of metals by some gastropods. *Mar. Pollut. Bull.* **21**, 137-143.
14. Nicolaidou, A. and Nott, J.A. (1998) Metals in sediment, seagrass and gastropods near a nickel smelter in Greece: Possible interactions. *Mar. Pollut. Bull.* **36**, 360-365.
15. Nott, J.A. and Nicolaidou, A. (1990) Transfer of metal detoxification along marine food chains. *J. mar. biol. Assoc., U.K.*, **70**, 905-912.
16. Bouquegneau, J., Martoja, M. and Truchet, M. (1984) Heavy metal storage in marine animals under various environmental conditions , in: J. Boles, J. Zedunaisky and R. Giles, eds., *Toxins, drugs and pollutants in marine animals*, Springer- Verlag, Berlin, pp 147-160.
17. Bebianno, M.J. and Langston, W.J. (1991) Metallothionein induction in *Mytilus edulis* exposed to cadmium. *Mar. Biol.* **108**, 91-96.

18. Bebianno, M.J. and Langston, W.J. (1992) Cadmium induction of metallothionein synthesis in *Mytilus galloprovincialis. Comp. Biochem. Physiol* **103C**, 79-85.
19. Soster, F.M., Harvey, D.T., Troska, M.R. and Grooms, T. (1992) The effect of tubificid oligochaetes on the uptake of zinc by Lake Erie sediments. *Hydrobiologia,* **248**, 249-258.
20. Lasenby, D.C. and Van Duyn, J. (1992). Zinc and cadmium accumulation by the opposum shrimp *Mysis relicta. Arcch. Environ. Contam. Toxicol.* **23**, 179-183.
21. Aller, R.C. (1978) The effect of animal-sediment interactions on geochemical processes near the sediment-water interface, in: M.L. Wiley , eds, *Estuarine Interactions*. Academic Press, N.Y. pp. 157-172.
22. Aller, R.C. (1980) Diagenetic processes near the sediment-water interface of Long Island Sound, II Fe and Mn. *Adv. Geophys.* **22**, 351-415.
23. Piper, D.Z., Rude, P.D. and Monteith, S. (1987) The chemistry and mineralogy of haloed burrows in pelagic sediment at Domes site A: the Equatorial North Pacific. *Mar. Geol.* **74**, 41-55.
24. Kersten, M. (1988) Geobiological effects on the mobility of contaminants in marine sediments, in: W. Salomons, B.L. Bayne, E.K. Duursma, U. Foerstner, *Pollution of the North Sea. An assessment.* pp, 36-58.
25. Lee, H.II. and Swartz, R.C. (1980) Biological processes affecting the distribution of pollutants in marine sediments. Part II. Biodeposition and bioturbation, in: R.A. Baker *Contaminants and sediments. Vol.2 Analysis, Chemistry, Biology.* Pp. 555-606.
26. Simkiss, K. and Mason, A.A. (1984) Cellular responses of molluscan tissues to environmental metals. *Mar. Envir. Res.* **14**, 103-118.
27. Taylor, M.G.and Simkiss, K. (1984) Inorganic deposits in invertebrate tissues. *Envir. Chem.* 3, 102-138.
28. Gibbs, P.E., Nott, J.A., Nicolaidou, A. and Bebianno, M.J. (1998) The composition of phosphate granules in the digestive gland of marine prosobranch gastropods: variation in relation to taxonomy. *J. mollusc. Stud.* In press.
29. Carmichael, N.G., Squibb, K.S. and Fowler, B.A. (1979) Metals in the molluscan kidney: a comparison of two closely related bivalve species (*Argopecten*), using X-ray microanalysis and atomic absorption spectroscopy. *J. Fish. Res. Bd Can.* **36**, 1149-1155.
30. Hopkin, S.P. and Nott, J.A. (1979) Some observations on concentrically structured, intracellular granules in the hepatopancreas of the shore crab *Carcinus maenas (L.). J. Mar. Biol. Ass. U.K.* **59**, 867-877.
31. Al-Mohanna, S.Y. and Nott, A.J. (1989) Functional cytology of the hypatopancreas of *Penaeus semisculatus (Crustacea: Decapoda)* during the molt cycle. *Mar. Biol. 101*, 535-544.
32. Bjerregaard, P. (1990) Influence of physiological condition on cadmium transport from haemolymph to hepatopancreas in *Carcinus maenas (L.).Mar. Biol.***106**, 199-209.
33. Vogt, G. and Quintino, E.T. (1994) Accumulation and excretion of metal granules in the prawn, *Paeneus monodon,* exposed to water-born copper, lead, iron and calcium. *Aquat. Toxicol.* **28**, 223-241.
34. Nott, J.A. and Nicolaidou, A. (1993) Bioreduction of zinc and manganese along a molluscan food chain. *Comp. Biochem. Physiol.* **104a**, 235-238.
35. George, S.G. (1982) Subcellular accumulation and detoxification of metals in aquatic animals, in: W.B. Vernberg *et al. Physiological Mechanisms of Marine Pollutant Toxicity.* Academic Press. pp 3-52.
36. Walker, G., Rainbow, P.S., Foster P., and Holland, D.L., (1975) Zinc phosphate granules in tissue surrounding the midgut of the barnacle *Balanus balanoides. Mar. Biol.* **33**, 161-166.
37. Rosijadi, G. (1992) Metallothioneins in metal regulation and toxicity in aquatic animals. *Aquat. Toxic.* **22**, 81-114.
38. Nott, J.A. Metals in food chains, in: W. Langston and M. Bebianno (eds.) *Metals in aquatic environments,* Chapman and Hall, London, pp. 387-414.
39. Frankenberg, D. and Smith, K.L.JR (1967) Coprophagy in marine animals. *Limnol. Oceanogr.* **12**, 443-450.
40. Booth, P.N. and Knauer, G.A. (1972) The possible importance of fecal material in the biological amplification of trace and heavy metals. *Limnol. Oceanogr.* **17**, 270-274.
41. Emerson, S., Janke, R. and Heggie, D. (1984) Sediment-water exchange in shallow water estuarine sediments. *J. mar. Res.* **42**, 709-730.
42. Alongi, D.M., Boyle, S.G., Tirendi, F. and Payn, C. (1996) Composition and behaviour of trace metals in Post-oxic sediments of the Gulf of Papua, Papua New Guinea. *Estuar. Coast. Shelf Sci.* 42, 197-211.

CHANGES IN MACROZOOBENTHOS COMMUNITIES INDUCED BY ANTHROPOGENIC EUTROPHICATION OF THE GULF OF GDANSK

A. SZANIAWSKA, U. JANAS, & M. NORMANT
Institute of Oceanography, Gdansk University, Al. Marszalka J. Pilsudskiego 46, 81-378 Gdynia, Poland

Abstract

Environmental changes in the Gulf of Gdansk (southern Baltic) due to eutrophication and toxic substances have many negative effects on communities of macrozoobenthos. The most significant changes are observed in the coastal areas and deeper parts of that region. The decline in bottom water oxygen concentration and hydrogen sulphide presence, observed during the last 50 years, caused an elimination of more sensitive macrofaunal species. The biomass of bivalves and polychaetes significantly increased, whereas that of crustacea decreased during this period. The bottom fauna mainly consists of facultative suspension-deposit feeders. The bivalve *Macoma balthica* is the most dominant species in the deeper part of the Gulf of Gdansk.

1. Introduction

Over a period of several decades an increase of eutrophication in the Baltic Sea has been observed [1]. This increase has resulted in many adverse changes in the Baltic waters, such as oxygen deficiency and hydrogen sulfide presence in the near bottom water layers and bottom sediment. Such phenomena have been observed in many areas such as the coastal waters of south-east Kattegat [2], the Bornholm Deep and the Gdansk Deep [3] [4]. Since the 1960's, due to oxygen deficiency and the increased presence of hydrogen sulfide faunal *deserts* in the deep parts of the Baltic have been observed. The Baltic, which has a well-developed halocline, is exposed to regular periods of oxygen deficiencies in deep waters [5] [6] [7]. During cold winters, due to ice cover, oxygen deficiency and the occurrence of hydrogen sulfide have been recorded even in shallow waters. Such phenomena, due to higher water temperatures, have also occurred during hot summers, especially at night, in the coastal waters of the Gulf of Gdansk, (Fig.1). It is explained by increased anthropogenic eutrophication of the water [8] [9].

The presence of hydrogen sulfide has been observed in practically all kinds of marine sediment [10] [11] [12] [13] [14] [15]. Hydrogen sulfide is a toxic compound which is created under natural conditions. It causes inhibition of cytochrome-c oxidase, which leads to the blocking of oxygen metabolism [16] [17] [18]. It is one of the most important ecological factors influencing the distribution and life processes of benthic organisms, [19] [20] [21]. It is known that stress caused by adverse oxygen conditions and high concentrations of hydrogen sulfide (several dozen micromoles) results in the migration of many species from endangered areas to new locations. However, some animals which cannot escape and are not able to adapt to unfavourable environmental conditions, die. The deficiency and lack of oxygen combined with the presence of hydrogen sulfide in the water strongly influences species composition, species behaviour and their metabolic and reproductive processes [22] [23] [24] [25] [26].

J.S. Gray et al. (eds.), Biogeochemical Cycling and Sediment Ecology, 147–152.

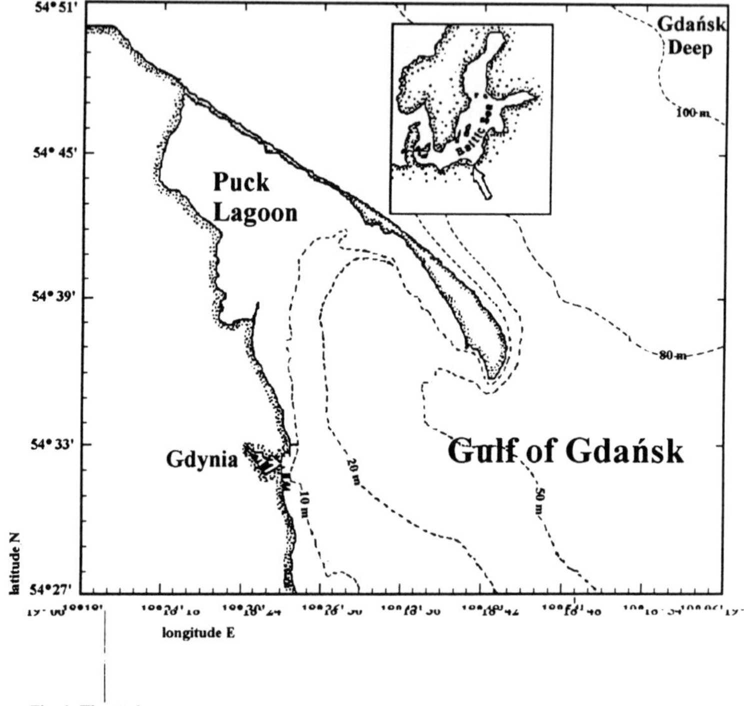

Fig. 1. The study area.

2. Characteristics of environmental conditions

2.1 Temperature

Water temperature in the Gulf of Gdansk changes seasonally. In the surface layer these changes are relatively high ranging from 2°C during winter to about 18°C in summer [27]. The highest temperatures are observed in August and the lowest in February. In deeper water layers the seasonal temperature amplitude is much smaller [28].

2.2 SALINITY

As with temperature, the salinity of surface and deep waters varies significantly. The salinity of the surface layers is lower due to fresh water input from rivers. In the Gulf of Gdansk the minimum annual salinity is 7.25 psu. The deep water layers, located below the halocline, are characterized by a higher salinity of about 13 psu [28].

2.3 OXYGEN CONDITIONS

The amount of oxygen in the waters varies seasonally. In the summer oxygen content in near bottom water layer drops below the annual average due to, the increase of water temperature, the increased consumption of oxygen by organisms and the mineralisation of organic matter. An increase in oxygen levels is observed in November and December [29]. In deep water layers, especially below the halocline, the amount of oxygen is lower and oxygen deficiency is a common problem.

2.4 SEDIMENTS

At the bottom of the basin, sediments of various particle sizes occur, from coarse-grained to fine-grained. The most common kind of sediment in shallow areas is sand of various granulation levels, while in deeper areas it consists of silt with a large amount of organic matter [30]. This large amount of organic matter on the sea bottom causes an increase of mineralisation intensity, resulting in oxygen consumption. This stagnation and the stratification below the halocline helps to create good conditions for the production of hydrogen sulfide.

2.5 HYDROGEN SULFIDE

In the deeper parts of the Gulf of Gdansk hydrogen sulfide occurs throughout the year in surface layers of sediment, however, the concentration varies seasonally. This compound can periodically occur in the near bottom waters [31] [32]. In surface layers of sediment an increase of hydrogen sulfide ihas been observed over time for deep regions, [26]. Concentrations of hydrogen sulfide in the 0-8 cm sediment layer vary from several to over 1000 µM (Fig. 2)

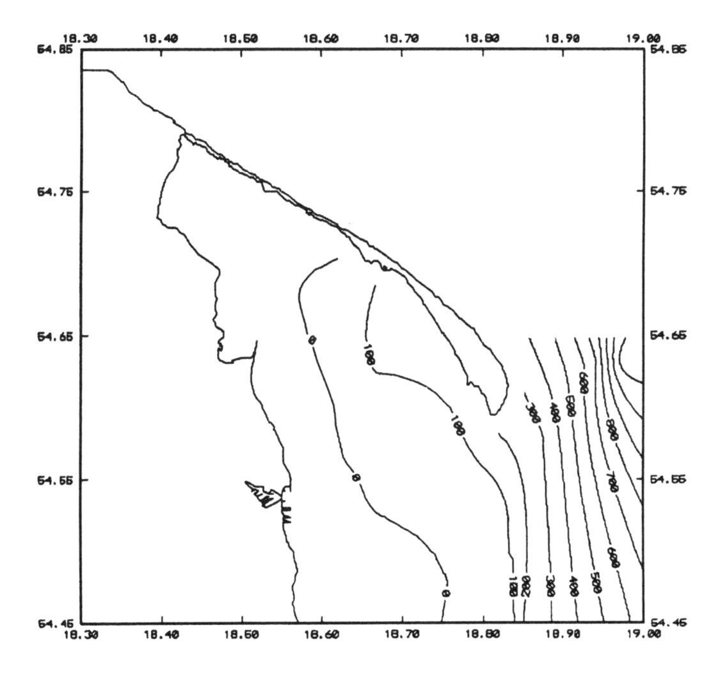

Fig. 2. Hydrogen sulfide concentration [µmol*dm^{-3}] in the pore waters of bottom sediments in the 0-8 cm layer.

3. Changes in macrozoobenthos

For many years investigations of macrozoobenthos community structure have been carried out in the southern Baltic Sea and the Gulf of Gdansk [33] [34] [35] [36] [37] [38] [39]. The smallest biomasses of macrobenthonic organisms are observed in shallow waters of up to 5 m in depth in the Gulf of Gdansk. This area is not well inhabited due to constant motion of water masses and river influence, which lead to variable temperatures, salinities and oxidation. In the 5-20m layer the biomasses of macrozoobenthos are higher. Below 20m the number of species decreases due to different environmental conditions, while their abundances and individual biomasses remain at high levels. Below 50m depth only a few species are observed, which is due to much lower food availability, as well as by much worse oxygen conditions and the presence of hydrogen sulfide in the sediment. Bottom fauna in this area mainly consist of facultative suspension-deposit-feeders with small numbers of predators and omnivorous species [40] (Fig. 3). Animals which live on the bottom of the Gulf of Gdansk, at depths >50m are exposed to hydrogen sulfide concentrations of several hundred micromoles. Such areas are inhabited by species with very low environmental requirements, such as *Macoma balthica*, *Pontoporeia femorata* and *Saduria entomon*. The absolute dominant species in this area of the Gulf of Gdansk is *M. balthica*, whose proportion of total biomass sometimes exceeds 90 %. In areas around and below the halocline, where oxygen conditions are the worst, two species occur namely, *M. balthica* and *Harmothoe sarsi*.

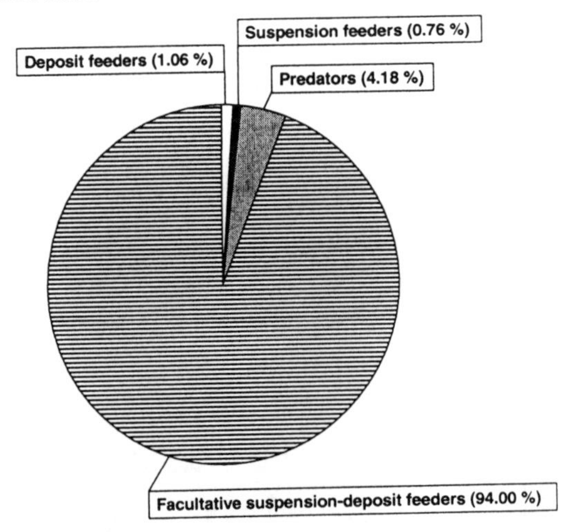

Fig. 3. Trophic structure of bottom fauna in the Gulf of Gdansk.

Over several decades many changes have been observed in the macrofauna at the bottom of the Gulf of Gdansk, especially in the coastal areas and deep regions. One of the main reasons for the changes in the biocoenosis is the strong eutrophication of the basin [39] [41] [42] [37]. The negative effects of such processes, especially inferior oxygen conditions and the presence of toxic hydrogen sulfide at the bottom cause the elimination of the most sensitive species and sometimes even whole groups of the macrofauna. Conversely, this allows for the expansion of the most resistant species.

The most drastic changes have occurred in the deep regions. At the beginning of the century the Gdansk Deep area was inhabited by a macrofauna set which consisted of 9 species with a combined biomass of 25.6 g m^{-2}, which included *Macoma calcarea, Scolops armiger, S. entomon, P. femorata, Halicryptus spinulosus* and *Harmothoe sarsi,* among others [43]. In recent years a catastrophic decrease in the number of species and the total macrozoobenthos biomass has occurred. Between 1984 and 1994 the only species periodically present in the waters of the Gdansk Deep was the semipelagical polychaete *H. sarsi* [44] [28] [45].

In the 1980's the ecological disaster spread outside the Gdansk Deep to other areas below the 80m isobath [41] [45]. Due to worsening oxygen conditions, the macrofauna died out almost completely, and only periodically could the highly tolerant polychaete *H. sarsi* be found [45]. By the end of the 1970s, those areas were inhabited by a set of five species, with *Scoloplos armiger* as the dominate species. The next mass repopulation of the Gulf appeared in 1990 when oxygen conditions rapidly improved, and *M. balthica* was practically the only species.

Long-term research results reveal that the species composition and biomass of the macrozoobenthos in the area above the halocline did not change very drastically. [46]

[3]. At depths of 30-60 m changes result from the total domination of *M. balthica* which is relatively resistant to oxygen deficiency or the presence of hydrogen sulfide and no fauna.

Significant changes can be observed in shallow waters, especially in the Puck Lagoon. Between 1962 and 1987 an increase of biomass of macrozoobenthos inhabiting this part of the Bay from 100 g m^{-2} to 300 g m^{-2} of bottom area and increase of areas where biomass exceeded 600 g m^{-2} were recorded [39] [44] [37]. During this period, bivalve biomass tripled while that of polychaetes doubled. At the same time crustacea biomass decreased five times. The process of eutrophication resulted in the disappearance of the underwater meadows which caused the local extinction of the the gastropod *Lymnea peregra* and bivalve *Parvicardium hauniense*, and crustacea such as *Crangon crangon* and *Palaemon adspersus* became less abundant.

If the oxygen conditions in the waters of the Gulf of Gdansk worsen even more and the areas of high hydrogen sulfide concentrations in sediment increase, changes in macrozoobenthos will be more significant.

4. References

1 Cederwall, H., Elmgren, R., 1990. Biological Effects of Eutrophication in the Baltic Sea, Particulary the Coastal Zone. AMBIO, **19**: 109-112.

2 Baden, S. P., Loo, L. O., Phil, L., Rosenberg, R., 1990. Effect of eutrophication on benthic communities including fish: Swedish west coast. Ambio, **19**: 113-122.

3 HELCOM, 1990. Second Periodic Assessment of the State of the Marine Environment of the Baltic Sea, 1984-1988, Background document. Baltic Sea Environment Proceedings **35** B, Helsinki.

4 HELCOM, 1996. Third Periodic Assessment of the State of the Marine Environment of the Baltic Sea, 1989-93, Background document. Baltic Sea Environment Proceedings **64** B, Helsinki.

5 Weigelt, M., 1990. Oxygen conditions in the deeper water of Kiel Bay and the impact of inflowing salt-rich water from the Kattegat. Meeresforschung, **33**: 1-22.

6 Weigelt, M., 1991. Short- and long-term changes in the benthic community of deeper parts of Kiel Bay (Western Baltic) due to oxygen depletion and eutrophication. Meeresforschung, **33**: 197-224.

7 Babenerd, B., 1991. Increasing oxygen deficiency in Kiel Bay (Western Baltic). A paradigm of progressing coastal eutrophication. Meeresforschung, **33**: 121-140.

8 Prena, J. (1994) Oxygen depletion in Wismar Bay (Western Baltic Sea) 1988. *Arch. Fish. Mar. Res.* **42 (1)**, 77-87.

9 Norkko, A. and Bonsdorff, E. (1996) Altered benthic prat - availability due to episodic oxygen deficiency caused by drifting algal mats. *P. Z. N. I.: Mar. Ecol.* **17 (1-3)**, 355-372.

10 Bagarinao, T., 1992. Sulphide as an environmental factor and toxicant: tolerance and adaptations in aquatic organisms. Aquat. Toxicol., **24**: 21-62.

11 Fenchel, T. M., Riedel, R. J., 1970. The sulfide system: a new biotic community underneath the oxidized layer of marine sand bottoms. Mar.Biol., **7**: 255-268.

12 Jørgensen, B. B., 1977. The sulfur cycle of coastal marine sediment (Limfjorden, Denmark). Limnol. Oceanogr., **22**: 814- 831.

13 J•rgensen, B. B. (1988) Ecology of sulphur cycle: Oxidative pathways in sediments. in: J. M. Cole, S. J. Ferguson (eds), *The nitrogen and sulfur cycles*, 42nd Symposium Soc. Gen. Microbiol. Univ. of Southampton , Cambridge University Press, Cambridge, pp. 31-63.

14 Jørgensen, B. B., Bang, M., Blackburn, H., 1990. Anaerobic mineralization in marine sediments from the Baltc Sea - North Sea transition . Mar. Ecol. Prog. Ser., **59**: 39-54.

15 Millero, F. J., 1991. The oxidation of H2S in Chesapeake. Estuar. Coast. Shelf. Sci., **33**: 521-527.

16 Nicholls, P.1975

17 Nicholls, P., Kim, J. K., 1981. Oxidation of sulphide by cytochrome aa3. Biochim. Biophys. Acta, **637**: 312-320.

18 Nichols, K., Kim, J. K., 1982. Sulphide as an inhibitor and electron donor for the cytochrome c oxidase system. Can .J Biochem., **60**: 613-623.

19 Vismann, B., 1991. Sulfide tolerance: physiological mechanisms and ecological implications. Ophelia, **34**: 1-27.

20 Giere, O., 1992. Benthic life in sulfide zones of the sea-ecological and structural adaptations to a toxic environmental. Verh. Dtsch. Zool. Ges., **85** (2): 77-93.

21 Grieshaber, M. K., Hardewig, I., Kreutzer, U., Schneider, A., Volkel, S., 1992. Hypoxia and sulfide tolerance in some marine invertebrates. Verh. Dtsch. Zool. Ges., **85** (2): 55-76.

22 Hagerman, L., Szaniawska, A., 1988. Respiration, ventilation and circulation under hypoxia in the glacial relict Saduria (Mesidotea) entomon. Mar. Ecol. Prog. Ser., **47**: 55-63.

23 Hagerman, L., Szaniawska, A., 1990. Anaerobic metabolic strategy of the glacial relict isopod *Saduria (Mesidotea) entomon*. Mar. Ecol. Prog. Ser., **59**: 91-96.

24 Hagerman, L., Szaniawska, A., 1991. Ion regulation under anoxia in the brackish water isopod *Saduria (Mesidotea) entomon*. Ophelia, **33**: 97-104.

25 Wiktor, K., Plinski, M., 1992. Long term changes in the biocenosis of the Gulf of Gdansk. Oceanologia, **32**: 69-79.

26 Janas, U., Szaniawska, A., 1996. The influence of hydrogen sulphide on macrofaunal biodiversity in the Gulf of Gdansk. Oceanologia, **38** (1): 127-142.

27 Nowacki, J., 1984. Ogólna charakterystyka warunków termiczno-zasoleniowych w Zatoce Puckiej. Zesz. Nauk. UG Oceanogr., **10**: 67-93.

28 Cyberska, B., Lauer, Z., Trzosinska, A., (eds), 1995. Warunki srodowiskowe polskiej strefy poludniowego Baltyku w 1994 roku, Inst. Meteorol. i Gosp. Wodnej., Materialy Oddzialu Morskiego, Gdynia.

29 Korzeniewski, K., (ed), 1993. *Zatoka Pucka. Fundacja Rozwoju Uniwersytetu Gdanskiego* p. 532.

30 Jankowska, H., Leczynski, L., 1993. Osady denne. In: Zatoka Pucka, Korzeniewski, K., (ed.), Fundacja Rozwoju Uniwersytetu Gdanskiego: 320-327.

31 Cyberska, B., Lauer, Z., Trzosinska, A., (eds), 1989. Warunki srodowiskowe polskiej strefy poludniowego Baltyku w 1988 roku. Inst. Meteorol. i Gosp. Wodnej, Materialy Oddzialu Morskiego, Gdynia.

32 Cyberska, B., Lauer, Z., Trzosinska, A., (eds), 1990. Warunki •rodowiskowe polskiej strefy poludniowego Baltyku w roku 1989. Inst. Meteorol. i Gosp. Wodnej, Materialy Oddzialu Morskiego, Gdynia.

33 Demel, K., Mulicki, W., 1954. Ilosciowe studia nad fauna denna Baltyku poludniowego. Prace MIR, **7**: 75-126

34 Herra, T., Wiktor, K., 1985. Composition and distribution of bottom fauna in coastal zone of the Gulf of Gdansk. SiMO, 46: 115-142.

35 Legiezynska, E., Wiktor, K., 1981. Fauna denna Zatoki Puckiej wlalciwej. Zeszyty Nauk. Wydz. BiNOZ UG, Oceanografia, **8**: 63-77.

36 Wiktor, K., Skóra, K., Wolowicz, M, 1980. Zasoby skorupiaków przydennych w przybrze•nych wodach Zatoki Gdanskiej. Zeszyty Nauk. Wydz. BiNOZ UG, Oceanografia, 7: 135-160.

37 Wolowicz, M., 1994. Impact de l'activite humaine sur les variations inter-annuelles de la malacofaune benthique en baie de Puck (S. Baltique). Haliotis, **23**: 43-50.

38 Warzocha, J., 1995. Classification and structure of macrofaunal communities in the southern Baltic. Arch. Fish. Mar. Res., **43** (3): 225-237.

39 Zmudzinski, L., 1976. Fauna denna wskalnikiem postepujacej eutrofizacji Baltyku. Studia i Materialy Oceanologiczne, **15**: 297-306.

40 Wiktor, K., 1985. An attempt to determine trophic structure of the bottom fauna in coastal waters of the Gulf of Gdansk. Oceanologia, **21**: 109-121.

41 Okolotowicz, W., 1985. Biomasa makrozoobentosu polskiej strefy Baltyku wskaznikiem jej zanieczyszczenia. Biuletyn MIR, XVI **(5-6)**: 27-40.

42 Wiktor, K., 1992. Fauna. In: *Studia i Materialy Oceanologiczne no 61 Marine Pollution* (2), Korzeniewski, K., (ed), Polish Academy of Science, National Scientific Comittee on Oceanic Research, Gdansk: 191-203.

43 Hegmeier, A., 1926. Die Arbaiten mit dem Petersenschen Bodengreifer auf der Ostseefahrt April und Juli 1926. Ber. Dtsch. Wiss. Komm. Meeresforsch. N. F., **5**: 154-173.

44 Zmudzinski, L., Osowiecki, A., 1991. Long-term in Macrozoobenthos of the Gdansk Deep. Int. Revue ges. Hydrobiol.,**76** (3): 465-471.

45 Osowiecki, A., Warzocha, J., 1996. Macrozoobenthos of the Gdansk, Gotland and Borholm Basins in 1978-1993. Oceanological Studies, **1-2**: 137-149.

46 Witek, Z., Bralewska, J., Chmielowski, H., Drgas, A., Gostkowska, J., Kopacz, M., Knurowski, J., Krajewska-Soltys, A., Lorenz, Z., Maciejewska, K., Mackiewicz, T., Nakonieczny, J., Ochocki, S., Warzocha, J., Piechura, J., Renk, H., Stopinski, M., Witek, B., 1993. Structure and function of marine ecosystem in the Gdansk Basin on the basis of studies performed in 1997. SiMO, 63, Marine Biology, **9**: 124pp.

DO BENTHIC ANIMALS CONTROL THE PARTICLE EXCHANGE BETWEEN BIOTURBATED SEDIMENTS AND BENTHIC TURBIDITY ZONES?

G. GRAF
University of Rostock, Department of Marine Biology, Freiligrathstrasse 7/8, 18055 Rostock, Germany

Abstract
Direct and indirect deposition and resuspension induced by benthic animals are compared with physical effects. Although some individual sub-processes, e.g. bioresuspension, are not yet sufficiently studied, it is concluded that in most cases animals increase the particle exchange between water and sediment by a factor of 2 to 10 and that these findings have to be incorporated into models of sediment cycling in coastal waters as well as in the deep-sea.

1. Introduction
Increasing particle concentrations close to the seafloor are a widespread phenomenon in marine shallow water systems as well as in the deep-sea. Only recently it was demonstrated that these benthic turbidity zones, which in the deep-sea are called nepheloid layers, provide a habitat for microorganisms benefiting from the hydrodynamic regime close to the sea floor [1] [2] [3]. The increased microbial activity and production enhances the food supply for bentho-pelagic fauna but also contributes to the formation of bottom near aggregates [4] [5].

Physical and biological processes contribute to the exchange of particles between the sediment and the overlaying water. Particles settling from the productive zone of the ocean and entering the turbidity zones are either deposited on the sediment surface, especially if they are incorporated in large and fast sinking aggregates, or kept in suspension, if the hydrodynamic regime balances the gravity component of a settling particle. Once incorporated into the sediment particles are redistributed downward by animals to finite depth called bioturbation depth. However, bioturbation does not imply only downward transport of particles into the sediment, but also includes the reversal, i.e. redistribution of deeper layers to the sediment surface and eventually subsequently biological induced resuspension. Thus, a comprehensive description of this resuspension loop (cf. Graf, [6]), should include all processes from the clear water minimum, (by definition the upper limit of the nepheloid layer or turbidity zone), to the bioturbation depth, taking into account the wide range of the vertical extensions and the physical discontinuities between the different compartments.

Near-bottom currents can possibly resuspend, transport and redistribute previously settled particles laterally over wide distances. On a few occasions particles may not even get in contact with the sediment surface and stay as "rebound" particles in suspension over a long period of time and distance. The understanding of these processes is essential for constructtion of quantitative budgets of organic or inorganic matter, e.g. carbon, for most benthic ecosystems. This holds especially true for shallow water ecosystems, such as Kiel Bight, where sediments are focused in topographic features like channels, or on continental slopes where high deposition areas have been observed at various continental margins. In these environments lateral advection may increase the input of organic carbon by a factor of 2 to 8 compared to vertical sedimentation [5] [7]. A completely different application of this topic is the near-

J.S. Gray et al. (eds.), Biogeochemical Cycling and Sediment Ecology, 153–159.

bottom distribution and transport of organic or inorganic pollutants. Many harmful substances like heavy metals or organochlorines are adsorbed onto particles and thus their dynamics are related to the same cycling as particles [8]. In addition the problem of resuspension is due to many technical problems in coastal waters (e.g. port construction and coastal protection) and major disturbances created by activities such as trawling and dumping. The aim of this paper is to evaluate the role of animals within the resuspension and deposition process.

2. Particle transport in the absence of animals

At a sea-floor bare of animals, near bottom currents will be the major factor to cause resuspension or mobilisation of particles from the seabed. A basic empirical approach by Hjulstrøm [9] predicts that down to a grain size of about 100 μm current velocities and consequently shear velocities required to mobilise a particle from the bed are decreasing. For smaller size classes of particles, representing in most cases cohesive and more consolidated sediments, the threshold shear stress and consequently the current velocity to mobilise particles from a hydrodynamically smooth bed increases again [20]. This empirical approach is still used in many models for sediment transport although so far no mathematical equation is available which would describe the whole range of grain sizes (c.f. Haupt et al., [11]). According to Hjulstrøm a particle of 100 μm diameter would be mobilised at a free stream velocity of about 15 cms^{-1}. However, it has to be kept in mind, that this relation has been established for well-defined clean grainsize fractions in laboratory experiments, which are significantly different from conditions at the sediment surface in the field.

In shallow waters, where oscillatory wave motion imposes high shear energy onto the sea floor, erosion of particles is much more rapid. Metha [12] demonstrated, that in wave influenced environments that are mainly dominated by porous or permeable sediments, shear resistance to erosion is significantly reduced. In about 20m water depth at two locations in the Kiel Bight the concentration of suspended matter in the benthic turbidity zone was significantly increased during a winter storm event compared to usual autumn situation [1]. These observations suggest that in energetically exposed shallow water environments physical processes like storms or waves are the major agents of sediment mobilisation.

However, also in deeper waters benthic storms have been observed at different occasions per year, with sufficiently high current velocities exceeding the required resuspension threshold reported for the deep sea [13].

For several reasons a well-defined widely applicable threshold between resuspension and non-resuspension is not available. Different sizes and properties of grain or particles result in varying mobilisation behaviour. The energy required to mobilise a grain or particle from the sea bed is in general higher than the amount of energy to keep a particle in suspension. Individual grains of larger size classes, like sands, tend to be mobilised into a saltatory motion at critical shear stress. Different sedimentological parameters such as chemical composition and the shape of the grains will effect the critical shear velocity. In natural systems, however, biological effects will strongly influence the critical shear even excluding animal activities. The sediment surface is generally covered by microbial biofilms [14]. The surfaces of light exposed sediments in shallow water environments are mostly dominated by microalgae. These algal biofilms tend to glue individual grains together and will significantly increase the critical shear velocity [15][16][17]. In such a case an isolated grain is not resuspended, but rather a piece of a mat will be swept away [18].

3. Indirect effects of biogenic structures

In shallow waters also waves can modify the smooth sea floor towards ripple formation, and at high current velocities pock marks may even be formed. In deeper water, however, animals will be the major factor modifying the sediment surface. In the literature of terrestrial ecology animals, which modify their environment by all sort of constructions, have been termed *engineers* [19]. Figure 1 summarises the dominant constructions of benthic *engineers*, potentially effecting the particle exchange between sediment and near bottom water.

Graf and Rosenberg [20] (1997) have summarised the effects of biodeposition and bioresuspension by benthic animals and distinguished between direct and indirect effects. A simple pit in the sea floor collects organic rich particles, a process frequently observed in deep-sea photographs (c.f. Billett et al., [21]) and apparently independent of whether the pit is inhabited by an animal or not (Fig. 1). Yager et al. [16] presented a systematic study of artificial pits in flume experiments. These authors demonstrated that the pit causes a downward movement of the lateral particle path, hitting the pit close to the downstream edge and then turning into a vortex within the pit. Consequently the residence time of particles within the pit is extended, which is a great advantage in terms of successful foraging for animals living at the bottom of such a pit, or animals living in burrows beneath the pit.

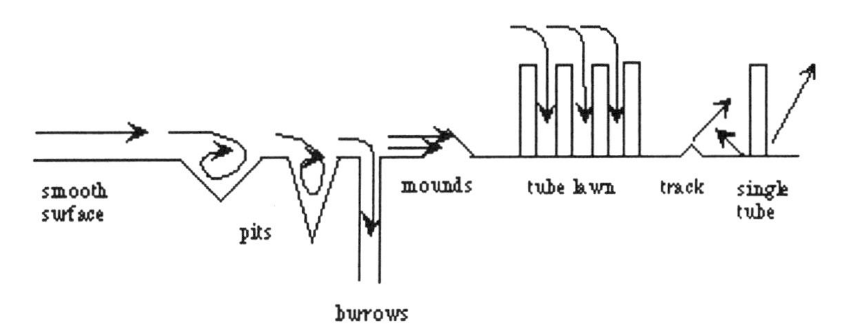

Figure 1: Schematic illustration of the indirect effect of biogenic structures on deposition and resuspension

An increased aspect ratio, (depth/diameter), of the pit will increase the amount of collected particies (Fig. 1). Only at very rough flow conditions the pits were washed out.

The opening of a burrow may be regarded as a pit with extremely high aspect ratio (Fig. 1). For the priapulid *Halicryptus spinulosus* Powilleit et al. [22] demonstrated that the amount of burrow openings is strongly correlated with downward particle transport. Although the sum of burrow openings represented only 0.3% of the sea floor this effect caused an enhanced biological mixing of the top 10 cm of the sediment. This bioturbation by a single species mixes the upper 10 cm of the sediment entirely over a period of 22 years, i.e. statistically a freshly settled particle will be transported from the sediment surface to the bioturbation depth and back to the sediment water interface within this period.

Different indirect depositional processes are involved in the case of obstacles protruding from the sea floor into the water. A clear example was recently presented by Huettel et al. [23], who investigated the effects of mounds on indirect deposition. For permeable, i.e. sandy sediments, they demonstrated that significant amounts of particulate tracers down to 1 μm in size are incorporated into the sediment. A linear relation between the height of these mounds and the penetration depth of particles was observed.

The effect of indirect deposition is more obvious in the case of dense accumulations of obstacles such as animal tubes, which can protrude several centimetres to decimetres into the water. Well-studied examples are tubes of spionid polychaets, forming lawns of tubes with abundances of up to several millions per square meter. Rhoads et al. [24] and Frithsen and Doering [25] have estimated the indirect depositional effect of such lawns to be 50% and 75 % respectively, of the overall deposition. A more systematic approach to this problem was recently presented by Friedrichs [26], who investigated the reduction of current velocities in artificial tube lawns. He was able to demonstrate that the reduction in flow speed was related to the abundance and diameter of tubes. In experiments with tube densities similar to those reported for the field the current velocity in the tube lawn was reduced to 0 cms^{-1} within 5 to 20 cm distance from the leading edge of the biological feature (Fig. 1).

Another type of biologically produced disturbances are isolated obstacles and smaller modifications of the sediment surface such as tracks created by animals crawling on the surface which may cause indirect resuspension (Fig. 1). Davis [27] reported that *Nucula annulata* and a mixed benthic population increased indirect resuspension of particles from the sediment surface by a factor of 3 to 8. A detailed study of the latter phenomenon was already presented by Nowell et al. [28] who reported that the bottom roughness height z_o was doubled by tracks of *Transenella tantilla* and that at the crest of the track the critical shear velocity was reduced from 1.74 cms^{-1} to 1.39 cms^{-1}.

Individual protruding tubes also lead to resuspension of sediment. The results of Eckmann et al.[29], Carey [30], and Eckmann and Nowell [31] suggest that a vortex-like resuspension occurs at the basis of the obstacle on the lee side of the tubes. In the case of the polychaete *Lanice conchilega* the animal can even benefit from this resuspension. The same effect has recently been shown for single arms of suspension feeding ophiurides [32].

4. Direct Biodeposition and Bioresuspension

Small particles are often intercepted, incorporated and subsequently deposited directly as faecal pellets or pseudofeces by suspension feeding animals. Especially sponges are known to take up particles < 8μm diameter [33]. Recently this finding was also confirmed for the deep-sea sponge *Thenea abyssorum* [34]. It was estimated that this species alone is responsible for a deposition of 7 - 10 mg suspended matter d^{-1} g^{-1} ash free dry weight, which was highly significant for this deep-sea community.

Sponges seem to spend only a small amount of energy for active suspension feeding. Riisgard et al. [35], estimated 0.85% of the total metabolism is used for pumping work of *Halichondria panicea.* Although only minor amounts of metabolic energy are required to intercept particles at sufficient current speeds, active suspension feeders like many bivalves may reach extremely high efficiencies in terms of absorption of seston from the lateral flux. Loo and Rosenberg [36] reported that bivalves are capable to actively remove 60% of the suspended matter from the horizontal particle flux, with even higher values during autumn.

Benthic animals can modify their behaviour and adapt to different current velocities. Bivalves, such as *Macoma balthica* turn the opening of their ingestion siphon towards the current (Fig. 2). Barnacles sweep their thoracic appendages through the water at low current speed, but switch to passive suspension feeding holding their cirri into the flow at current velocities above 3.1 cm s^{-1} [37]. An excellent example was recently published by Loo et al.[32], who demonstrated that the brittle star *Amphiura filiformis*, a passive suspension feeder, stretches its arm up to 2 cm into the water at low current speeds. In contrast at higher current speeds the animal is holding its arms perpendicular to the current close to the sediment.

Passive suspension feeders depend on a direct interception of the drifting particles with their body appendages. In the above example of *A. filiformis* it appears that the hydrodynamic regime may be more complex than previously assumed and thus poorly understood. In this context the capture of larger particles such as aggregates may appear as a new important topic in this sort of investigation. For spionid polychaetes it was shown that they spend most of their time suspension feeding using their two helically coiled tentacles [38]. Although the area for direct interceptions is very small, their enormous abundance can lead to significant deposition of pellets.

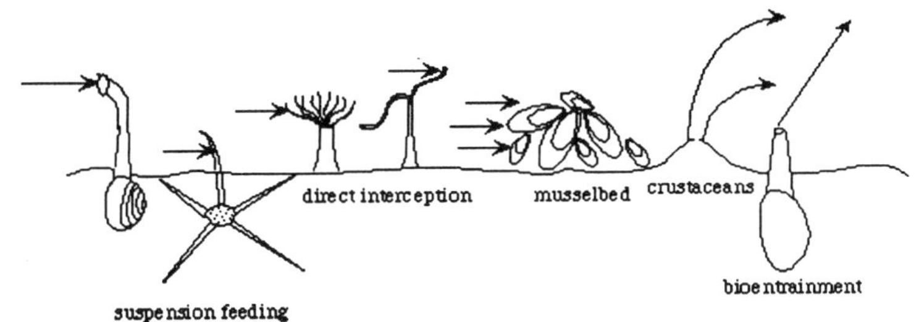

Figure 2: Schematic illustration of direct deposition and resuspension

The most impressive biodepositional fluxes were reported for mussel beds. In the Dutch Wadden Sea *Cardium edule* and *Mytilus edulis* were reported to produce 100 000 t dr. wt yr^{-1} and 175 000 t dr. wt yr^{-1} of biodeposits respectively [39]. These old estimates were supported by more recent studies of Asmus and Asmus [40] and Muschenheim and Newell [41], who reported biodeposition fluxes between 1 to 70 g C m^{-2} d^{-1}.

A very key study on the effect of biodeposition was recently presented by Thomsen and Flach [42]. These authors investigated a benthic community from the Oslofjord in a flume experiment and observed that even under super critical flow conditions the benthic community could compensate for the erosional losses of the sediment, by selectively incorporating fresh phytodetritus.

Direct bioresuspension is one of the fluxes between the sediment and the bottom water. Nevertheless the few existing examples suggest that this process can be significant. Especially deep-dwelling crustaceans like the thallasionid *Callianassa subterranea*, which have to remove the sediment material from their burrow systems onto the sediment surface adjacent to their burrow exit or into the bottom water [43].

Bivalves tend to reject their faecal pellets and pseudofeces into the water column. A well -studied example was presented by Bender and Davis [44], who investigated *Yoldia limulata* from the Narragansett Bay. A median size animal is able to resuspend 440 g dry sediment per year, considering their abundance could this value translates into a bioresuspension rate of 15.8 - 24.6 kg dry sediment m^{-2} yr^{-1}. For other benthic animals it is known that regular bioresuspension events do occur frequently, but they are not yet quantified [27].

5. Conclusions

In high-energy environments, e.g. tidal channels, periods of storms and high waves, physical processes will dominate resuspension of sediments. However, the resuspended particles at the sediment surface are the result of biological activities in the sediment. In most cases animals directly or indirectly control the transport of particles across the sediment water interface, especially the non-local transport between deeper sediment strata and the bottom water. Yet not all the processes involved are sufficiently studied. Nevertheless it is already clear that the activity of animals can be responsible for higher biodeposition or bioresuspension rates, enhanced by a factor of 2 and may in some cases reach a factor of 10 compared to the purely physical processes. This finding should be included into models describing the distribution of suspended matter in coastal waters as well as for models of the nepheloid layers. While particle mixing in the sediment is theoretically well described, (a recent excellent example was given by Sotaert et al., [45]), the numerical modelling of near bottom particle dynamics, including the near bed hydrodynamics and particle interaction of various size classes like aggregate formation and destruction, has only just begun [2] [46].

So far there is only one analytical model by Boudreau [47] which includes a non-local mixing between sediment and water. With this model the author could successfully fit a near bottom seston profile of a continental slope published by

Newberger and Caldwell [48] using reasonable values for the parameter settling velocity (2.6 m d^{-1}), injection height, i.e. height of bioresuspension (3 cm), and frequency of resuspension (0.0038 d^{-1}). He came to the conclusion that the effect of bioresuspension may enhance particle exchange between the sediment and the water column by a factor of 10.

6. Acknowledgements

The author thanks Dr. W. Ritzrau and Dr. S. Jähmlich for helpful comments on the manuscript.

7. References

1. Ritzrau, W. and Graf, G. (1992) Increase of microbial biomass in the benthic turbidity zone of Kiel Bight after resuspension by a storm event. *Limnol. Oceanogr.* **37**, 1081-1086.
2. Ritzrau, W. (1996) Microbial activity in the benthic boundary layer (BBL): Small scale distribution and its relationship to the hydrodynamic regime. *J. Sea Res.* **36**, 171-180.
3. Ritzrau, W., Thomsen, L., Lara, R.J. and Graf, G. (1997) Enhanced microbial utilisation of dissolved amino acids indicates rapid modification of organic matter in the benthic boundary layer. *Mar. Ecol. Prog. Ser.* **156**, 43-50.
4. Thomsen, L. and Graf, G. (1995) Benthic boundary layer characteristics of the continental margin of the western Barents Sea. *Oceanologica Acta* **17**, 597-607.
5. Thomsen, L., Graf, G., Juterzenka, K. von and Witte U. (1995) An in situ experiment to investigate the depletion of seston above an interface feeder field on the continental slope of the western Barents Sea. *Mar. Ecol. Prog. Ser.* **123**, 295-300.
6. Thomsen, L. and Ritzrau, W. (1996) Aggregate studies in the benthic boundary layer at a continental margin. *J. Sea Res.* **36**, 143-146.
7. Graf, G. (1992) Benthic-Pelagic Coupling: A Benthic View. *Oceanogr. Mar. Biol. Annu. Rev.* **30**, 149-190.
8. Kersten, M., Dicke, M., Kriews, M., Naumann, K., Schmidt, D., Schulz, M., Schwikowski, M. and Steiger, M. (1988) Distribution and Fate of Heavy Metals in the North Sea. In B.L. Bayne, W. Salomons, E.K. Duursma and U. Förstner, (eds) *Pollution of the North Sea An Assessment Part II: Input and Behavior of Pollutants*. Springer-Verlag, Berlin. pp. 300-347.
9. Hjulstrøm, F. (1935) Studies of the morphological activity of rivers as illustrated by the river Fyris, *Bull. Geol. Inst.* Univerity Uppsala.
10. Werner, F., Erlenkeuser, H. Grafenstein, von U., McLean, S. Sarnthein, M., Schauer, U., Unsöld, G., Walger, E. and Wittstock, R. (1987) Sedimentary records of benthic processes. In J. Rumohr, E. Walger and B. Zeitzschel (eds), *Seawater-Sediment interactions in coastal waters.* Springer-Verlag, Heidelberg, pp. 162-262.
11. Haupt, B. J., Schäfer- Neth, C. and Stattegger, K. (1994) Modeling sediment drifts: *A coupled oceanic circulation- sedimentation model of the northern North Atlantic.* Paleoceanography **9**, 897-916.
12. Mehta, A.J. (1988) Laboratory Studies on Cohesive Sediment Deposition and Erosion. In J. Dankers and W. v. Leussen, (eds) *Physical Processes in Estuaries.* Springer Verlag, Heidelberg. pp. 427-445.
13 Aller, J.Y. (1989) Quantifying sediment disturbance by bottom currents and its effect on benthic communities in a deep-sea western boundary zone. Deep-Sea Res. **36**, 901-934.
14. Meyer-Reil, L.-A. (1994) Microbial life in sedimentary biofilms - the challenge to microbial ecologists. *Mar. Ecol. Prog. Ser.* **112** 303-311.
15. Stolzenbach, K.D. (1989) Particle transport and attachment. In W.G. Characklis and P.A. Wilderer (eds), *Structure and Function of biofilms.* Wiley, New York, pp. 33-47.
16. Yager, P.L., Nowell, A.R.M. and Jumars, P. A. (1993) Enhanced deposition to pits: A local food source for benthos. *J. Mar. Res.* **51** 209-236.
17. Grant, J. and Emmerson, C. (1994) Resuspension and stabilization of sediments with microbial biofilms: Implications for benthic-pelagic coupling. in W.E. Krumbein, D.M. Paterson and L.J. Stal (eds), *Biostabilization of sediments.* BIS - Verlag, Oldenburg, pp. 121-134.
18. Grant, J., and Bathmann, U. V. (1987) Swept away: resuspension of bacterial mats regulates benthic-pelagic exchange of sulfur. *Science* **236** 1472-1474.
19. Lawton. J.H. (1994) What do species do in ecosystems? *Oikos* **71**, 367-374.
20. Graf, G. and Rosenberg, R.. (1997) Bioresuspension and Biodeposition: A Review. *J. Mar. Sys.* **11**, 269-278.
21. Billett, D.S., Lampitt, M.,R.S. and Rice, A.L.M. (1983) Seasonal sedimentation of phytoplankton to the deep-sea benthos. *Nature* **302**, 520-522.
22. Powilleit, M., Kitlar, J. and Graf, G. (1994) Particle and fluid bioturbation caused by the priapulid worm Halicryptus spinulosus (v. Seibold). *Sarsia* **79**, 109-117.
23. Huettel, M., Ziebis, W. and Forster, S. (1996) Flow-induced uptake of particulate matter in permeable sediments. *Limnol.Oceanogr.* **41**, 309-322.
24. Rhoads, D.C., Yingst, J.Y. and Ullmann, W.J. (1978) Sea floor stability in Lomng Island Sound. Part I. Temporal changes in erodibility of fine grained sediments. In M.L. Wiley (ed.), *Estuarine Interactions.* Academic Press, New York, pp. 603.

25. Frithsen, J.B. and Doering, P.H. (1986) Active enhancement of particle removal from the water column by tentaculate benthic polychaetes. *Ophelia* **25**, 169-182.
26. Friedrichs, M. (1996) Auswirkungen von Polychaetenröhren auf die Wasser-Sediment-Grenzschicht. Ms. Thesis, Kiel University, pp. 82.
27. Davis, W.R. (1993) The role of bioturbation in sediment resuspension and its interaction with physical shearing. *J. Exp. Mar. Biol. Ecol.* **171**, 187-200.
28. Nowell, A.R.M., Jumars, P.A. and Eckman, J.E. (1981) Effects of biological activity on the entertainment of marine sediments. *Mar. Geol.* **42**, 133-153.
29. Eckmann, J.E., Nowell, A.R.M. and Jumars, P.A. (19981) Sediment destabilization by animal tubes. *J. Mar. Res.* **39**, 361-374.
30. Carey, A.D. (1983) Particle resuspension in the benthic boundary layer induced by flow around polychaete tubes. *Can. J. Fish. Aquat. Sci.* **40**, 301-308.
31. Eckmann, J.E. and Nowell, A.R.M. (1994) Boundary skin friction and sediment transport about an animal-tube mimic. *Sedimentology*, **31**, 851-862.
32. Loo, L.-O., Jonsson, P.R., Sköld, M. and Karlsson, Ö. (1996) Passive suspension feeding in Amphiura filiformis (Echinodermata: Ophiuroidea): feeding behaviour in flume flow and potential feeding rate of field populations. *Mar. Ecol. Prog. Ser.* **139**, 143 - 155.
33. Reiswig, H.M. (1971) Particle Feeding in Natural Populations of Three Marine Demosponges. *Biol. Bull.* **141**, 568-591.
34. Witte, U., Brattegard, T., Graf, G. and Springer, B. (1997) Particle capture and deposition by deep-sea sponges from the Norwegian-Greenland Sea. *Mar. Ecol. Prog. Ser.* **154**, 241-252.
35. Riisgard, H.U, Thomassen, S., Jakobsen, H., Weeks, J.M. and Larsen, P.S. (1993) Suspension feeding in marine sponges Halichondria panicea and Haliclona urceolus: effects of temperature on filtration rate and energy cost of pumping. *Mar. Ecol. Prog. Ser.* **96**, 177-188.
36. Loo, L.-O., and Rosenberg, R. (1989) Bivalve suspension-feeding dynamics and benthic-pelagic coupling in an eutrophicated marine bay. J. Exp.Mar. Biol. Ecol. **130**, 253-276.
37. Trager, G.C., Hwang, J.-S. and Strickler, J.R. (1990) Barnacle suspension-feeding in variable flow. *Mar. Biol.* **105**, 117-127.
38. Taghon, G.L. and Greene, R.R. (1992) Utilization of deposited and suspended particulate matter by benthic "interface" feeders. *Limnol. Oceanogr.* **37**, 1370-1391.
39. Verwey, J. (1952) On the ecology of distribution of cockle and mussel in the Dutch Waddenzee, their role in sedimentation, and source of their food supply, with a short review of the feeding behaviour of bivalve mollusks. *Arch. Neerl. Zool.* **10**, 172-239.
40. Asmus, R.M. and Asmus, H. (1991) Mussel beds: limiting or promoting phytoplankton? *J. Exp. Mar. Biol. Ecol.* **148**, 215-232.
41. Muschenheim, D.K. and Newell, C.R. (1992) Utilization of seston flux over a mussel bed. *Mar. Ecol. Prog. Ser.* **85**, 131-136.
42. Thomsen, L. and Flach, E. (1997) Mesocosm observations of fluxes of particulate matter within the benthic boundary layer. *J. Sea Res.* **37**, 67-79.
43. Rowden, A.A. and Jones, M.B. (1994) A contribution to the biology of the burrowing mud shrimp, Callianassa subterranea (Decapoda: Thalassinidea). *J. Mar. Biol. Assoc. U.K.* **74**, 623-635.
44. Bender, K., and Davis, W. R.. (1984) The Effect of Feeding by Yoldia Limatula on Bioturbation. *Ophelia* **23**, 91-100.
45. Sotaert, K.,Hermann, P.M.J., Heip, C. deStigter, H.S., van Weering, T.C.E., Epping, E., und Helder, W. (1996) Modelling 210Pb-derived activity in ocean margin sediments: Diffusive versus nonlocal mixing. *J. Mar. Res.* **54**, 1207-1227.
46. Ritzrau, W. and Fohrmann, H. (in press). Field and numerical studies of near bed aggregate dynamics. In J. Harff and K. Stattegger, (eds), *Computerised modelling of sedimentary systems.* Springer Verlag, Heidelberg.
47. Boudreau, B.P. (1997) A one-dimensional model for bed-boundary layer particle exchange. *J. Mar. Sys.* **11**, 279 - 303.
48. Newberger, P.A. and Caldwell, D.R. (1981) Mixing and the bottom nepheloid layer, *Mar. Geol.* **41**, 321-336.

IMPACT OF CATCHMENT LAND-USE ON AN ESTUARINE BENTHIC FOOD WEB

D. RAFFAELLI
Culterty Field Station, University of Aberdeen, Newburgh, Ellon, Aberdeenshire, Scotland, AB41 6AA

Abstract

Long-term changes in agricultural land-use has resulted in enhanced nutrient (N) concentrations in the Ythan estuary, Aberdeenshire. Over the same time period there has been an increase in mat-forming macroalgae (*Enteromorpha, Chaetomorpha, Ulva*). These weed mats have both direct effects and indirect effects on the ecology of the estuary. Macro-algal biomasses in excess of 1 $kg.m^{-2}$ generate a highly reduced sediment environment through smothering of the sediment surface and local organic enrichment. The densities of many invertebrate species are significantly reduced under these weed mats, locally reducing the availability of prey to shorebirds. In contrast, the abundance of invertebrates and shorebirds may be enhanced in areas distant from the mats through subsidies of organic material from the mats themselves.

1. Introduction

Considerable attention has been paid recently to the importance of allochthonous material for the dynamics of assemblages [1]. The idea that such "subsidies" may be important is hardly novel: a huge effort was made under the International Biological Programme in the 1970s to document the flows of material within and between major assemblage types. However, the importance of these external subsidies for the dynamics and organization of the receiving assemblages is not at all well documented for most systems [1]. In this respect, marine benthic ecologists have progressed further than most. The significance of the open nature of benthic marine systems has long been recognized for carbon inputs, recruitment processes and community organization and this has permitted a better understanding of the likely effects on the benthos of anthropogenic enhancements of subsidy materials. However, only relatively recently have the importance of terrestrial, catchment-scale processes for marine benthic assemblages been acknowledged (e.g. [2] [3] [4] [5] [6]). Here, I describe the results of long-term studies carried out on the relatively small catchment and estuary of the Ythan river, Aberdeenshire, Scotland, to illustrate how catchment land-use practices have altered nutrient subsidies to the estuary with cascading effects on the benthos and higher trophic levels.

2. The Ythan river and its catchment

The Ythan river (Figure 1) is 63 km in length, draining just under 700 km^2 of gently rolling lowland characterized by glacial deposits of silt and clay interspersed with stones and sand. Most of the catchment is under 260m in altitude ([7]).

J.S. Gray et al. (eds.), Biogeochemical Cycling and Sediment Ecology, 161–171.

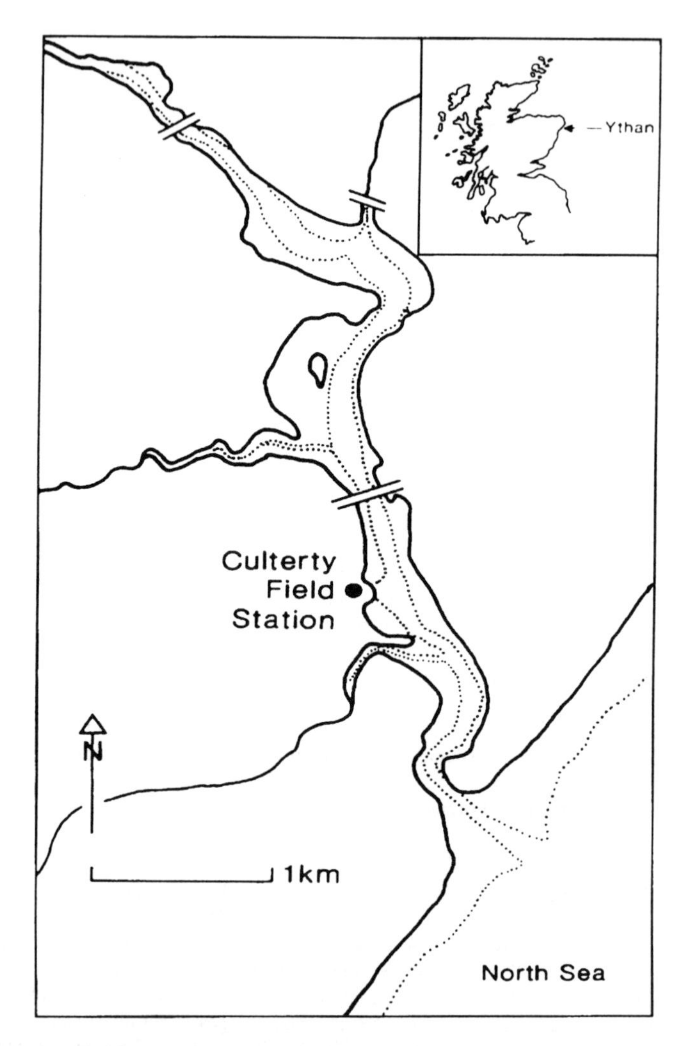

Figure 1. Location of the Ythan estuary, Aberdeenshire, Scotland.

Given these attributes it is not surprising that 95% of the catchment land is currently under arable crops and grassland. Like many other parts of northern Europe, the pattern of land-use within the Ythan catchment has seen dramatic changes in the last 40 years (Figure 2), reflecting changing national and European agricultural policies and incentives. Most strikingly is the increase in the amount of land under intensive cereal and oil-seed rape production (Figure 2). These crops are nitrogen-demanding and the application rate (kg/ha) of nitrogenous fertilizers has also increased, so that the overall nutrient inputs into the catchment have increased markedly [8] [9].

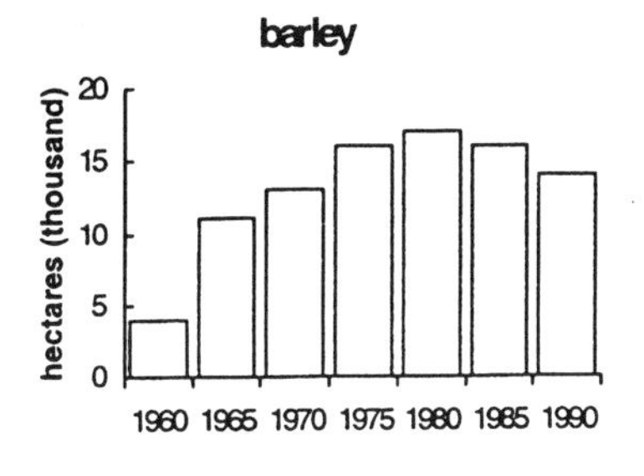

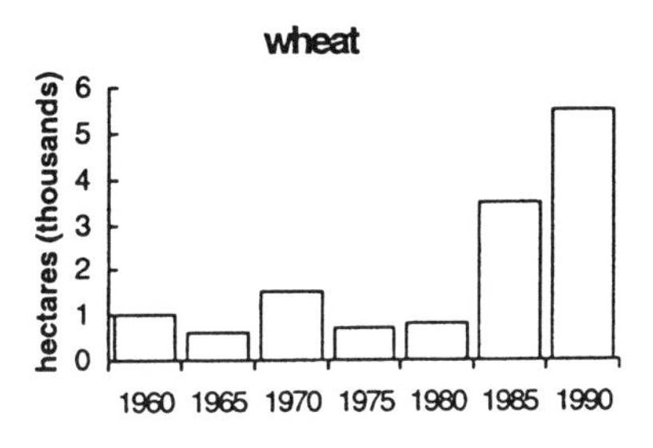

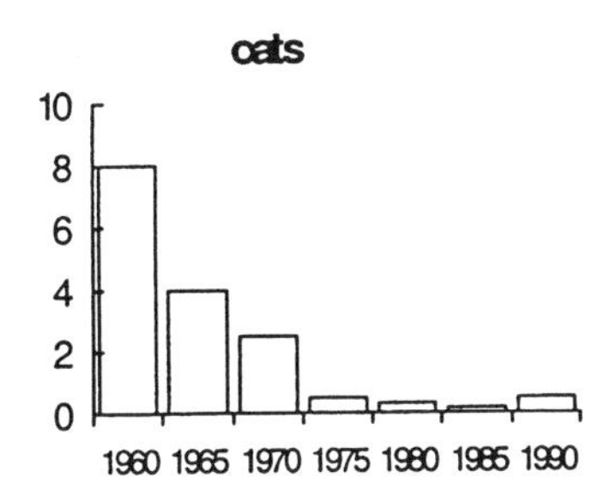

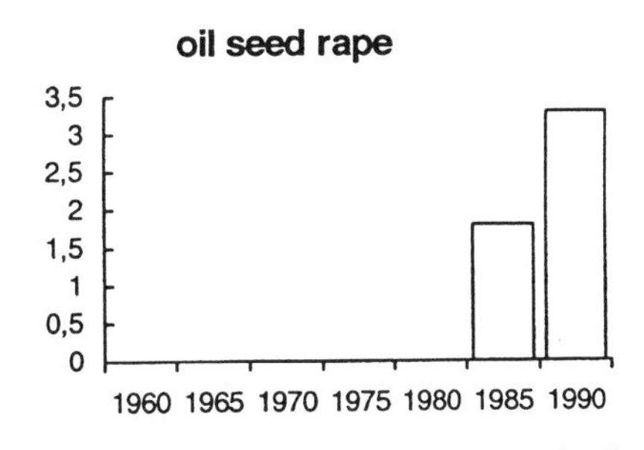

Figure 2. Changes in aspects of land-use in the Ythan catchment. From Scottish Office records for all parishes totally included within the catchment.

These changes in land-use are reflected in the nutrient status of the river [9] [5] and it is estimated that concentrations of total oxidised nitrogen (TON) in the river have increased from ca. 100-150μM in the late 1960s to ca. 500-550μM in the 1990s [8]; Figure 3. These changes are also reflected in the estuary itself (Figure 3). The phosphorus concentrations in the river are more variable and there is a much weaker trend over time [8][9]. Clearly, the estuary is now receiving much more nitrogen from catchment sources than before.

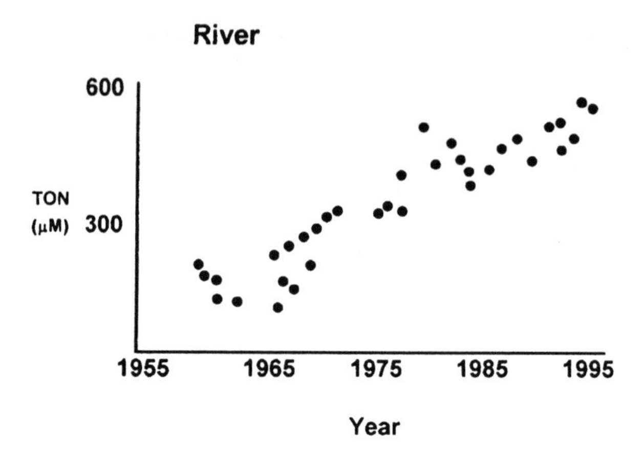

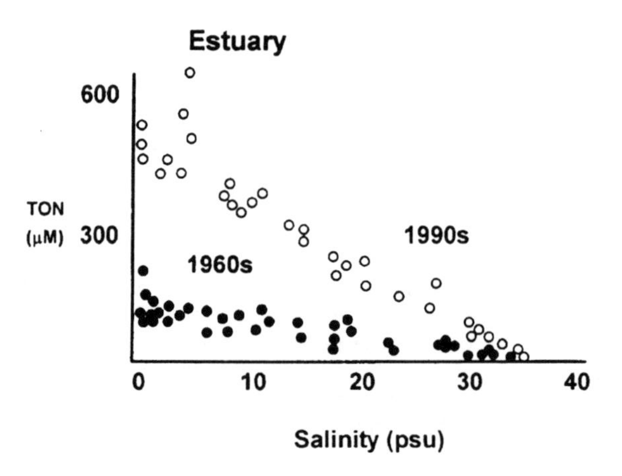

Figure 3. Long-term changes in nitrogen concentrations in the river Ythan (upper) and the estuary (lower). Data from Ball *et al* [8].

3. The Ythan estuary

The estuary itself (57°N, 2°W)is one of the smallest in the UK, but probably one of the best documented in terms of its biology [10]. The Ythan is tidal for about 8 km from the mouth and averages about 300m in width. The tidal range is up to about 3m and the flushing time is relatively short, estimates ranging 1.15 tidal periods [11] to several days (Balls, pers. comm.). The low water channel is about 71 ha and there are 115 ha of mudflat and 70 ha of mussel beds and sand [12]. The extensive flats support high densities of invertebrates (Table 1) large amounts of which are utilized by shorebirds and fish (Figure 4).

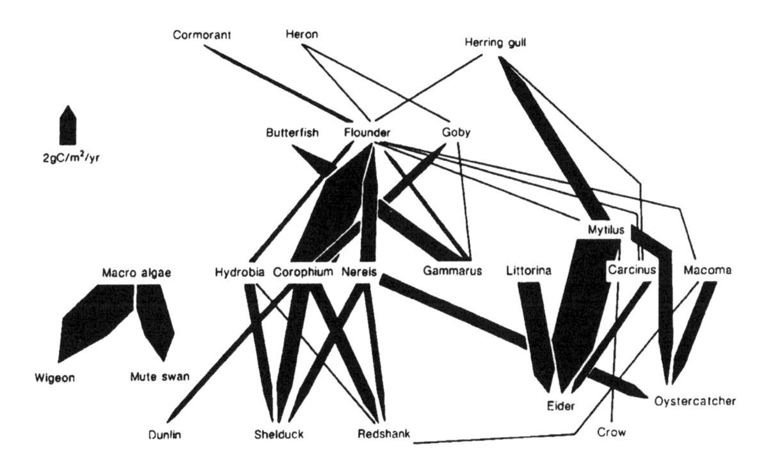

Figure 4. Major energy flows ($gC.^{-2}.yr^{-1}$) between producers and consumers in the Ythan estuary. Data from Baird and Milne [12].

The dynamics of the interactions between these predators and their invertebrate prey have been studied extensively since the late 1960s (when the Ythan was a IBP site) to date [12] [13] [14] [15] [16]. A by-product of these studies has been the accumulation of substantial data sets on a variety of ecological parameters relevant to the nutrient subsidy changes described above.

TABLE 1. Typical densities of the main invertebrate species on the Ythan mudflats. These species comprise most of the macroinvertebrates likely to be encountered in sediment samples on the Ythan.

	Numbers in m^2
Corophium volutator	20,000 - 50,000
Hydrobia ulvae	100,000 - 500,000
Nereis diversicolor	400 - 600
Macoma balthica	500 - 800
Manayunkia aestuarina	10,000 - 15,000
Pygospio elegans	5,000 - 8,000
Tubificoides (Peloscolex) benedini	10,000 - 15,000
Tubifex costatus	3,000 - 8,000

The most relevant data sets relate to the distribution of green opportunistic benthic macroalgae (*Enteromorpha*, *Chaetomorpha*, *Ulva* and *Cladophora*), macroinvertebrates (especially *Corophium volutator*, *Nereis diversicolor*, *Hydrobia ulva* and *Macoma balthica*) and shorebirds (mainly shelduck *Tadorna tadorna*, redshank *Tringa totanus*, curlew *Numenius arquatus*, bar-tailed godwit *Limosa lapponica*, dunlin *Calidris alpina*, oystercatcher *Haematopus ostralegus*, turnstone

Arenaria interpres and knot *Calidris canutus*).

4. Long-term changes in weed mats, invertebrates and shorebirds

Benthic green macroalgae form extensive mats during the summer on the Ythan, declining again over the autumn and winter. The peak of weed growth is from late July to early September, depending on weather conditions and species (*Ulva* usually reaches its peak somewhat later than *Enteromorpha*). Ground-truth surveys indicate that weed mats in excess of 1kg wet weight/m² appear on aerial photographs and we have mapped the extent of these mats from photographs taken between 1954 and 1997 [17]. These data show that:

- weed mats were not as extensive in the 1950s and 1960s as they are in the 1980s and 1990s (Figure 5).
- the middle section of the estuary has always supported some weed cover, but mats have spread upstream (and possibly downstream) in recent years (Figure 6) in 1996 there was a marked decline in weed cover.

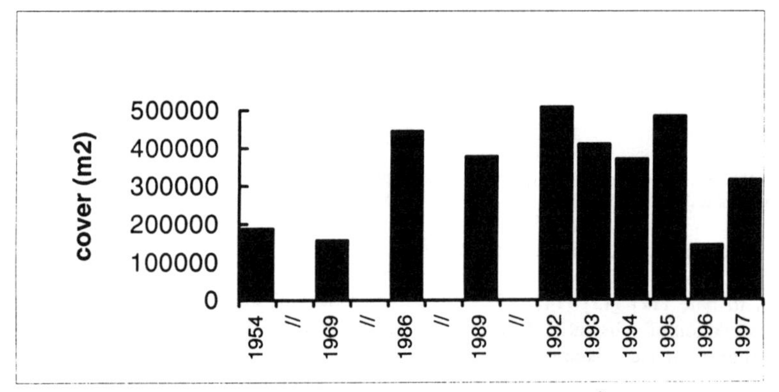

Figure 5. Long-term changes in coverage by macro-algal mats (>1 kg.m^{-2}) of the Ythan estuary's mudflats. From Raffaelli *et al* [17].

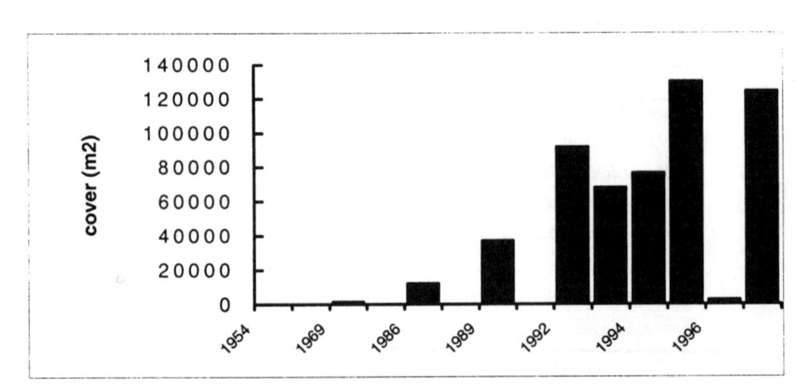

Figure 6. Changes in the cover of macro-algal mats (>1 kg.m^{-2}) at Forvie burn, in the upper reaches of the Ythan estuary.

Our planimetric analysis of weed mats (>1kg/m²) is consistent with pixel-by-pixel spectral analysis of the same photographs (G. Ferrier, pers. comm.) as well as with historical biomass estimates at specific locations [5] [17]. The decline in weed cover in 1996 is attributed to a major river hydraulic event in October 1995 which removed the overwintering biomass on which the following (1996) spring bloom normally builds [17].

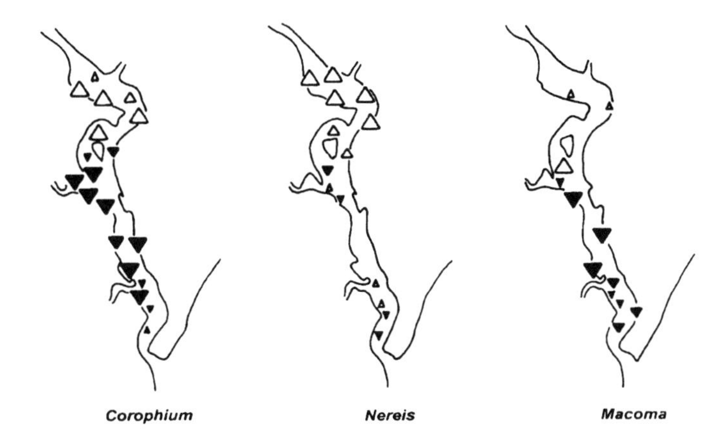

Figure 7. Changes in abundance of invertebrates 1964-1990. Largest triangles represent an increase (open) or decrease (closed) over this period of 30 ind/100cm² for *Corophium* and 10 ind/100cm² for *Nereis* and *Macoma*.

The distribution and abundance of those invertebrates which are major prey species for shorebirds has similarly changed since the 1960s. Surveys using the same procedures and protocols carried out in 1964 [18] and 1990 [19] reveal that areas known to be affected by weed mats during the summer supported far fewer individuals of the amphipod *Corophium volutator* in 1990 compared to 1964, this species virtually having disappeared from several areas where it was once abundant (Figure 7). A similar trend is seen for the bivalve *Macoma balthica* (Figure 7). Both these species, as well as the polychaete *Nereis diversicolor*, appear to have increased in abundance in the upper reaches of the estuary, which at that time (1990) was not severely affected by weed mats (Figure 7).

The decline in *Corophium* and *Macoma* are almost certainly due to the spread of weed mats over this period. Invertebrate populations and sediment physico-chemistry in weed mat-affected and adjacent weed-free areas of mudflat are strikingly and consistently different (Figure 8).

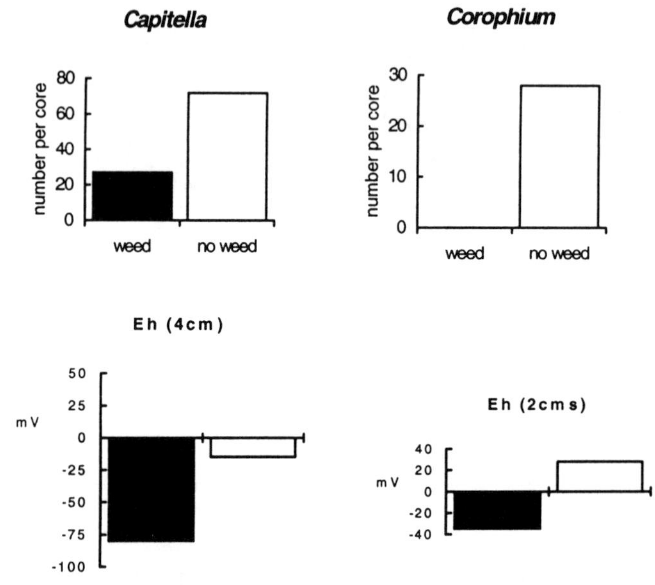

Figure 8. Invertebrate abundance and redox potential at 2 and 4 cm under weed mats and in weed-free patches on South Quay mudflat, Ythan estuary. Error bars are 95% confidence limits. From Raffaelli *et al* [19].

Also, experimental addition of different amounts of macroalgae to previously unaffected mudflats [20] demonstrated a dose-dependant response in invertebrate abundance and sediment properties (Figure 9). The presence of weed mats generates reducing conditions in the sediment through reduction in oxygen exchange with the overlying water and with the air at low tide (a smothering effect)

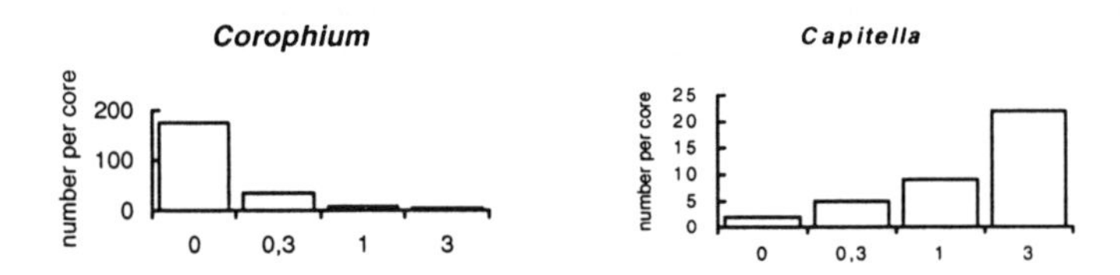

Figure 9. Effects on *Corophium* and *Capitella* of adding different amounts of weed mats to sediments. Data from Hull [20].

interference by the weed with its feeding behaviour [19]. Adults of bivalves, such as *Cerastoderma* and *Macoma,* move to the sediment-water interface, where they become vulnerable to predation by shorebirds and fish. Paradoxically, high densities of juveniles (spat and post-larvae) of several species, including *Nereis* and *Macoma*, are often associated with the mats, probably because of hydrodynamic entrainment by the filaments. The interactions between the algal mats, invertebrates and sediments are summarized in Figure 10. We believe that these weed-invertebrate interactions account for the historical changes in invertebrate distributions shown in Figure 7. The increases in invertebrate densities in the upper reaches probably reflect a general eutrophication of the estuary and export of the organic material fixed by the weed mats in downstream sections.

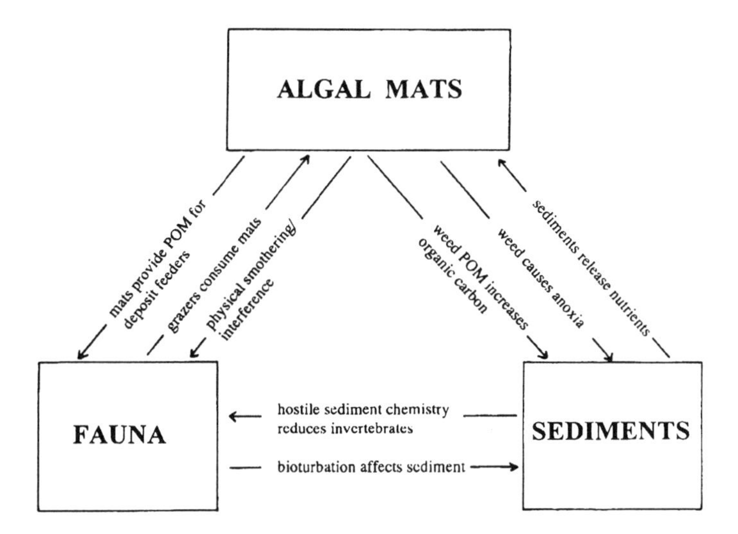

Figure 10. Interactions between weed mats, invertebrates and sediments.

Changes in the abundance and distribution of shorebirds since the 1960s are consistent with those of their major invertebrate food species as well as changes in weed mats. Winter counts of several shorebird species increased significantly between the 1960s and the early 1980s (Figure 11), in many cases against national trends. There is evidence of a subsequent decline in at least four species, oystercatcher, redshank, dunlin and bar-tailed godwit This pattern of change of an increase, followed by a decrease in abundance has been anticipated [5]. There seems to have been a general increase in the carrying capacity of the estuary in the earlier stages of eutrophication with higher invertebrate densities in upstream areas and in weed-free patches in the downstream sections. Areas supporting weed mats during the summer rapidly decline in the winter, and are quickly colonized by invertebrates from weed-free areas probably because these areas are then organically enriched. As long as a mosaic of weed-affected and weed-free patches exists, winter invertebrate densities will remain high. However, if weed mats continue to spread in extent then the area of weed-free refugia will inevitably decline, along with invertebrates and their shorebird predators. The patterns of change seen in Figure 11 are consistent with this hypothesis. Also, for redshank and shelduck there have been shifts in abundance to sites furthest upstream and downstream and less use of the middle reaches where weed growth is greatest [17].

Shelduck have continued to decline significantly since the 1960s against national trends, as well as shifting their centre of distribution away from the middle reaches [17]. Shelduck feed by pushing their raked beak through the surface mud and winnow the sediment from small invertebrates. Buried weed interferes with this behaviour [21] making this species highly sensitive to the presence of weed mats. Recent studies involving experimental manipulation of large areas of weed mats by Dayawansa [22] have confirmed that the foraging success of redshank and other waders is reduced by the presence of the mats.

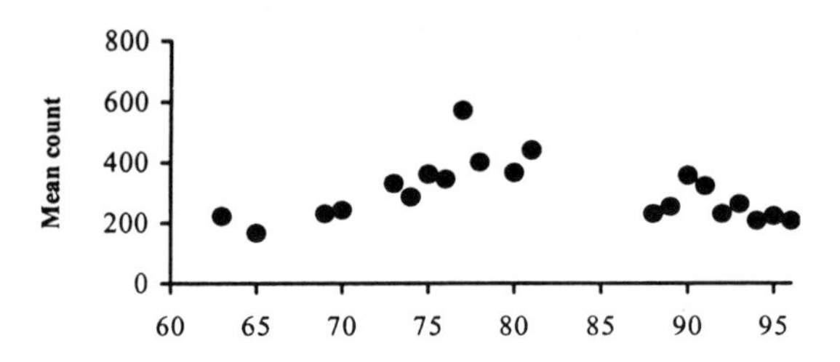

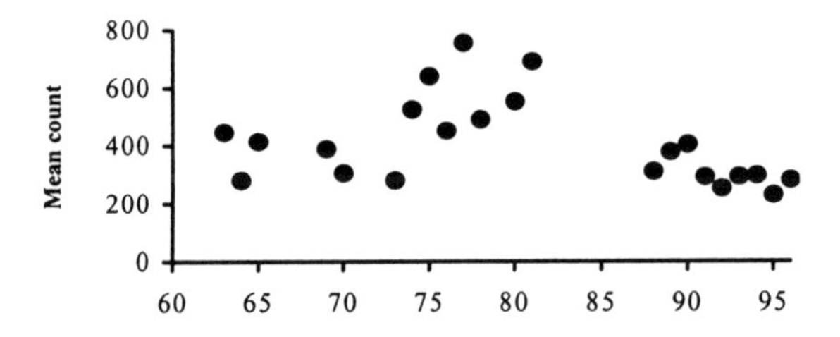

Figure 11. Winter counts of oystercatcher and redshank on the Ythan estuary. Data from Raffaelli *et al* [17].

5. Conclusions

Our analysis of the Ythan provides persuasive evidence that macroalgal blooms can have major impacts on estuarine food web organization and structure. We are confident of the mechanisms underlying the interactions between weed mats and invertebrates and between weed mats and shorebirds, because we have been able to experimentally manipulate the components of the system in order to rigorously test the hypotheses. However, explanations for the spread of weed mats are more controversial. Experimental tests of the link between nitrogen in river water and weed growth is not possible at the estuary scale, but the evidence is compelling and alternative competing hypotheses (such as changes in hydrography) have been falsified [17].

There remains the final question as to whether the ecological changes we have documented are reversible, and if so over what time-scale? The loss of overwintering weed biomass in October 1995 through a major storm event provided a natural experiment. The 1964 and 1990 large-scale surveys were repeated in February 1997, following a season with little weed growth. Several of the locations where *Corophium* had declined between 1964 and 1990 showed recovery, several to 1964 densities [17], indicating that if weed blooms declined then the invertebrate assemblages and shorebird populations might recover to their pre-eutrophication abundances and distributions. Assuming that nutrients are responsible for the macroalgal growth, the effects of reducing nutrient run-off from the catchment on weed blooms in the estuary are more difficult to predict. The period of time between nitrogen being applied to the soil and its arrival in the estuary is unknown and could be in the order of years, so that the effects of any remedial measures in the catchment might not be manifest in the ecology of the estuary for some time. This response lag is likely to be extended further by nutrients stored in the Ythan's sediments (Figure 10) which are expected to continue to be released thereby initiating and maintaining blooms long after any remedial measures had been applied to the catchment land.

6. References

1. Polis, G.A., Holt, R.D., Menge, B.A. and Winemiller, K.O. (1996) Time, space, and life history: influences on food webs, in G.A. Polis and Winemiller, K.O. (eds.), Food webs: integration of patterns & dynamics, Chapman & Hall, London, pp.435-460.
2. Giller, P.S., Hildrew, A.G. and Raffaelli, D. (1994) Aquatic ecology: scale, pattern and process, Blackwell Science, Oxford.
3. McCoomb, A.J. and Humphries, R. (1992) Loss of nutrients from catchments and their ecological impacts in the Peel-Harvey estuarine system, Western Australia, Estuaries, 15, 529-537.
4. Raffaelli, D. (1992) Conservation of Scottish estuaries, Proceedings of the Royal Society of Edinburgh, 100B, 55-76.
5. Raffaelli, D.G., Hull, S.C. and Milne, H. (1989) Long-term changes in nutrients, weed mats and shorebirds in an estuarine system, Cahiers de Biologie Marine, 30, 259-270.
6. Schramm, W. and Nienhuis P.H. (1996) Marine benthic vegetation. Recent changes and the effects of eutrophication, Springer, Berlin.
7. NERPB (1994) River Ythan catchment review. Report of the North East River Purification Board.
8. Balls, P.W., MacDonald, A., Pugh, K. and Edwards, A.C. (1995) Long-term nutrient enrichment of an estuarine system: Ythan, Scotland (1958-1993), Environmental Pollution, 90, 311-321.
9. MacDonald, A.M., Edwards, A.C., Pugh, K.B. and Balls, P.W. (1995) Soluble nitrogen and phosphorus in the river Ythan system, U.K.: annual and seasonal trends, Water Research, 29, 837-846.
10. Prater, A.J. (1981) Estuary birds of Britain and Ireland, T. & A.D. Poyser, Carlton, Staffordshire.
11. Leach, J.H. (1969) Hydrography of the Ythan estuary, with special reference to detritus in the production of benthic microflora, PhD Thesis, University of Aberdeen.
12. Baird, D. and Milne, H. (1981) Energy flow in the Ythan estuary, Aberdeenshire, Scotland, Estuarine, Coastal and Shelf Science, 13, 455-472.
13. Hall, S.J. and Raffaelli, D. (1991) Food web patterns: lessons from a species-rich web, Journal of Animal Ecology, 60, 823-824.
14. Milne, H. and Dunnet, G.M. (1972) Standing crop, productivity and trophic relations of the fauna of the Ythan estuary, in R.S.K. Barnes and J. Green (eds.), The Estuarine Environment, Applied Science Publishers, London, pp. 86-106.
15. Raffaelli, D. and Hall, S.J. (1992) Compartments and predation in an estuarine food web, Journal of Animal Ecology, 61, 551-560.
16. Raffaelli, D.G. and Hall, S.J. (1996) Assessing the relative importance of trophic links in food webs, in G.A. Polis and Winemiller, K.O. (eds.), Food webs: integration of patterns & dynamics, Chapman & Hall, London, pp. 185-191.
17. Raffaelli, D., Way, S., Balls, P., Paterson, P., Hohman, S. and Corp, N. (1999) Major changes in the ecology of the Ythan estuary: how important a re physical factors? Aquatic Conservation (in press)
18. Goss-Custard, J.D. (1966) The feeding ecology of redshank, Tringa totanus L., in winter, on the Ythan estuary, Aberdeenshire, PhD Thesis, University of Aberdeen.
19. Raffaelli, D., Limia, J., Hull, S. and Pont, S. (1991) Interactions between the amphipod Corophium volutator and macroalgal mats on estuarine mudflats, Journal of the Marine Biological Association of the United Kingdom, 71, 899-908.
20. Hull, S.C. (1987) Macroalgal mats and species abundance: a field experiment, Estuarine, Coastal and Shelf Science, 25, 519-532.
21. Atkinson-Willes, G.L. (1976) The numerical distribution of ducks, swans and geese as a guide to assessing the importance of wetlands, Proceedings of the International Conference on Wetland and Waterfowl, Heilingenhafen, 1974, 199-254.
22. Dayawansa, P.N. (1995) The distribution and foraging behaviour of wading birds on the Ythan estuary, Aberdeenshire, in relation to macroalgal mats, PhD Thesis, University of Aberdeen.

NATURAL VARIABILITY AND THE EFFECTS OF FISHERIES IN THE NORTH SEA: TOWARDS AN INTEGRATED FISHERIES AND ECOSYSTEM MANAGEMENT?

MAGDA J.N. BERGMAN & HAN J. LINDEBOOM
Netherlands Institute for Sea Research, P.O. Box 59 1790 AB Den Burg, Texel, The Netherlands

Abstract
Marine ecosystems are not in steady state, but exhibit continuous changes in production and species composition of different trophic levels. Sudden changes in biomasses or species composition, a reversal of trends, increased seasonal variation and cyclic behaviour all seem to contribute to the interannual and decadal variability of the North Sea ecosystem. An array of possible causes is introduced and the need for more long-term data sets is stressed. From this highly variable ecosystem men tries to extract a maximum quantum of fish using more and more efficient fishing methods, resulting in an overexploitation of fish stocks and notable effects on other species. In this paper an estimate is given for the fishing mortality, *i.e.* the total direct mortality, in the populations of non-target invertebrate species generated by the trawl fisheries in the Dutch sector of the North Sea in 1994. For the species studied the fishing mortality appeared to range from 7 to 48%. Long term observations indicate that stocks of several species declined and some species disappeared from the southern North Sea.

So far, attempts to control overfishing failed. Measures to regulate fisheries should be integrated into a general policy for the ecosystem. A sustainable North Sea ecosystem has to be a common objective for fisheries and nature management, also taking into account the high natural variability of the ecosystem. In this paper, several measures are proposed: an overall reduction in fishing effort leading to a moratorium in case of steep declines in fish stocks, stimulation of new designs of more selective and less damaging gears, and designation of areas closed for fishing activities to protect species and habitats that can not be protected sufficiently otherwise.

1. Introduction

Marine ecosystems are not in steady state but show large interannual and decadal variations. Periods with large algal blooms alternate with years or decades with relatively low algal biomasses. Biomasses or reproductive success may double or halve between two consecutive years. Periods with decreasing trends are followed by periods with increasing trends. Sometimes the ecosystem seems more or less stable for a certain period, but then the system seems to flip to another semi-steady state [1].

Fishery with bottom trawls started in these highly variable northwestern European seas as early as in the 13th century. As concern raised about the sustainability of the stocks, the International Council for the Exploration of the Sea (ICES) was founded in 1904 to coordinate the international fishery policy in the North Sea. Results of management related scientific research, mainly studies on stock assessment, formed the basis for this policy. Despite all research effort and the subsequent regulations in fisheries management during this century, many commercial fishstocks became heavily overfished during the last 30 years. Next to being the source for this overexploitation, trawling is certainly also one of the key factors in the changes in the benthic ecosystem

J.S. Gray et al. (eds.), Biogeochemical Cycling and Sediment Ecology, 173–184.

in the south-eastern part of the North Sea [2].

In this paper an overview is given of the variability in the North Sea ecosystem. The potential impact of fisheries on the benthic ecosystem is illustrated by calculating the fishing mortality in a number of invertebrate species. Long term impact of fisheries on the benthic ecosystem is indicated. Arguments are given to initiate a policy for sustainable use of the North Sea and possible consequences are presented for fisheries policy.

2. Variability in the North Sea ecosystem

2.1 Phenonema observed

Several long-term data sets on phytoplankton, zooplankton, macrofauna, fish and birds have been collected in the Wadden Sea and North Sea. When the various data sets are combined, it is striking that certain changes are very sudden and not gradual, as one would expect, for example, from gradual increasing human impact. Field data (Fig. 1) indicate that the algal biomass in the western Wadden Sea doubled between 1976 and 1978, followed by the macrobenthos in 1980, and in 1978, the breeding success of Eider ducks (*Somateria mollissima*) increased by several orders of magnitude [3][4][5]. Similar phenomena were observed in the North Sea. In the vicinity of Helgoland, changes in the

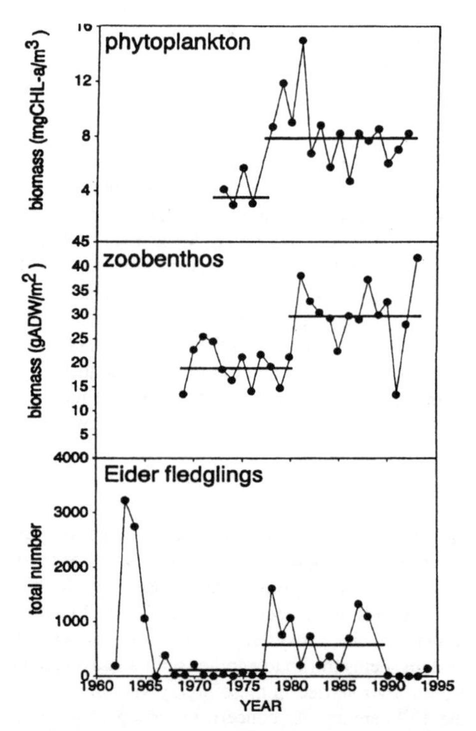

Fig. 1. Average chlorophyll-a concentration (after Cadée and Hegeman, [3], Macrofauna biomass (after Beukema, [4]) and Eider fledglings (after Swennen, [5]) in the western Wadden Sea. The different levels observed in the data set are indicated.

phytoplankton species composition were observed and the biomass of flagellates <10 μm went from 3 μg/l in the period '62-'77 to 15 μg l^{-1}in the period '78-'91 [6]. Analyses of the Continuous Plankton Recorder (CPR) data indicated a minimum abundance in zooplankton and a simultaneous shift in species composition in the late 70s [7]. Shifts in macrofauna assemblages [8] and benthic respiration rates were reported, all hinting at a rapid major change of the marine ecosystem in the entire North Sea in the late seventies. A major shift in 40 environmental variables measured in the marine ecosystem in the second half of the seventies has also been recorded in the North-Eastern Pacific [9]. Another rapid change in the CPR data was recently reported for the mid 80s [10], whereas in the Wadden Sea the breeding success of eider ducks once again collapsed in the early 90s (see Fig. 1). The latter coincided with a dramatic decrease in standing stocks of mussels and cockles in the Dutch Wadden Sea. The collapse of the North Sea fish stocks at the end of the previous century and in the thirties were also rather sudden changes, indicating that also the target species of fisheries may be subject to this phenomenon.

In addition to these rapid changes, other striking phenomena have been observed. The CPR data indicate that the monthly variation of the net colour index was about twice as high in the 80s as compared to the 70s (see [11]).

In several data sets, cyclic changes with different frequencies have been observed. Neudecker and Purps [12] found indications for a 6-year cycle in the amount of shrimps landed around the North Sea. Between 1969 and 1996, the density of *Macoma balthica* recruits showed a maximum every 6 years, starting in 1973, with only two other maxima in other years (Dekker, pers.comm.). In the same period, cockles (*Cerastoderma edule*) also showed periodical recruitment patterns. However, their recruitment failed in the 90's. Gray and Christie [13] analysed plankton data from the North Atlantic and found evidence of 3-4, 6-7 and 10-11 years cycles, whereas benthic data suggested 6-7 and 10-11 years cycles. Witbaard [14] found a 25-30 year cycle in the variation in annual shell growth of *Arctica islandica* collected at the Fladen grounds in the North Sea. For the CPR data a 30-year cycle has been suggested.

However, up till now most data sets are too short for a reliable analyses of these longer cycles. Analyses of phenomena observed in marine sediments may help to solve this problem. Pike and Kemp [15] found significant periodicities in the deposition of diatom mat laminae of ~11 years, 22-24 years and ~50 years in Gulf of California sediments deposited in the early holocene. Cyclic deposition has been observed in many sediments and their paper gives a selection of observations all indicating similar periodicities.

In conclusion, many longer data series indicate interannual and decadal changes among which the following phenomena may be observed: sudden rapid changes, gradual changes (*e.g.* in the direction of trends), changes in (seasonal) variability, changes in dominance of species and cyclic variation. What causes this behaviour of the marine ecosystem?

2.2 *POSSIBLE CAUSES*

Links with changes of short-term or large-scale weather patterns have been suggested in several papers. Wind, winter and/or summer temperatures or rainfall are mentioned as possible causes. A shift in storm frequencies or wind directions might cause changes in sediment water exchange in shallow areas or mixing of stratified waters in deeper areas [1]. Beukema [16] found that the occurrence of cold winters strongly influenced the species composition of intertidal benthic communities. He demonstrated that the winter of 1979 affected the macrofauna in the Balgzand area. Other strong winters, but not all, led to similar effects.

High temperatures in summertime may influence the distribution of species adapted to cooler waters. Cadée and Hegeman [3] showed that there is a positive correlation between the freshwater input from Lake IJssel, and thus rainfall, and the length of algal blooms in springtime. Therefore, specific differences in weather phenomena may explain changes in the marine ecosystem. The long-term changes observed in the CPR data have been attributed to changes in westerly weather patterns or related to the North Atlantic Oscillation (NAO; [7]).

Burroughs [17] mentions cycles of 2, 3-4, 5-7, 10-12, 18.6, 22, 80-90 and 200 years observed with a relatively high frequency in meteorological records, all of which might have a relationship with certain weather and astronomical features. Are these cycles reflected in the marine ecosystem? Although the data serie is still short, the fledgling success of the eiderducks has periods with optima and minima of approximately 11 years. Recently, it has been argumented that storm patterns in the temporate zone may be influenced by the occurrence of sunspots [18]. Cycles of 11 or 22 years have been observed in several sedimentary records and data sets mentioned before, which hint at a possible influence of the 11 year sunspots cycle or the 22 year Hale cycle on marine systems. So far, many authors have critisized the possible relationship with sunspots, because the difference in solar irradiation is too small, others have pointed at the possibility of the effect of the changing magnetic field [17] [18].

On the other hand, the 18.6-year period in the regression of the longitude of the node - the line joining the points where the Moon's orbit crosses the ecliptic [17] - may also have an influence on biota. Peaks in tidal forces due to this phenomenon occurred in 1950, 1969 and 1988. In the Wadden Sea an 18.6 year cycle was found in sedimentation rates [19]. And organisms living in the intertidal area, such as the macrofauna shown in Figure 1, may be subject to influences of this cycle.

For the sudden changes at the end of the 70s, a relationship with the great salinity anomaly which entered the North Sea in 1977/78 was proposed [1]. Not the salinity itself but a difference in macro- or micro-nutrient concentrations in this distinguishable amount of water could have led to changes in the ecosystem. Fromentin and Planque [20] found a close relationship between the North Atlantic Oscillation (NAO) and the abundance of the copepods *Calanus finmarchicus* and *C. helgolandicus* in the eastern North Atlantic. Whereas Corten [21] pointed at the relationship between hydrography and the occurrence of herring. In their analyses Gray and Christie [13] showed evidence of cycles in hydrographic data for the North Atlantic Ocean with periods of 3-4, 6-7, 10-11, 18-20 and 100 years. They indicate a relationship between hydrographic and biotic cycles.

Up till now, we have looked for the possible cause of the observed variations outside the system, but it could also be caused by internal cycles in the marine ecosystem. High biomasses or high reproduction rates of one species may alternate with similar phenomena in other species. These sort of changes have been observed, *e.g.* in fish stocks [22], and using model calculations it has been shown that sudden declines or increases may be explained by internal cycles [23].

However, we will never be able to answer the cause-effect questions if no longer real time dataseries are collected, and the continuous collection of data, often hampered by limiting funding, should be strongly supported.

Apart from the natural variability of the ecosystem, pollution, eutrofiering and offshore activities generally cause effects on environmental quality and productivity on a local scale (coastal zone; around riggs); fishery acts as a large scale activity that exploits the living resources and thus depends entirely on and affects the ecosystem directly. Impact of human activities must be seen against the background of natural variation in the system. Conversely, the observed changes in the ecosystem represent the integrated effects of natural and man induced variation. In the following the magnitude and effects of the beam trawl fisheries on the North Sea benthic system will be discussed.

3. Impact of North Sea fisheries

3.1 COMMERCIAL FISH STOCKS

During this century fishing mortality shows an almost continuous increase with exceptions only during the world wars. Between 1945 and 1975 total North Sea catches doubled up to $3*10^6$ tonnes per year [24]. For most commercial fish species, that normally exhibit a natural mortality of about 20%, up to 40 to 60% of the standing stock is caught every year [25]. Pelagic and demersal fisheries severely overexploited the stocks of herring, mackerel, cod, haddock, whiting, sole and plaice [26]. As a result, populations are build upon only a few yearclasses and fishery catches exist almost exclusively of immature fish. Some spawning stocks have reached an all time low level, other have reduced to around or even beyond the level where the numbers of adults can be expected to become a limiting factor to subsequent recruitment [27, 28]. Stocks of several ray species declined and disappeared from the Dutch coastal zone (stingray *Dasyatis pastinaca*) and Dutch continental shelf (common skate *Raja batis*, thornback ray *Raja clavata*) after the second world war [29, 30]. Rays and sharks respond readily with decreasing spawning stocks to increasing fishing mortality, due to a relative belated age of reproduction, a minor reproduction capacity and the fact that their egg-capsules are bound to the seabed, causing an increased sensibility to demersal fishing methods.

3.2 NON-TARGET INVERTEBRATE SPECIES

Invertebrate bottom fauna without commercial value (*e.g.* bivalves, gastropods, starfish, anemones) is also caught in the trawl net, or hit by the gear but escaping underneath or via the meshes. The direct mortality due to a single trawling among these categories of fauna was measured in experimental fieldstudies. The total direct mortality for different species appeared to range from 10 to 80% of the initial densities in the trawl path [31] [32]. As only a small fraction was actually caught in the net, this mortality is predominantly due to the contact with the gear (groundrope and tickler chains). Direct mortality mainly occurred among animals, mortally damaged or exposed to predators, in the trawl path after the passage of the trawl. Since species sensitive to fishing mortality or to reduction of appropiate habitats have disappeared from the ecosystem long ago, fieldstudies to measure the direct mortality of invertebrates had to be performed with species that managed to survive during 30 years of intensive trawling, *i.e.* species relatively resistent to beam trawling.

To illustrate the potential impact of trawling on invertebrate bottom fauna, the fishing mortality for a number of these infauna species was calculated in the Dutch sector of the North Sea in 1994. This mortality was calculated using three variables: the trawling frequency of the different demersal fleets in 1994, the spatial distribution of these species, and the estimates of the direct mortality due to a single experimental trawling with commercial trawls. For detailed information on calculations and direct mortality estimates, see [33]. In short, the following data-sets and assumptions were used in the calculations.

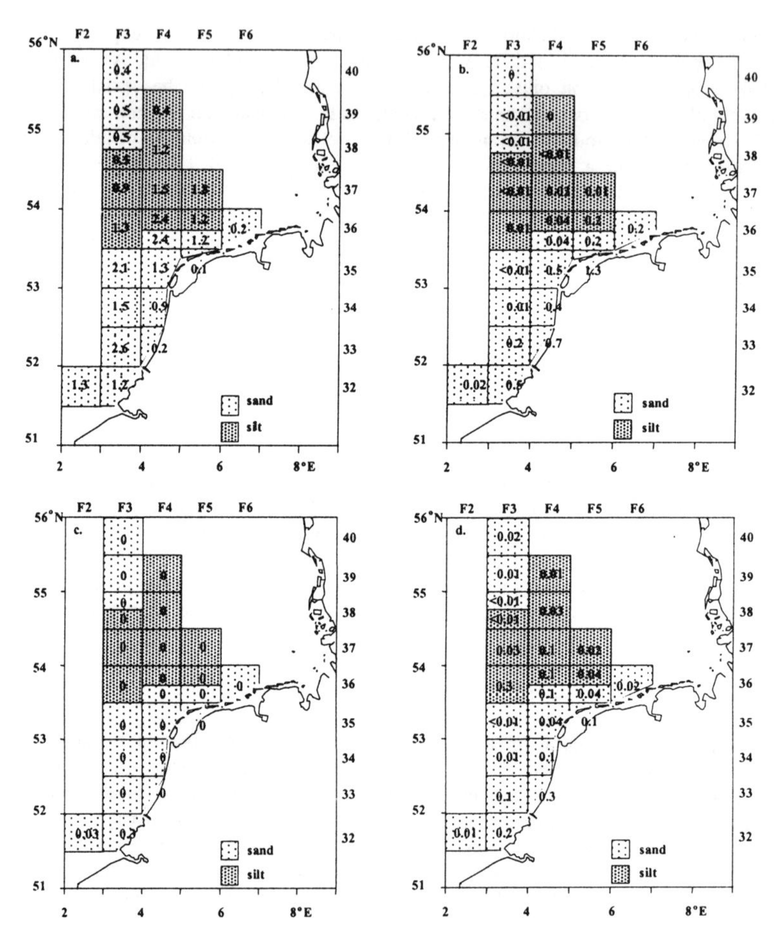

Fig. 2. Distribution of the trawling effort (mean trawling frequency *i.e.* trawled area / total area) per ICES rectangle in 1994, for four types of trawl fisheries, over the Dutch sector (data from Polet *et al.*, [34]). a. 12 m beam trawl fisheries using tickler chains; b. 4 m beam trawl fisheries using tickler chains; c. 4 m beam trawl fisheries using chain matrices; d. otter trawl fisheries (flatfish and roundfish).

The trawling frequencies per ICES rectangle (*i.e.* trawled area / total area) by the Dutch, Belgium, German and British fleets in 1994 were calculated from the numbers of fishing hours [34] and the surface area of the ICES quadrants. This was done for 4 m and 12 m beam trawls with ticklers, for 4 m beam trawls with chain matrices, and for otter trawls, respectively (Fig. 2). The 12 m beam trawl fishery appeared the dominant type of trawling, with an average frequency for the Dutch sector of 1.23, in contrast to mean frequencies of 0.13, 0.01 and 0.06 for the 4 m beam trawl fishery with tickler chains, with chain matrices and the otter trawl fishery, respectively. The 12 m beam trawl fishery occurred predominantly offshore. The 4 m beam trawl fishery was mainly restricted to the coastal zone, in which the trawling intensity was approximately as high as the 12 m beam trawl fishery offshore. To simulate a clustered distribution of trawling activities in the calculation of the fishing mortality, as was indicated by Rijnsdorp et al. [35], each ICES rectangle was divided in nine subrectangles, each representing a certain percentage of the total trawling effort in that retangle. Population densities of a number of larger sized invertebrates were estimated in 1996 [36], with a benthos dredge (Triple-D) designed to estimate reliably low abundant infauna species

[37]. It was assumed that the distribution patterns in 1996 were roughly similar with those in 1994 and that species were homogenuously distributed within an ICES statistical rectangle. The direct mortality in invertebrates species due to a single experimental trawling with the different types of trawls was calculated in a recent EU-project [31]. As the mortality estimate for 12 m and 4 m beam trawls with ticklers did not differ greatly, the mean estimate was used.

For all invertebrate species considered, the 12 m beam trawl fishery caused the highest fishing mortality in the Dutch sector (Table 1). Fishing mortality in invertebrate populations largely depends on the spatial distribution of both the effort of trawling fleets and of the species. Especially for species living predominantly in silty offshore areas (*e.g.* the bivalve *Dosinia lupinus*) and for those occurring in all types of sediments (*e.g.* the sea urchin *Echinocardium cordatum*), the fishing mortality due to the 12 m beam trawl fisheries was much higher than due to the 4 m beam trawl fisheries, mainly because of the coastal distribution of the latter. On the contrary, for species that are restricted to sandy areas, in which the highest efforts of the 4 m beam trawl fisheries are found, this difference in fishing mortality was less pronounced (*e.g.* the bivalves *Spisula solida* and *Ensis* spp., dominated by the coastal species *Ensis americanus*).

The fishing mortality in the invertebrate populations in the Dutch sector due to the total effort of trawl fisheries in 1994 varied from 7 to 48%. For half the number of the species considered, annual fishing mortality was more than 25%. Although species for which the calculations could be performed were able to maintain a certain population density despite this fishery induced mortality, others, showing either higher direct mortalities or population characteristics less suited to resist this pressure, may not be able to withstand this high fishing mortality and may have become rare.

3.3 LONG TERM IMPACT ON INVERTEBRATE POPULATIONS

Whether fishing mortality affects the population sizes and distribution of non-target invertebrates depends on the life cycle characteristics of a particular species. Age of maturation, numbers and survival of eggs and larvae, and recruitment to the adult population are the main parameters. Long term impact of fisheries on invertebrate species could not be measured by comparing the species composition in trawled and non-trawled areas, as non-fished reference areas do not exist in the southern North Sea. Analyses of historical datasets on occurrence, abundance and catchability of invertebrate species gave indications about possible long term impacts of fisheries. Demersal fisheries is thought to be responsible for the decreasing abundances of a number of bivalve species, like *Ostera edulis*, *Modiolus modiolus*, *Arctica islandica*, edible crab (*Cancer pagurus*), lobster (*Homarus gammarus*), gastropods (e.g. *Buccinum undatum, Neptunea autiqua*), anemones, sponges and tube building worms [38, 39, 40, 41].

Modelling results indicate that the observed decline in stocks of crabs, lobsters and gastropods coincide with increasing trawling intensities and the change in demersal trawl type [41].

Table 1. Fishing mortality (%) in the total population of invertebrate species in the Dutch sector due to different bottom trawl fisheries in 1994. - = no total direct mortality estimate availabe for this species. [blank] = no overlap in trawling and species distribution. 12m BT = 12 m beam trawl fishery; 4m BT = 4 m beam trawl fishery; OT = otter trawl fishery.

Species	Fishing mortality (%) in the Dutch sector in 1994				
	12m BT ticklers	4m BT ticklers	4m BT chain matrix	OT	All trawl fisheries
ECHINODERMS					
Astropecten irregularis	18	0		0	18
Echinocardium cordatum	25	3	0	4	30
Ophiura texturata	7	1	0	1	9
BIVALVES					
Abra alba	30	2		0	31
Arctica islandica	15	0		1	15
Chamelea gallina (1-2 cm)	7	0	0	0	7
Chamelea gallina (2-4 cm)	28	1	0	2	30
Dosinia lupinus	29	1		1	31
Ensis spp.	10	5	0	1	16
Gari fervensis	39	0		4	42
Mactra corallina	17	2	-	1	20
Phaxas pellucidus	21	0		2	22
Spisula solida	21	11	1	-	31
Spisula subtruncata	16	8	0	3	25
GASTROPODS					
Lunatia catena	33	22	-	-	48
Turritella communis	16	2		0	18
CRUSTACEANS					
Corystes cassivelaunus - female	21	1	0	1	23
Corystes cassivelaunus - juvenile	35	1		2	37
Corystes cassivelaunus - male	32	1		3	35
Thia scutellata	21	4	0	-	25
OTHER GROUPS					
Aphrodita aculeata	25	1		1	27
Pelonaia corrugata	18	0		0	19

Fishery causes shifts in species composition in the ecosystem by favouring the short-lived species at the cost of longlived species. In heavily trawled areas, larger numbers of opportunistic (small sized, fast reproducing, short lived) species were found than in less intensively trawled areas, *i.e.* around wrecks and in areas closed for fisheries, where sensitive, large sized, and long lived species occur more frequently. Such shifts in species composition have been found both in experimental fieldstudies [42] and in the North Sea (German Bight) [39]. Indications were found that the complexity of habitats for benthic fauna has been reduced by the elimination of species that create holes or structures in/at the seabed [43], and that complete habitats have been removed such as certain gravel areas [44], and areas with boulders (e.g. the Texel Stones).

Due to the relative scarcity of older yearclasses in the commercial fish populations, prey-predator relationships within the ecosystem may indirectly be affected. And particularly the predation on small sized fish is correspondingly lower. This feed-back mechanism not only leads to an increased recruitment of certain commercial species, but indications were found of higher abundances of non-target fish species like solenette, scaldfish, lesser weever and dragonet [45]. As small sized fish exploit a very different food source compared to adult fish, predation pressure on benthos might have changed. The enlarged populations of crabs in the North Sea [39] might be the result of the relative absence of large sized predator fish.

Although changes in the benthic ecosystem have to be considered against the background of high natural variability, results of many studies (*e.g.* [2]) showed that trawl fisheries contributed considerably to the shifts in species composition observed.

4. A sustainable fishery in a variable North Sea ecosystem

4.1 INTEGRATED FISHERIES AND NATURE MANAGEMENT

In conclusion, many longer data series indicate interannual and decadal variability in the marine ecosystem. The observed fluctuations seem to be a combination of *e.g.* sudden rapid changes, gradual changes, changes in (seasonal) variability, changes in dominance of species and cyclic variations. For certain periods ecosystems might be more or less stable, in other periods systems flip to another semi-steady state.

In this variable system, the intensive North Sea trawl fisheries has major implications: commercial fish stocks are heavily overfished and many of them have reached an all time low level. A number of invertebrate populations declined and several species almost disappeared from the Dutch sector, the complexity of habitats has been reduced, and specific relationships in the system changed.

Attempts to control the exploitation of fish stocks, since 1983 based on technical maesures and a quota system, has been unsuccesfully for almost 100 years, mainly due to lack of control and enforcement and because of its inacceptability by the fishing industry. Obviously, measures to regulate fisheries should be integrated into a general policy for the marine system. Fisheries management should also take into account the natural variability of the marine ecosystem with possible sudden changes, changing composition of fauna, and periods with low recruitment and biomasses of the preferred target species. Or to put it simply: even in an ideal system there will always be "rich and poor years". To underpin this management policy, more knowledge about the occurrence and causes of these ecosystem phenomena is urgently needed.

A sustainable North Sea ecosystem has to be a common objective for fisheries and nature management. This objective should be aimed to serve fisheries perspectives, *i.e.* a permanent availability of a variety of fish and shellfish stocks with (high) economic value. While, for nature perspectives, extinction of species should be prevented and (parts of) the natural ecosystem should be conserved. For an integrated management, general aims have to be formulated. As the ultimate effectivity of an integrated North Sea management depends primarily on mutual committment, resulting in control and enforcement, it might be worthwhile to introduce a system of shared responsibility *i.e.* a new conception of rights and duties of both government and fishery sector [46].

4.2 MEASURES TO SUPPORT AN INTEGRATED MANAGEMENT

At present, there is consensus about the view that the North Sea is heavily overfished [28] [46]. A reduction in fishing effort of about 20 to 40% will enable the commercial stocks to extent towards a more natural population structure (numeric as well as *qua* age distribution), that is less vulnerable to natural fluctuations *i.e.* extreme cold/warm winters or shifts in ocean currents. Spawning stocks enlarged to above biologically safe levels are less vulnerable to natural fluctuations that may affect future recruitment. More stable fish populations showing higher resilience will result in less sudden changes in licensed quota and thus in higher economical rendements. Due to a normalised age distribution in the stocks, the relative proportion of undersized fish in the catch will decline and, in a quota system, the aimless destruction of juvenile discards may be forced back. Because proper fisheries management, has to account for many relationships in the complex ecosystem, impact of natural (climate, wind, currents) and antropogenic factors (eutrophication, pollution, fisheries) that generate gradual and short term, as well as large scale shifts in the populations, fishery has to be further reduced or even temporarily banned in case of too strong declines in stocksize. To cope with this economically, joint ventures of different types of fisheries (fish, shellfish, and shrimps) might be a solution. A method to gradually increase the fishstocks could be found in lowering the quota in years with large stocks. As all subsidies to fisheries only aggravete the overexploitation problem, an arteficial increase in costs of fishing by *e.g.* putting levies on oil, annual fees on fishing rights and general environmental use, limited pay for decommissioned ships, may have some steering capacity [28].

A reduction in fishing effort will lead to a reduction of the impact. Reduced trawling effort generates a reduction in the disturbed surface of the seabed and therefore in fishing mortality of invertebrates. As a result of which *e.g.* bivalve populations may enlarge. Decreasing activities of the industrial fishing fleet will lead to increasing abundancies of (semi)pelagic fish e.g.sandeel and sprat, creating more food for predatory fish, seabirds and seamammals. In general, a reduction in fishing effort fits in the nature management perspective with respect to conservation, restoration and development of ecological values (diversity, natural processes, protection of species and habitats).

Apart from a major reduction in fishery effort, alternative gears have to be designed to catch target fish more selectively and to minimize the mortality in undersized and non-target fish and invertebrates. Larger meshed nets generally are more selective as relatively small sized animals are able to escape. If *e.g.* the legalized minimum size for sole, a species with the smallest marketable size, the largest body flexibility, and the minimum mesh size, are enlarged, the amounts of undersized flatfish (plaice, dab, flounder) caught and discarded would be much lower. Reduced amounts of discards can also be achieved by anticipating on the behaviour of fish: the insertion of large meshed separation panels in the back of a trawl enables juvenile gadoids to escape. Obligatory fishing with specific gears (*i.e* more selective gears) in defined fishing grounds offers an instrument to steer ecological developments in the fishery sector. To lower the mortality in invertebrates in the trawl path, fysical contact between the groundrope and the fauna has to be minimalised. Solutions may be found in other arrangements of tickler chains, or chasing the target fish by electrical shocks or water jets.

As even the most selective trawling gears will have bottom contact by means of the groundrope, direct mortality will still be induced in epifauna species living at the seabed (*e.g.* bivalves, sponges), slowly swimming fish, and egg capsules of rays and whelks (*Buccinum undatum*). Habitat structures build by tube building worms or bivalves will be destroyed as well. Therefore, a significant reduction in fishing effort and the development of selective gears will not result in a sufficient low fishing mortality to enable the recovery of populations of sensible animals like rays, long lived bivalves, sedentary epifauna species (sponges, anenomes, hydroids), whelks, and structure building fauna species. For the conservation of these species the designation of areas closed to harmful fisheries is needed [47]. Size and location of the closed areas depend on the characteristics of the species considered. To protect typical southern North Sea species, an area of 140*90 km north of the Wadden isles, that includes several types of seabeds and depth zones, seems to be appropiate [40]. In this area, specific habitat characteristics and ecological relationships can develop. Most invertebrate species showing seasonal migrations will be protected in such an area. For fish like rays probably even larger areas are needed, but the proposed area will at least give protection for local populations [30]. Closed areas in the coastal zone give juvenile fish an important protection against a premature death in fishing nets. Zonation in the North Sea contributes to restoration and conservation of the ecological values, as in the closed areas the ecosystem may experience a sustainable development without the continuous pressure of fishing activities.

5. References

1. Lindeboom, H.J., Raaphorst, W. van, Beukema, J.J., Cadée, G.C. and Swennen, C. (1995) (Sudden) changes in the North Sea and Wadden Sea: Oceanic influences underestimated? Dt. Hydrogr. Z. Suppl. **2**, 87-100.
2. Lindeboom, H.J. and Groot, S.J. de (eds) (1998) *The effects of different types of fisheries on the North Sea and Irish Sea benthic ecosystems.* NIOZ Rapport 1998-1 / RIVO-DLO Report C003/98, 404 pp.
3. Cadée, G.C. and Hegeman, J. (1993) Persisting high levels of primary production at declining phosphate concentration in the Dutch coastal area (Marsdiep). Neth. J. Sea Res. **31**: 147-152.
4. Beukema, J.J. (1992) Long-term and recent changes in the benthic macrofauna living in tidal flats in the western part of the Wadden Sea. NIOZ Publ. Ser. **20**, 135-141.
5. Swennen, C. (1991) Fledgling production of Eiders Somateria mollissima in The Netherlands. J. Orn. **132**, 427-437.
6. Hickel, W., Mangelsdorf, P. and Berg, J. (1993) The human impact in the German Bight: Eutrophication during three decades (1962-1991). Helgol. Meeresunters. **47**, 243-263.
7. Aebischer, N.J., Coulsen, J.C. and Colebrook, J.M. (1990) Parallel long-term trends across four marine trophic levels and weather. Nature **347**, 753-755.
8. Josefson, A.B., Jensen, J.N. and Ærtebjerg, G. (1993) The benthos community structure anomaly in the late 1970s and early 1980s - a result of a major food pulse? J. Exp. Mar. Biol. Ecol. **172**, 31-46.
9. Ebbesmeyer, C.C., Cayan, D.R., McLain, D.R., Nichols, F.H., Peterson, H.H. and Redmond, K.T. (1990) 1976 step in the pacific climate: forty environmental changes between 1968-1975 and 1977-1984. In J.L. Betancourt and V.L. Tharp (eds*), Proc. 7th Annual Pacific Climate (PACLIM) Workshop*. Calif. Dept. of Water Resources, Interagency Ecol. Stud. Prog. Rep. 26, pp. 115-126.
10. Reid, P.C., Edwards, M., Hunt, H.G. and Warner, A.J. (1998) Phytoplankton change in the North Atlantic. Nature **391**, 546.
11. Buchanan, J.B. (1993) Evidence of benthic pelagic coupling at a station off the Northumberland coast. J. exper. mar. Biol. Ecol. **172**, 1-10.
12. Neudecker, Th. and Purps, M. (1996) Zur Periodizität der Krabbenanlandungen-oder: Wie werd die nächste Krabbensaison? Das Fisherblad. 6: 159-161.
13. Gray, J.S. and Christie, H. (1983) Predicting long-term changes in marine benthic communities. Mar. Ecol. Prog. Ser. **13**, 87-94.
14. Witbaard, R. (1996) Growth variations in *Arctica islandica* L. (Bivalvia, Mollusca); A reflection of hydrography related food supply. ICES J. Mar. Sci. **53**, 981-987.
15. Pike, J. and Kemp, A.E.S. (1997) Early Holocene decadal-scale ocean variability recorded in Gulf of California laminated sediments. Paleoceanography **12**(2), 227-238.
16. Beukema, J.J. (1990) Expected effects of changes in winter temperatures on benthic animals living in soft sediments in coastal North Sea areas. In J.J. Beukema et al. (eds), *Expected effects of climatic change on marine coastal ecosystems*. Kluwer Academic Publishers, Dordrecht, pp. 83-92
17. Burroughs, W.J. (1992) *Weather cycles Real or Imaginary?* Cambridge University Press, Cambridge, UK, 207 pp.
18. Haigh, J.D. (1996) The impact of solar variability on climate. Science **272**, 981-984.
19. Oost, A.P., Haas, H. de, Ijnsen, F., Boogert, J.M. van den and Boer, P.L. de (1993) The 18.6 year nodal cycle and its impact on tidal sedimentation. Sedimentary Geology **87**, 1-11.
20. Fromentin, J-M. and Planque, B. (1996) *Calanus* and environment in the eastern North Atlantic. II. Influence of the North Atlantic Oscillation *on C. finmarchicus* and *C. helgolandicus*. Mar. Ecol. Prog. Ser. **134**, 111-118.
21. Corten, A., 1990. Long-term trends in pelagic fish stocks of the North Sea and adjacent waters and their possible connection to hydrographic changes. Neth. J. Sea Res. **25** (1/2), 227-235.
22. Hempel, G. (1978) North Sea fisheries and fish stocks - A review of recent changes. Rapp. P.-v. Réun. Cons. int. Explor. Mer **173**, 145-167.
23. Silvert, W. (1993) Size-structured models of continental shelf food webs. In V. Christensen and D. Pauly (eds), Trophic models of aquatic ecosystems. ICLARM Conf. Proc. **26**, pp. 40-43.
24. Holden, M.J. (1978) Long term changes in landings of fish from the North Sea. Rapp. P.-v. Reun. Cons. int. Explor. Mer **172**, 11-27.
25. Jones, R. (1984) Some observations on energy transfer through the North Sea and Georges Bank food webs. Rapp. P.-v. Reun. Cons. int. Explor. Mer **183**, 204-217.
26. Daan, N. (1989) The ecological setting of North sea fisheries. Dana 8, 17-31.
27. Daan, N., Bromley, P.J., Hislop, J.R.G. and Nielsen, N.A. (1990) Ecology of North Sea fish. Neth. J. Sea Res. **26**, 343-386.
28. Daan, N. (1996) *Desk study on medium term researech requirements in relation to the development of integrated fisheries management objectives for the North Sea.* RIVO-DLO Report C054/96, 83 pp.
29. Walker, P. (1996) *Ecoprofile rays and skates on the Dutch continental shelf and North Sea.* Rapport RIKZ 96.005, 76 pp.
30. Walker, P. (1998) *Fleeting Images. Dynamics of North Sea ray populations.* Thesis, University of Amsterdam, 145 pp.
31. Bergman, M.J.N., Ball, B., Bijleveld, C., Craeymeersch, J.A., Munday, B.W., Rumohr, H. and Santbrink, J.W. van (1998a) Direct mortality due to trawling. In H.J. Lindeboom and S.J. de Groot (eds), *The effects of different types of fisheries on the North Sea and Irish Sea benthic ecosystems.* NIOZ Rapport 1998-1 / RIVO-DLO Report C003/98, pp. 167-185.
32. Craeymeersch, J.C., Ball, B., Bergman, M.J.N., Damm, U., Fonds, M., Munday, B.W. and Santbrink,

J.W. van (1998) Catch efficiency of commercial trawls. In H.J. Lindeboom and S.J. de Groot (eds), *The effects of different types of fisheries on the North Sea and Irish Sea benthic ecosystems*. NIOZ Rapport 1998-1 / RIVO-DLO Report C003/98, pp. 157-167.

33. Bergman, M.J.N., Craeymeersch, J.A., Polet, H. and Santbrink, J.W. van (1998b) Fishing mortality in invertebrate populations due to different types of trawl fisheries in the Dutch sector of the North Sea in 1994. In H.J. Lindeboom and S.J. de Groot (eds), *The effects of different types of fisheries on the North Sea and Irish Sea benthic ecosystems*. NIOZ Rapport 1998-1 / RIVO-DLO Report C003/98, pp. 353-359.
34. Polet, H., Ball, B., Blom, W., Ehrich, S., Ramsay, K. and Tuck, I. (1998) Fishing gears used by different fishing fleets. In H.J. Lindeboom and S.J. de Groot (eds), *The effects of different types of fisheries on the North Sea and Irish Sea benthic ecosystems*. NIOZ Rapport 1998-1 / RIVO-DLO Report C003/98, pp. 83-120.
35. Rijnsdorp, A.D., Buijs, A.M., Storbeck, F. and Visser, E. (1997) *De micro-verspreiding van de Nederlandse boomkorvisserij gedurende de periode van 1 april 1993 tot en met 31 maart 1996*. RIVO-DLO Rapport C006/97, pp 39.
36. Bergman, M.J.N. and Santbrink, J.W. van (1997) *Verspreiding en abundantie van grotere (epi)bentische macrofaunasoorten op het NCP en de invloed van boom-korvisserij op de verspreiding van deze fauna*. BEON data report: NIOZ 96 V.
37. Bergman, M.J.N. and Santbrink, J.W. van (1994) A new benthos dredge (Triple-D) for quantitative sampling of infauna species of low abundance. Neth. J. Sea Res. **33**, 129-133.
38. Vooys, C.G.N. de, Witte, J.IJ., Dapper, R., Meer, J. van der and Veer, H.W. van der (1993) *Lange termijn veranderingen op het Nederlands continentaal plat van de Noordzee: trends in evertebraten van 1931-1990* NIOZ Rapport 1993-17, 68 pp.
39. Rumohr, H., Ehrich, S., Knust, R., Kujawski, T., Philippart, C.J.M. and Schroeder, A. (1998) Long term trends in demersal fish and benthic invertebrates. In H.J. Lindeboom and S.J. de Groot (eds), *The effects of different types of fisheries on the North Sea and Irish Sea benthic ecosystems*. NIOZ Rapport 1998-1 / RIVO-DLO Report C003/98, pp. 280-353.
40. Bergman, M.J.N., Lindeboom, H.J., Peet, G., Nelissen, P.H.M., Nijkamp, H. and Leopold, M.F. (1991) *Beschermde gebieden Noordzee, noodzaak en mogelijk-heden*. NIOZ Rapport 1991-3, 195 pp.
41. Philippart, C.J.M. (1998) Long term impact of bottom fisheries on several by-catch species of demersal fish and benthic invertebrates in the south-eastern North Sea. ICES J. Mar. Sci. **55**, 342-352.
42. Tuck, I., Ball, B. and Schroeder, A. (1998) Comparison of undisturbed and disturbed areas. In H.J. Lindeboom and S.J. de Groot (eds), *The effects of different types of fisheries on the North Sea and Irish Sea benthic ecosystems*. NIOZ Rapport 1998-1 / RIVO-DLO Report C003/98, pp. 245-280.
43. Auster, P.J., Malalesta, R.J., Langton, R.W., Watling, L., Valentine, P.C., Lee, C., Donaldson, S., Langton, E.W., Shephard, A.N. and Babb, I.G. (1996) The impacts of mobile fishing gear on seafloor habitats in the Gulf of Maine (northwest Atlantic). Implications for conservation of fish populations. Review Fisheries Science **4**(2), 185-202.
44. Leth, J.O. and Kuipers, A. (1996) *Effects on the seabed sediment from beam trawling in the North Sea*. ICES CM 1996/Mini:3.
45. Fonds, M. and Blom, W. (1996) *Onderzoek naar mogelijkheden tot vermindering van discardproductie door technische aanpassing van boomkornetten*. BEON Rapport 96-15, 50 pp.
46. Bergman, M.J.N., Daan, N., Lanters, R.L.P., Salz, P., Smit, H., Vries, I. de and Wolff, W.J. (1997) *Kansen voor natuur en visserij in de Noordzee*. Werk-document IKC Natuurbeheer W-41, 33 pp.
47. Lindeboom, H.J. (1995) Protected areas in the North Sea: an absolute need for future marine research. Helgol. Meeresunters. **49**, 591-602.

COMMUNITY COMPOSITON OF TIDAL FLATS ON SPITSBERGEN: CONSEQUENCE OF DISTURBANCE?

J.M. WESLAWSKI*, M. SZYMELFENIG**
**Institute of Oceanology Polish Academy of Sciences, street Powstancow Warszawy 55, Sopot 81-712, Poland*
***Institute of Oceanography, University of Gdansk, street Pilsudskiego 46, Gdynia 81-370, Poland*

Abstract
Physical characteristics of six tidal flats on the island of Spitsbergen in the Svalbard archipelago and their macro-infaunal abundance and biomass, and meiofaunal biomass were compared. All sites are covered by fast ice for 6-8 months and experience low temperatures that kill all macrofauna by December. Only nematodes were observed to survive winter under the fast ice cover. New colonisation occurs in late May. All sites also experience massive freshwater discharge and sedimentation during summer. Flats are of similar size (1 to 4 km^2) and characterised by fine-grained surface sediments, salinity below 5 PSU, and 40 to 120 mg/l of particulates in the water. Despite these physical similarities, the fauna inhabiting these tidal flats are very different. Of the 39-macrofauna species sampled, only one (*Lumbricillus* spp.) was found at more than 3 of the sampled areas. Most of the species were collected from only one or two areas. Macrofaunal biomass ranged from 0.9 to 10g dw/m^2 while density ranged from 100 to 8000 ind/m^2. Meiofauna abundance ranged from 100 to 8000 ind/10cm^2, and biomass from 0.1 to 1.7g dw/m^2. It is proposed that the suite of taxa occupying intertidal flats on Spitsbergen depends on the pool of colonists in the nearby subtidal.

1. Introduction

Physically stressed marine environments like exposed rocky littoral, sandy beaches, and mudflats often host similar, well-defined animal communities [1] [2]. Intertidal mudflats from a wide geographic range are commonly inhabited by a suite of species of which *Macoma balthica, Nereis diversicolor, Scoloplos armiger, Arenicola marina,* and *Hydrobia spp.* are the most common. Communities with this composition have been recorded from the North Sea [3], White Sea [4], Northern Norway [5], New Foundland [6], Greenland [7] and Iceland [8]. Mudflats are also common in the Arctic. On Spitsbergen, a number of intertidal flats have been studied [9] [10] [11], Weslawski unpub. data), revealing a rich fauna. Surprisingly, species composition, density and biomass vary tremendously among these flats. The aim of the present study is to examine the differences among these flats and determine if differences can be explained by physical factors.

2. Materials and methods

Data for this paper come from published and unpublished studies of six intertidal flats on Spitsbergen (Figure 1). Data from Adventfjord were collected in July 1996 and 1997 on the tidal flat at the mouth of the Advent River. Macrobenthos were collected during low tide, with a 21-cm diameter core, inserted 10cm into the sediment. Three replicate cores were taken at each of 10 stations, combined, sieved through a 0.5 mm mesh sieve, and the material retained on the sieve preserved in 4% buffered formaldehyde. Meiofauna were collected from the same areas as the macrofauna, with a 2-cm diameter core, inserted 5cm into the sediment. Samples were stained with Rose Bengal

J.S. Gray et al. (eds.), Biogeochemical Cycling and Sediment Ecology, 185–193.

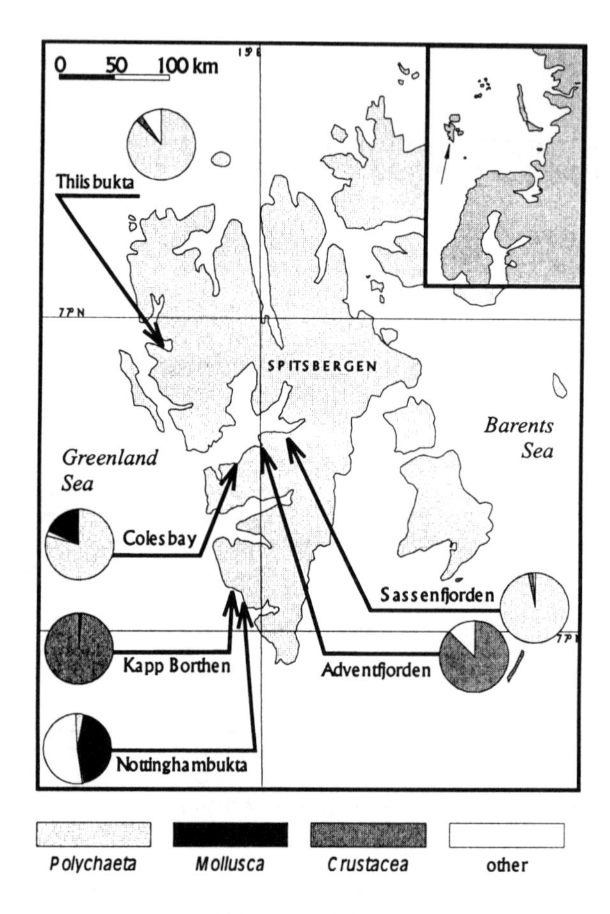

Fig.1 Percent share of major taxonomic groups in macrobenthos biomass in the investigated localities

in 4% formaldehyde, and sieved according to standard procedures [12]. We calculated meiofaunal biomass by converting linear measurements to volume according to Feller and Warwick [13]. Macrofaunal dry-weight was determined by drying animals at 60°C for 24 hours. For each macrofaunal species, the mean specimen size and weight were established and used for biomass calculations based on established relationships [14].

Data from Nottingham Bay and Kapp Borthen were collected by us and have been partly published [12] [15]. Other data were taken from the literature (Nottingham Bay, [9] [11], Sassenfjord, Colesbay, and Thiis Bay, [10]. All studies used a 0.5mm mesh sieve for collection of macrofauna. The first author visited each of the tidal flats considered in this study during the summers 1996 and 1997.

3. Results and discussion

3.1 Physical settings

Despite being located in different regions of Spitsbergen, the 6 flats have similar physical characteristics (Table 1). The flats are typified by riverine discharge resulting in a salinity as low at 5 PSU at low tide, very fine surface sediment, large suspension loads and consequently large sedimentation rates. Probably the most important physical stress for the infaunal community is the 6-8 months coverage by fast ice. When the ice melts in spring, it removes the upper part (10-cm) of sediment.

4. Faunal characteristics

Thirty-nine species of macrofauna and 6 taxa of meiofauna were collected from Spitsbergen tidal flats during summer (Table 2). No taxon was found in all localities, and only one (*Lubbricillus* spp.) was found at more than 3 locations. Density of macrofauna ranged from 40 to 6000 ind/m^2, and biomass from 0.9 to 10 g dw/m^2 (Table 3). The contribution of meiofauna to total benthic infaunal biomass ranged from 10 to 50% (Table 3). Size frequency distribution of benthic species, shows almost equal contribution to biomass from meio and macrofauna (Fig. 2). Such size distributions have been recorded from temperate tidal flats [16] and from Svalbard in the deep sublittoral [17].

Table 1. Physical conditions of 6 intertidal soft-sediment habitats on Spitsbergen Island. The authors collected data for all sites. Fast ice covers all sites from December until June.

	Nottingham Bay	Adventfjord	Sassenfjord	Colesbay	Kapp Borthen	Thiis Bay
Area exposed during low tide (km^2)	2	3	0.5	0.25	2	0.2
Length of supporting river (km)	3	21	15	4	2	1
Initial sediment load in supporting river (mg/dm^3)	200-400	150-500				
Suspended matter during high tide (mg/dm^3)	30 to 100	40 to 130				
Summer maximal temperature at low water	14	8	6		8	
Sedimentation ($g/m^2/day$)		130-260				26
Percent of silt-clay in surf. sediments	55 to 60	40 to 60	24	46	40	15 to 32
Daily salinity range (HW to LW)	0.5 to 30	2 to 28	5 to 30	15 to 30	10 to 30	
Organic matter content (%)	3	2 to 5				

Table 2. Mean macrofaunal density (individuals/ m^2) from 6 intertidal soft-sediment areas and meiofuanal density (individuals/ 10 cm^2) f rom 3 intertidal soft-sediment areas on Spitsbergen. Feeding types: D= deposit feeder, C=carnivore, S=suspension feeder.

Macrofaunal Taxa	Feeding Type	Mean Indiv. Dw (g)	Nottingham Bay[1]	Adventfjord[2]	Sassenfjord[3]	Colesbay[1]	Kapp Borthen[4]	Thiis Bay[3]
Ampharete acutifrons	*D*	1	0.1					
Ampharete finmarchica	*D*	1	0.1					
Anonyx sarsi	*C*	20	0.1					
Antinoella sarsi	*C*	30	1.5					
Astyris rosacea	*D*	30	0.1					
Brada granulata	*D*	20	7					
Calliopius laeviusculus	*C*	10	0.1					
Capitella sp	*D*	1			1130	212		4028
Chaetozone setosa	*D*	2						71
Chironomidae n.det	*D*	1		8				
Chone duneri	*S*	2	20					
Cylichna alba	*C*	20	2					
Cylichna scalpta	*C*	20	0.1					
Dendrodoa grossularia	*S*	25	41					
Eteone longa	*C*	2	0.1		565			
Eteone spetsbergensis	*C*	2	0.1					
Euchone rubrocincta	*S*	2						71
Fabricia sabella	*S*	2	0.1					
Gammarellus homari	*C*	5	0.1					71
Gammarus oceanicus	*C*	5	0.2				1500	
Gammarus setosus	*C*	5	0.1	6	71			
Halicryptus spinulosus	*D*	20	32					
Ischyrocerus anguipes	*C*	5	0.1					
Liocyma fluctuosa	*S*	5	190			70	50	
Lumbricillus spp	*D*	1		200	141	35		990
Lumbrinereis fragilis	*C*	5						
Mya truncata	*S*	50	2					
Nemertea n det	*C*	10	0.1					

Onisimus littoralis	*C*	2	4	600			2750	
Orchomene minuta	*C*	5	0.1					
Phascolosoma spp	*D*	10	0.1					
Polydora caulleryi	*D*	2						71
Polydora quadrilobata	*D*	2						353
Priapulus caudatus	*D*	15	23					
Pygospio elegans	*D*	1	0.1					
Scoloplos armiger	*D*	2			1356	600		1060
Spio filicornis	*D*	1						2120
Spionidae n.det.	*D*	1			5159	35	50	
Travisia forbesi	*D*	1	2					
Meiofaunal Taxa								
Nematoda			6459	1640	-	-	466	-
Harpacticoida			112	30	-	-		-
Copepoda nauplii			6869	267	-	-	3	-
Turbellaria				0.8	-	-		-
Oligochaeta			3	0.3	-	-		-
Tardigrada				0.9	-	-		-
Ostracoda			304	0.4	-	-		-

1 [9] [10] 2 [27] 3 [10] 4 [15]

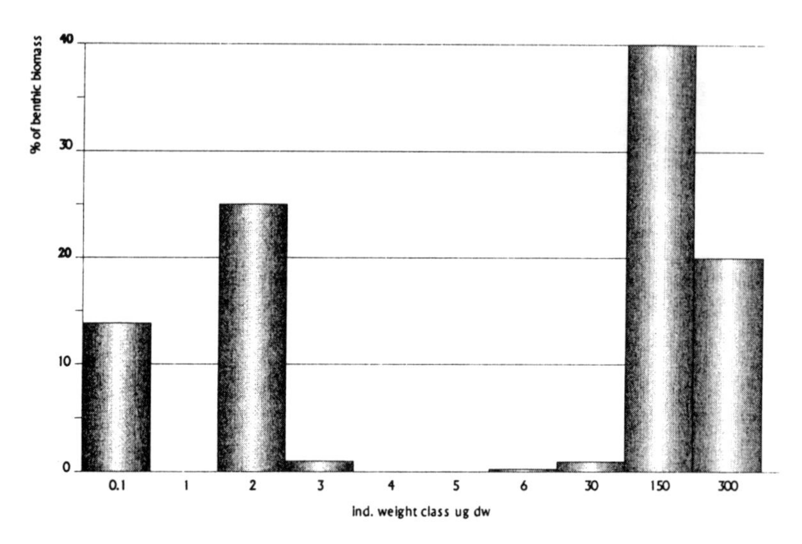

Fig.2 Percent share of organisms in different size (weight) classes in zoobenthos biomass, Adventfjorden tidal flat, July 1996

4.1 IS THERE A TYPICAL ARCTIC MUDFLAT COMMUNITY?

Infaunal abundances and species composition vary tremendously among the intertidal areas examined. Even at taxonomic levels higher than species, very different groups dominate localities: Polychaetes (Sassenfjord, Colesbay), Crustacea (Adventfjord), Priapulida (Nottingham Bay) (Figure 1). Intertidal soft-sediments exhibit very little similarity in faunal composition among locations compared with other marine habitats (*Laminaria* beds, subtidal soft-sediments, and rocky intertidal) on Svalbard that can exhibit a 60-70% similarity in taxonomic composition among sites (Weslawski unpub. data). A Low level of similarity is observed among glacier bays on Spitsbergen, where less than half of the species occurred in all localities [18].

Macrofauna and most of the meiofauna appear to die or leave intertidal areas during the period of fast ice. No animals were recovered from several macrofaunal samples taken from under fast ice cover in winter revealed no animals. Only nematodes, at one tenth of summer density, have been found under fast ice (Weslawski unpub. data). Wisniewska (pers. comm.) found Harpacticoids under the fast ice in Nottingham Bay. Other taxa appear after fast ice melt. Meiofauna recover quickly following disturbance [19] and meiofaunal taxonomic composition and density are similar among Spitsbergen mudflats [12].

Effects of ice on intertidal communities have received little attention. Intertidal mud flats in the Bay of Fundy (Canada) are consistently covered by ice in winter [20]. The top few centimetres of sediment may be frozen in ice [21] and blocks of ice moved by tidal currents leave tracks visible from the air [20]. Some taxa, particularly in the upper intertidal, experience winter mortality, presumably due to the combined effects of ice and low temperature [22]. Many taxa, however, are not influenced by winter conditions [21]. Where severe, ice scour can remove late successional species and allow the colonisation of opportunists [23] [24]. The intertidal of Svalbard certainly experiences sufficiently severe conditions that the seasonal removal of macrofauna can be expected.

Table 3. General characteristics of macro and meiobenthos from 6 intertidal soft-sediment habitats on Spitsbergen. Data were collected during the summer and from literature cited in Table 2.

	Nottingham Bay	Adventfjord	Sassenfjord	Colesbay	Kapp Borthen	Thiis Bay
Average biomass of macrofauna (g dw/m^2)	3.3	1.4	10.6	1.8	13.3	10.7
Average biomass of meiofauna (g dw/m^2)	1.7	0.7			1.2	
Average macrofauna density (indiv./m^2)	325	814	8422	952	4350	8834
Average meiofaunal density (indiv./10cm^2)	6000	2500			800	
Taxon dominating macrofaunal biomass	Priapulida	Crustacea	Polychaeta	Polychaeta	Crustacea	Polychaeta
Neighbouring marine benthic assemblage	Phytal zone	Deep sublittoral	Shallow soft sublittoral	Shallow soft sublittoral	Shallow hard bottom	Shallow soft sublittoral

All species listed in Table 1 have been recorded in the sublittoral. No exclusively intertidal species occur on Spitsbergen mudflats. This contrasts with Spitsbergen's rocky intertidal which is inhabited by *Littorina saxatilis, Semibalanus balanoides,* and *Gammarus setosus* that have only been recorded from the intertidal [15] [25].

The differences in macrofaunal abundance and composition among flats are likely caused by different patterns in yearly recolonisation. Most benthic species from latitudes as high as Svalbard might be expected to have direct development or lecitotrophic larvae [26]. There has been no study of the reproductive behaviour of Svalbard polychaetes and bivalves. Most of the larvae of benthic animals are present in coastal plankton from April to early June before the flats are ice-free [27]. While some larval colonisation of intertidal areas certainly occurs, many of the colonists may be juveniles transported from the subtidal. If this is true, than the nearby subtidal is likely an important source area for intertidal recolonization.

The subtidal adjacent to each mudflat is different among areas (Table 3). Nottingham Bay, with the richest faunal list, is adjacent to extensive, shallows composed of a mixture of hard- and soft-bottom, overgrown with dense *Laminaria* meadows [9].

The mudflat at Adventfjord is very poor in macrofaunal species diversity. It is flanked by a steep muddy slope, dropping from 0 to 40m in less than 500m distance. The shallow sublittoral zone (2-20m) adjacent to the Adventfjord intertidal is very narrow and lacks any macrophytes [15]. Tidal flats examined by Ambrose and Leinaas [10] (Colesbay, Thiis Bay, Sassenfjord) are probably recolonized from the nearby shallow sublittoral which supports a high diversity of polychaetes [28]. Consequently, the recolonisation of the flats examined, that takes place each year following ice melt, relies on source areas with very different community composition and very different macrofaunal assemblages develop on the adjacent tidal flats. A similar explanation has been offered to explain the composition of intertidal rock pools. Astles [29] found considerable faunistic differences among examined ponds and related each pool's faunal composition to the history of each locality. The importance of history in determining patterns of species distribution and abundance on sublittorial hard [30] and intertidal soft [31] substrates has long been recognised.

Other factors might contribute to the structure of intertidal communities on Spitsbergen. Shore birds are conspicuous predators on some Spitsbergen mudflats [32] and they can influence the distribution and abundance of intertidal fauna [33] [34] [35] [36]. Despite similarities in grain size among flats, flats may differ in sediment organic content and the abundance of benthic diatoms. The quantity and quality of food has been demonstrated to influence the structure of arctic subtidal soft-sediment communities [37] [38]. The impact of these and other factors in explaining patterns of distribution and abundance of infauna on arctic intertidal flats need to be examined.

In summary, tidal flats on Spitsbergen support seasonal communities inhabited by opportunistic species. Annual colonisation of the flats may be largely dependent on recruits from the adjacent sublittoral community that differ among locations in their physical and biological characteristics. Consequently, tidal flats on Spitsbergen support very different macrofaunal communities. Seasonally ephemeral communities are probably typical of arctic intertidal soft-sediments. Macrofaunal species community structure on these flats, like on Spitsbergen, are dependent on sublittoral communities for recruits. To the extent that arctic sublittoral source communities vary in composition, intertidal flats will harbour different suites of species.

5. Acknowledgements

We wish to express our thanks to Drs. Alexander Keck and Ole Jorgen Lönne for their help with the University Studies on Svalbard (UNIS) and IOPAS Co-operative Project "Arctic Tidal Flats" and to Dr. William G. Ambrose, Jr. and two referees for their comments on earlier drafts of this manuscript.

6. References

1 Ellis, D.V., Wilce, R.T. (1961) Arctic and subarctic examples of intertidal zonation, Arctic **14**: 224-235.

2 Brown, A.C. McLachlan, A. 1990. *Ecology of Sandy Seashores* Elsevier.328 pp.

3 Thorson, G. (1957) Bottom communities, Mem. Geol. Soc. Amer. **76**: 461-534.

4 Bek, T.A. (1990) Coastal zone in the White Sea ecosystem, Zhurnal Obscej Biologii, **51**: 116-124.

5 Sneli, J.-A. (1968) The intertidal distribution of polychaetes and molluscs on a muddy shore in Nord-More, Norway, Sarsia **31**: 63-68.

6 Steele, D.H. (1983) Marine ecology and zoogeography, in G. R. South (ed.) *Biogeography and ecology of the Island of New Foundland*, Dr W. Junk Publ. The Hague, pp. 421-465.

7 Steven, D.(1938) The shore fauna of Amerdloq fjord, West Greenland, Journ. Animal Ecology **7**: 53-70.

8 Ingolfsson, A. (1996) The distribution of intertidal macrofauna on the coasts of Iceland in relation to temperature, Sarsia **81**: 29-44.

9 Legezynska, E., Moskal, W., Weslawski, J.M., Legezynski, P. (1984) The influence of environmental factors on the distribution of bottom fauna in Nottingham Bay (Spitsbergen), Oceanografia **10**: 157-172.

10 Ambrose, W.G., Leinaas, H.P. (1989) Intertidal soft-bottom communities on the West Coast of Spitsbergen, Polar Biology **8**:393-395.

11 Rozycki, O., Gruzczynski, M. (1991) Quantitative studies on the infauna of an Arctic estuary, Nottinghambukta, Svalbard, Polish Polar Research **12**: 433-444.

12 Szymelfenig, M., Kwasniewski, S., Weslawski, J.M. (1995) Intertidal zone of Svalbard. 2 Meiobenthos density and occurrence, Polar Biology **15**: 137-141.

13 Feller, R.J., Warwick, R.M. (1998) Energetics, in R.P. Higgins, H. Thiel (eds.), *Introduction to the study of meiofauna*, Smith. Int. Press, Washington- London, pp. 181-196.

14 Berestovskij, E.G., N.A. Anisimova, S.G. Denisenko, E.N. Lunnova, W.M. Savinov, and S.F. Timofeev (1989) *The relation between size and weight of invertebrates and fish species from the North-East Atlantic*. Apatity, AN CCCP, 23 pp. (in Russian)

15 Weslawski, J.M., Wiktor, J., Zajaczkowski, M., Swerpel, S. (1993) Intertidal zone of Svalbard. 1. Macroorganisms distribution and biomass, Polar Biology **13**: 73-79.

16 Kendall, M.A., Warwick, R.M., Somerfield, P.J. (1997) Species size distributions in Arctic benthic communities, Polar Biology **17**: 389-392.

17 Sprung, M., Asmus, H. (1995) Does the energy equivalence rule apply to intertidal macrobenthic communities? Netherlands Journal of Aquatic Ecology **29**: 369-376.

18 Wlodarska-Kowalczuk, M., Weslawski, J.M., Kotwicki, L. (1998) Spitsbergen glacial bays macrobenthos - a comparative study, Polar Biology **20**: 66-73.

19 Christie, H., Berge, J.A. (1995) In situ experiments on recolonisation of intertidal mudflat fauna to sediment contaminated with different concentrations of oil, Sarsia **80**: 175-185.

20 Gordon, D.C. Jr., C. Desplanque (1983) Dynamics and environmental effects of ice in the Cumberland Basin of the Bay of Fundy. Canadian Journal of Fisheries and Aquatic Sciences. **40**:1331-1342.

21 Wilson, W.H. (1991) The importance of epibenthic predation and ice disturbance in a Bay of Fundy mudflat. Ophelia Supplement 5: 507-514.

22 Wilson, W.H. (1988) Shifting zones in a Bay of Fundy soft-sediment community: patterns and processes. Ophelia **29**: 227-245.

23 Lenihan, H.S. and J.S. Oliver 1995. Anthropogenic and natural disturbances to marine benthic communities in Antarctica. Ecological Applications **5**: 859-874.

24 Conlan, K.E., H.S. Lenihan, R.G. Kevitek, J.S. Oliver 1998. Ice scour disturbance to benthic communities in the Canadian High Arctic. Marine Ecology Progress Series. **166**:1-16.

25 Weslawski, J.M. (1990) Distribution and ecology of South Spitsbergen coastal marine amphipoda (Crustacea), Pol. Arch. Hydrobiol. **37**: 503-519.

26 Thorson, G. (1950) Reproductive and larval ecology of marine bottom invertebrates, Biological Reviews **25**:1-45

27 Weslawski, J.M., Kwasniewski, S., Wiktor, J., Zajaczkowski, M., Moskal, W. (1988) Seasonality of Spitsbergen fjord ecosystem, Polar Research **6**: 185-189.

28 Gromisz, S. (1993) Occurrence and species composition of Polychaeta (Annelida) in Hornsund fjord (South Spitsbergen), in K.W. Opalinski, R.Z. Klekowski (eds.) *Landscape, life world and Man in High Arctic*, Institute of Ecology Publishing Office, Warszawa, pp.199-206.

29 Astles, K.L. (1993) Patterns of abundance and distribution of species in intertidal rock pools, Journal Marine Biological Association United Kingdom **13**: 555-569.

30 Sutherland, J. (1981) The fouling community at Beaufort, North Carolina, a study in stability, American Naturalist **118**: 499-519.

31 Ambrose, W.G. Jr. (1984) Influence of residents on the development of a marine soft-bottom community. Journal of Marine Research **42**:633-654.

32 Leinaas, H.P., W.G. Ambrose, Jr. (1992) Foraging activity of the purple sandpiper, Calidris maritima, on a beach at Ny Ålesund, Spitsbergen. Cinclus **15**: 85-91.

33 Boates, J.S. and P.C. Smith (1979) Length-weight relationships, energy content and the effects of predation on Corophium volutator (Pallas) (Crustacean:Amphipoda). Proceedings Nova Scotian Institute Science **29**: 489-499.

34 Schneider, D.C. and B.A. Harrington (1981) Timing of shorebird migration in relation to prey depletion. Auk **98**: 801-811.

35 Quammen, M.L. (1984) Predation by shorebirds, fish, and crabs on invertebrates in intertidal mudflats: an experimental test. Ecology **65**: 529-537.

36 Wilson, W.H. (1994) The effects of episodic predation by migratory shorebirds in Grays Harbor, Washington. Journal of Experimental Marine Biology and Ecology. **177**: 15-25.

37 Grebmeier, J.M. and J.P. Barry (1991). The influence of oceanographic processes on pelagic-benthic coupling in polar regions: a benthic perspective. Journal of Marine Systems **2**: 495-518.

38 Ambrose, W.G., Jr. and P.E. Renaud (1995) Benthic response to water column productivity: evidence for benthic-pelagic coupling in the Northeast Water Polynya. Journal of Geophysical Research **100**:(C3):4411-4421

THE PROBLEM OF SCALE: UNCERTAINTIES AND IMPLICATIONS FOR SOFT-BOTTOM MARINE COMMUNITIES AND THE ASSESSMENT OF HUMAN IMPACTS

SIMON F. THRUSH, SARAH M. LAWRIE, JUDI E. HEWITT AND VONDA J. CUMMINGS
National Institute of Water and Atmospheric Research, P.O. Box 11 115, Hamilton, New Zealand.
E-mail: S.THRUSH@NIWA.CRI.NZ

Abstract

Natural ecological systems are heterogeneous in a manner that is rarely consistent from scale to scale. Patterns not apparent on a particular scale of observation may emerge when the sampling scales are varied. In this paper we focus on how the issue of scale relates to the generation and application of ecological information to help resolve environmental problems. We illustrate potential problems of matching the scale at which information is generated to that at which it is applied by discussing potential scale-dependence in the results of field experiments and the problems of identifying the large-scale effects of commercial fishing on marine benthic communities. We briefly describe the various approaches that can be used to assess scale effects and suggest the development of more integrative research programmes with studies conducted at different space or time scales. Inevitably for many large-scale environmental issues there is a lack of both appropriate controls and an ability to rigorously demonstrate effects. This has important implications for resource management concerning a trade-off between confidence and generality in environmental information and emphasises the need for a more integrative process of predicting and testing large-scale effects.

1. Introduction

An important goal for ecology is to develop an understanding of how physical and biological processes influence the distribution and abundance of individuals, populations and communities. Generating this understanding will lead to better predictions of future change to and risk assessments for ecological systems. In turn, this should lead to better environmental management. Ultimately the criteria for environmental health are changes in resident ecological systems. Thus developing an understanding of natural processes and how they are influenced by human activities is a significant component of environmental management, and needs considering along with social, economic and ethical issues.

Despite the pressing need to manage environmental change associated with human activity, it is often difficult to predict the details of change in marine soft-sediment ecosystems. These are complex systems driven by processes operating over a variety of space and time scales. We can document catastrophes adequately, but it is far more difficult to predict long-term, broad-scale changes. In many instances we simply do not have sufficient knowledge to adequately address environmental questions. For example, how do catchment management practices influence the ecology of a harbour? Are there functional ecological links between exploited finfish and their habitats? What are the impacts of removing many types of

J.S. Gray et al. (eds.), Biogeochemical Cycling and Sediment Ecology, 195–210.

predatory fish from coastal waters? Attempting to answer questions such as these raises problems with scaling-up from specific studies to address large-scale management issues [1] [2], or with isolating processes of interest from other sources of variation in large-scale studies [3] [4].

Identifying how the effects of ecological/environmental processes change with variation in spatial and temporal scale is one of the most important issues facing ecologists. In this paper we focus on how the issue of scale relates to the generation and application of ecological information to resolve environmental problems. However, the issue is also relevant to more theoretical aspects of ecology [5] [6] [7] [8].

We then highlight current limitations in ecological knowledge and implications of scale (particularly spatial scale) for future research. We also briefly address the implications of the scale issue for environmental management.

2. Why is scale an issue?

To address this question we must consider both the inherent complexity of natural systems and the limitations of both our knowledge and our methods of generating information. Scale is potentially an especially important issue when attempting to understand marine benthic ecosystems because of the importance of hydrodynamic processes over a variety of spatial and temporal scales [9] [10]. This is especially true for organisms in the semi-fluid soft-sediment habitats as a result of changes in their mobility associated with transitions between planktonic dispersal and settlement, post-settlement juvenile dispersal and more sedentary adult phases.

2.1.EFFECTS OF SCALE ON PERCEPTION.

A key feature of ecological systems is their heterogeneity. We generate hypotheses and theories as to the functioning of ecosystems on the basis of emergent patterns. If the structure and function of ecosystems were homogeneous in space and time, scale would not be so important an issue. But heterogeneity is an important functional component of ecological systems, not a nuisance that obscures underlying process [11] [12]. In many cases heterogeneity is the result of non-random processes, i.e., it is not white noise. Heterogeneity is rarely consistent from scale to scale; patterns not apparent on a particular scale of observations may emerge when the sampling scales are varied. A good example of how meaningful patterns may be revealed depending on scale of observation is provided by Schneider and Piatt [13]: correlations of abundance between the density of seabirds and schooling fish (their prey) only emerged at large spatial scales (2-6 km).

In Fig. 1 the effect of variations in spatial scale of sampling on the information we gain is shown by the types of patterns that emerge in the distribution of two common intertidal bivalves (*Macomona liliana* and *Austrovenus stutchburyi*). The spatial scales range from; 1. Suitable habitats in northern New Zealand; 2. density variation between sandflats in Manukau Harbour [14] ; 3. density variation within a 9000 m^2 plot (site AA in Thrush et al. [11]; 4 and 5. density variation in a 36 m^2, and 0.075 m^2 plots (sandy site in Hewitt et al. [15]). The information available changes as you move from one scale to another, e.g., at large scales the small-scale features of bivalve density distribution are unresolvable. Despite the different patterns that emerge at different levels of resolution, all are ultimately based on individual bivalves living in sandflats. From Fig. 1 it is clear that extrapolation from some sampling regimes, when knowledge of across and within scale variation is lacking, is potentially very limited.

If we are to recognise the limitations of our methods, determining how we perceive natural systems has to be considered in terms of the practical dimensions of the study design. To do this, it is helpful to separate three components of scale: *grain*, the area of an individual sampler (e.g., core or grab), *lag*, the inter-sample distance, and *extent*, the total area from which samples were collected (Fig. 2). These descriptions originate from the geostatistical literature (see [16]), but have been applied to ecological data (e.g. [5] [17] [18] [19]. Comparisons of these components lead to explicit description of the scope of the field study, i.e., the ratio of upper to lower limit of measurement [7]. Comparisons of scope enable a more precise comparison of one field study with another.

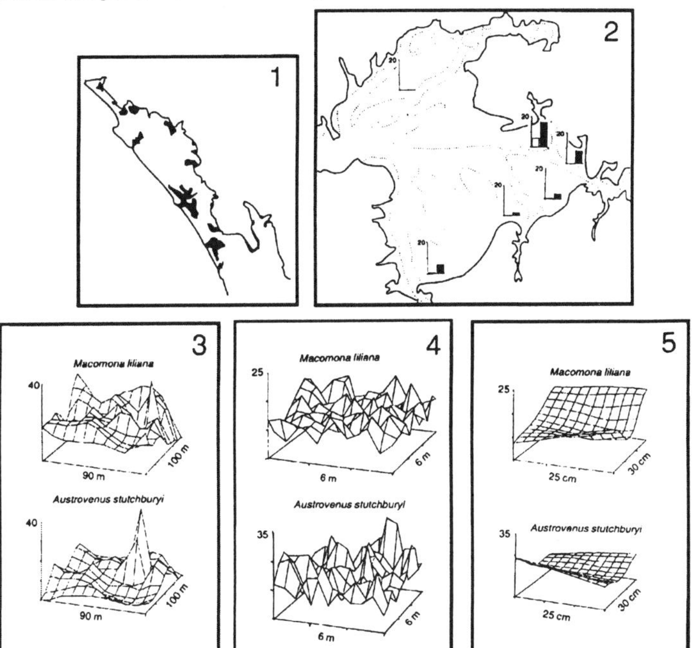

Fig. 1. Variations in pattern depend on scale of observation. 1. Northern New Zealand, shaded areas show the location of large sandflat habitats suitable for the common bivalves *Macomona liliana* and *Austrovenus stutchburyi*. 2. Density variation (mean number per 0.07 m^2) on six sandflats in Manukau Harbour, Auckland. 3. Density variation within a 9000 m^2 site. 4. Density variations within a 36 m^2 site. 5. Density variation within a 0.075 m^2 site.

Extent determines the amount and type of heterogeneity encompassed by a study. Thus seeking an increase in the generality of a study forces an increase in the extent. However, any field study involves the careful balance of confidence and generality [20]. We need to be confident in the information we generate, particularly as weak interactions, indirect and threshold effects are common, although often difficult to predict (e.g. [21] [22] [23] [24] [25]). To increase our confidence we increase replication and decrease lag. If the budget for the field work is fixed this inevitably leads to a decrease in extent. Making decisions about the trade-off between confidence and generality inevitably involve consideration of the spatial and temporal scales over which processes operate.

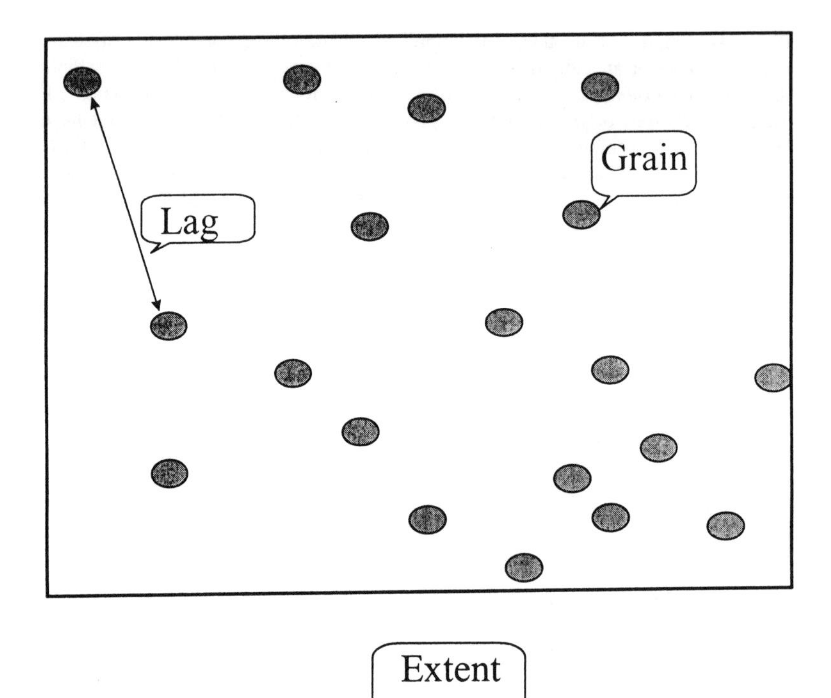

Fig. 2 Spatial sampling characteristics: grain, lag and extent.

2.2 SCALES OF PROCESSES

Variations in observed heterogeneity, within and across space (Fig. 1), emphasise the fact that ecological processes commonly act over a continuum. The demarcation of scales, either theoretically or by the practicalities of sampling, is merely a tool for categorisation and understanding. Also, while a particular process may act over different scales, its relative importance in contributing to emergent patterns can vary.

As the processes inferred in creating the patterns observed are dependant on the scale of observation, so the types of questions that can be addressed and the theory generated from the observations will also vary from scale to scale. Controversies in ecology often reflect comparisons of studies conducted over different scales. For example, the relative importance of active site selection vs passive deposition of larvae can at least partially be accounted for by the importance of these factors varying with spatial scale [26].

Although, physical and biological processes interact over many different scales, the influence of abiotic factors can usually be observed at large-scales (e.g., [19]). Generally biological interactions are thought to operate within the constraints imposed by large-scale abiotic factors. The upper spatial limits of biological processes are scaled to mobility, while the lower spatial limits are scaled by the organisms size [15] [27]. Both limits may change during the organisms life. For example, as larvae or juveniles, bivalves may be dispersed over large areas associated with currents or sediment transport, while as adults their mobility may be confined to comparatively small distances. Biological interactions such as competition, predation, disease, mutualism take place between individuals of differing size and mobility functioning under different physical conditions. This leads into the concept of functional heterogeneity [28] which acknowledges that organisms percieve/respond to their environment on multiple scales. Thus, while there may be scales which are inappropriate for the study of a particular process, there is no single correct scale at which to study [8] [29].

2.3 PREDICTING IMPACTS AND THE CONFOUNDING EFFECTS OF SCALE

Field experiments have proved to be a very useful technique in developing a mechanistic understanding of some ecological processes. There is an extensive body of literature on both their design (e.g., [30] [31] [32] [33]) and their application to marine environmental management (e.g., [34] [35] [36]). However, the experimental approach typifies the problem of accepting confidence at the expense of generality. All field studies, but particularly experiments, are invariably conducted over limited space and time scales (e.g., [29] [37] [38]) and we must question how far the results can be extrapolated both in terms of extent and grain. For experiments, the potential problems with transient dynamics, indirect effects, artefacts, environmental variability and site history must be considered (e.g, [39] [40] [41]. Even where processes are operating on scales amenable to experimental manipulation, individual experiments frequently are conducted at only one location and the results of similar experiments in different locations are not always consistent [42]. Furthermore, even small changes in experimental grain (plot size) can produce different results [43] [44] [45]. These potential problems are most likely to be exacerbated with experiments designed to directly examine large-scale impacts as these are often difficult to design stringently. In fact most experiments are likely to under-estimate or miss large-scale effects [46].

Nevertheless, using experiments to study the response of benthic communities to disturbance has made some important contributions to benthic ecology and our ability to assess environmental impacts. Such studies have provided insight into likely direct and indirect effects, the potential for recovery and the implications of disturbance-generated patchiness for the structure and stability of populations and communities. However, when the scale of disturbance used in experiments is different from that of the process potentially generating an impact, the information generated from the experiments is mis-applied if it is just multiplied-up to predict large-scale effects (Fig. 3). In the rest of this section we illustrate both this point, and some strengths and weaknesses of the experimental approach.

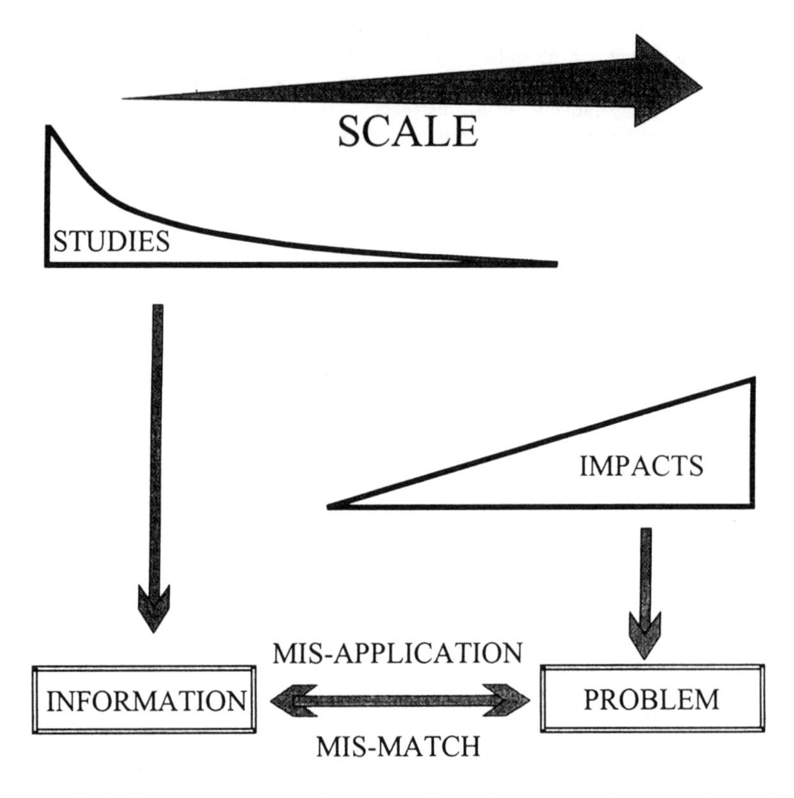

Fig. 3 Disparity between scales of research and human impacts can lead to mis-match and mis-application of information to problems.

Based on a review of disturbance/recovery experiments in soft sediments Hall et al. [38] concluded that defaunated areas of the order of 1 m^2 would be expected to recover in about 1 month. However, much slower rates of macrofaunal recolonisation into experimental patches were apparent in an experiment conducted in a dynamic sandflat habitat [44], with macrofaunal assemblages in 0.2 m^2, 0.8 m^2 and 3.2 m^2 plots still different (> 50% Bray-Curtis dissimilarity) from the ambient assemblage after 9 months. This was surprising given that, on sandflats, large numbers of post-larvae are often transported with sediment bedload [47] [48]. Differences were observed in the rates of recolonisation between different plot sizes, emphasising the problem of directly extrapolating from small-scale experiments to predict recovery times for larger disturbances. However, this experiment also demonstrated that sediment instability increased with increasing plot size and emphasised that emigration from recovering patches may significantly slow the rates of recolonisation. It thus elucidated some important factors (patch size, emigration, recovery time and interactions between hydrodynamic conditions and habitat stability) that need to be incorporated in models used to predict recovery processes in large-scale disturbances. In particular it demonstrated that the destruction of organisms that maintain habitat stability is likely to result in very slow recovery dynamics, especially in wave disturbed habitats.

Disturbance events may have both positive and negative effects on marine communities. While the specifics of individual events are important, the interpretation of the consequences of disturbance events will also depend on the spatial and temporal scale of concern (see [38] [49]). For example, ecological theory tells us that small-scale disturbance events at intermediate frequencies may have beneficial effects on the system at larger scales [50]. This is because spatial heterogeneity within communities is frequently viewed as a mosaic of patches that has the potential to increase diversity and lead to population, community and ecosystem stability over large scales (e.g., [28] [51] [52] [53]). However, disturbances on large scales (e.g., habitat disturbance by fishing which removes surface dwelling organisms and modifies sediment topography) are much more likely to reduce heterogeneity in benthic communities.

Problems with predicting ecosystem effects can be illustrated by looking at habitat disturbance by commercial fishing, possibly the largest direct human disturbance which impacts benthic communities, [54] [55]. Over time, repeated disturbance will select for species with facultative responses to disturbance and communities are likely to become dominated by juvenile stages, highly mobile species and rapid colonist/opportunist species. These impacts may also be long-term, resulting in ecosystem-wide changes (e.g., Wadden sea [56], 1982; English Channel [567; Australian North West Shelf [58]). Issues which need to be considered include: the damage or death of organisms other than the target species; changes in the density and type of predators; and habitat modifications (particularly removal of larval settlement sites, nursery areas and reduced heterogeneity). The broader context of modifications to sediment texture and microbial activity [59] [60] and resuspension of contaminants or increases in benthic/pelagic nutrient fluxes [61] also need to be considered. Thus, given its spatial extent, fishing has the potential to affect marine communities on a global scale. This is not only a conservation issue, exploited species are integral components of natural systems and the broader effects of their exploitation are ultimately likely to have important ramifications for the management and sustainability of industrial fisheries.

While the potential for negative effects of such impacts appear obvious, evaluating magnitudes of specific effects has been difficult and the evidence attributing changes to fishing has been considered equivocal. The large space and time scales over which this industry operates further complicate the situation. Generally it has not been possible to conduct studies with a very high potential for demonstrating cause and effect because, historically, management activity has not incorporated sufficient lead-in time to collect before impact data and appropriate reference areas have not been reserved.

With the lack of unifying theories concerning the functioning marine benthic systems, despite extensive experimentation particularly over the last 25 years, the implication is we should be quite cautious in making generalisations. This is particularly important in applied ecology where information generated from small scale experiments or through case studies often needs to be used in the assessment of human impacts over large space and time scales [62].

3. Ways of addressing scale effects

So far we have attempted to demonstrate the importance of recognising scale in terms of natural processes, study methods, anthropogenic impacts and the associated problems in addressing questions of change in the marine environment. Next we assess some of the tools used by ecologists in attempting to address issues of scale.

3.1 DESCRIPTIVE STUDIES AND THE ANALYSIS OF PATTERN

Studies that describe patterns in space or time are a very important first step, and lead to the development of better hypotheses. Direct analysis of variability has proved to be a useful tool in assessing the importance of processes operating within a system [63] [64] [65]. Identifying patterns at various spatial scales may provide a clue as to the kinds of processes that operate at that particular scale. This, in turn, may help develop more specific hypotheses for testing or may reduce the number of competing hypothesis. In the long-term this will allow the development of models of the action and interaction of processes operating on different scales [5] [66] [67]. Description of pattern by describing heterogeneity can also help in determining representative study areas [1] [3] [68] and appropriate sampling schemes [18][69] [70].

Fractal geometry is an increasingly popular tool for analysing the structure of spatial variance. The fractal dimension (D) is often used to indicate the relative importance of large-scale versus small-scale controlling factors [71] [72]. Low values of D are an indication of the dominance of large-scale processes, whereas high values suggest complex interacting processes and/or localised variations. One of the important concepts of fractal geometry is self-similarity, i.e., a structure examined at increasingly larger resolution shows details that display the same variations as were seen at smaller resolutions. If analysis of a structure (e.g., spatial) revealed this sort of behaviour we assume that it was caused by processes operating in the same way over all scales. If this is not the case, and it rarely is, the implication is that we can not make conclusions about one scale based on what happens at another. However, the value of the fractal dimension at different scales can be used to indicate the presence and relative importance of scales of variation [71]. Fractal geometry can also be used as a modelling tool to test predictions about interacting processes and their relative importance. Fractal geometry is one way that ecologists can view landscapes at multiple scales and thereby achieve predictability in the face of complexity [73] [74] [75].

3.2 MODELLING WITH SIMPLIFICATION.

When developing a model of current patterns within a harbour, hydrodynamic modellers usually measure current velocities at a few places for a period of time and then assume that processes are consistent over space and time. Current velocity, generated by two fundamental forces (gravity and friction), is assumed homogeneous within each model grid square to provide the means for larger scale predictions. Benthic ecologists with extensive knowledge of local variability would not claim such consistency for most ecological processes that operate at the population and community levels. Yet where patterns of distribution and abundance are strongly and directly coupled to physical processes, simple models can be effectively used within the bounds of our understanding of the physical processes themselves. Modelling can also be used to incorporate heterogeneity both within and across scales. However, predictive power diminishes as the potential for complex interactions increases.

3.3 LONG-TERM STUDIES

Long-term study sites and monitoring programmes can provide very important information, but there is usually a trade off in effort such that long-term studies are not generally well replicated in space. Lack of replication in space gives rise to the problem of generalising from single locations. However, identifying major controls on the ecology of the system can help to elucidate to what extent site specific information is generalizable [76]. Also, for applied monitoring programmes the need to generalise may be less of an issue because (by definition) the study is focused on an area of specific interest.

3.4 *SCALING EXPERIMENTS*

Another approach is to incorporate some component of scale, within experiments at the treatment level, to test questions about the scale dependence of processes (e.g., [44] [45] [77] [78]). As we have previously discussed, changes in grain have important implications for disturbance/recovery studies and provide information relevant to a number of ecological issues. Of course there are limitations to how far this approach can be taken before the experiment becomes logistically and ethically impossible. Sampling experimental plots over time provides information on the changes in ecological response to manipulations through time [79]. Combining the results of such field experiments with information generated by theoretical studies produces some important generalisations that enable us to make predictions of potential environmental effects.

3.5 *DIMENSIONAL ANALYSES AND SCALING-UP FROM PREDICTIONS*

Dimensionless ratios can be used to predict the relative importance of processes at various scales [80] [81] [82] [83]. Dimensionless ratios are plotted as a function of spatial and temporal scale to produce a summary diagram from which the spatial and temporal scales over which particular processes are important can be determined. However, the technique has not been commonly applied to questions in benthic ecology (but see [84]). Recently a method has been proposed to allow some objective determination of the limits to which the effect of processes can be scaled-up [27]. Critical to this process is natural history information, both in understanding cause and effect relationships and making predictions.

3.6 *THE ROLE OF NATURAL HISTORY INFORMATION*

Assembling available natural history information about the temporal and spatial scales over which some processes are likely to have a major influence is crucial; firstly to decide at what scales we may (or may not) study particular processes and secondly to decide how natural processes may interact with human disturbances.

We cannot determine one scale or even several separate scales that will be exactly appropriate for addressing a particular question, but we can recognise inappropriate scales. We can also recognise what types of questions can be addressed and what predictions can be made. For benthic systems dimensional analysis indicates that determining potential scales of movement of organisms is critical as these will provide clues as to whether birth or death processes will be apparent at particular scales or will be swamped by fluxes [27].

If reasonably detailed information on species/community aggregations, temporal dynamics, mobility and functioning is available, more effective studies can be designed. In particular this information can allow us to balance the experimental approach with studies on large spatial and long temporal scales that can provide the big picture. Fitting experiments within a basic understanding of the spatial and temporal dynamics of the system being studied can further the interpretation of experimental effects and help generalisations and predictions to be made (e.g., [24]). Good ecological information enables development of studies sensitive to scaling issues. It enables us to build a picture of ecological functioning by identify species distributions along key large-scale environmental gradients and identifying local variation associated with biotic interactions.

3.7 *GRADIENT STUDIES*

Although processes may interact over a variety of scales, the role of abiotic factors is especially important at larger scales (e.g., habitat structure, oceanography, primary productivity gradients, temperature). Ordering studies down large-scale abiotic gradients enables us to gain an understanding of the changing role of small scale processes with increasing extent [85]. When large-scale abiotic gradients can be isolated, generality can not only be assessed but the role of particular processes in accounting for site-to-site variation tested. This understanding could not be achieved by simple random sampling. Even where predictions can not be made *a priori*, sampling down spatial gradients enables indications of the scale of processes to be postulated [86].

4. Summary - synthesising studies from different scales.

The variability of ecological responses with scale emphasise that it is not always possible to do the one definitive, critical experiment. To provide information on how particular processes operate over various spatial and temporal scales, and how they influence the ecology of populations or communities, ideally we would study the process and its direct and indirect effects over all possible scales. But this is not possible in most situations. As is the case with most techniques applied to environmental problems, there is no one practical approach that is a panacea for all the potential problems. An integrative approach that uses a variety of techniques should be used whenever possible (Fig. 4). But an important question is, how do we formally integrate these different kinds of studies? We need to develop ways to integrate small-scale experimental studies into large-scale spatial mosaics [8] [87] [88]. Developing a quantitative synthesis of pattern and process will involve an iterative checking among pattern, process and prediction [7] [8] [27] [87] [89] [90]. Whatever approach or combination of approaches is taken, it is important to focus questions appropriately and to explicitly state the spatial and temporal scale of the studies and thus acknowledge their limitations.

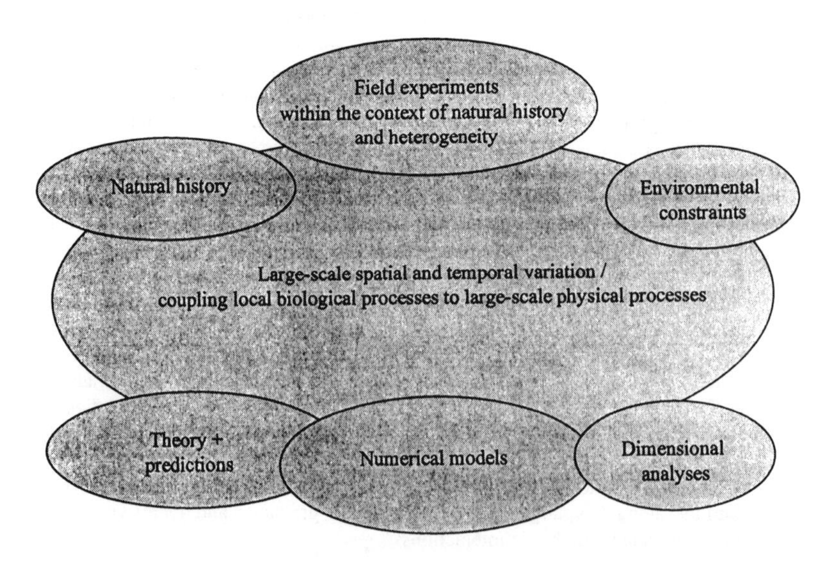

Fig. 4 Components of a multi-scaled study programme.

Study design is not only concerned with adequate sample size and replication but also relevance to the problem and a realisation of the limits to extrapolation.

5. Current and future implications for environmental management of coastal marine and estuarine resources

There are three basic ways in which scientific research and information can help in the wise management of coastal resources. Firstly there are the specific problems of today where focussed-research is needed to provide information. Secondly, information can be used to assess and prioritise risks to the environment. But, thirdly, there are also the environmental issues of the future. It is important we remember that many of the important environmental issues of today (e.g., pesticides, global warming) were not discovered by focused, outcome-orientated research programmes.

For many large-scale environmental problems, conclusions need to be reached despite the lack of adequate controls and demonstrated experimental responses. It is important to realise that conclusively identifying cause and effect relationships in the traditions of experimental science will not be feasible because the necessary comparisons will not be available. Studies that do not show significant results need to be interpreted cautiously; the lack of significance may be due to a lack of statistical power or to the fact that the study was conducted at the wrong scale for the questions asked. Explicit specification of the power and the scope of the study, together with the level of extrapolation would allow a balanced risk assessment of various options to be made. The issue then becomes how best to assess the risk that certain paths of interaction will produce deleterious ecological changes. The role of science is not only to provide information on the environment but also to inform management of the risks of different options [54] [61] [91].

Once predictions of large scale effects have been made (e.g., impact statements and environmental assessments), appropriately designed and cost-effective surveys should be used to test predictions about the large scale human impacts. In many cases, this may provide more useful and appropriate information than elegant but small-scale experiments. However, identifying and ultimately understanding large-scale effects is not simply a case of conducting large-scale surveys. Although large-scale abiotic factors set the boundary conditions under which ecological processes operate, with the exception of obvious disaster areas much of the information concerning large-scale effects comes from an understanding of natural history and fine-scale details [23]. Continuing to generate information on the natural history of organisms and individual system constraints are important. This information is difficult to obtain and often requires intensive and long-term research efforts, but it is essential for making predictions that are sensitive to issues of scale.

Little is known of how processes on one scale can initiate changes that cascade through ecological systems. For example, to address issues of global change we need to be able to answer questions of how large scale process will influence ecological processes on local scales and how these effects will vary from place to place within the system as a whole. If the simple situation that large-scale patterns were only controlled by physical processes were true, then we should be in a much better position to predict. However, emergent patterns generated by local processes are important, much of the large-scale variability can be driven by smaller scale processes [64] [92]. Ecological systems are complex systems with non-linear interactions, feedbacks, lags and priority effects [41] [93] [94].

Information generated from comparatively pristine systems is important to understanding effects in potentially degraded ones. Assessments of environmental effects centred only on the study of disturbed areas has been described by Loehle [95] as analogous to early psychologists defining human behaviour largely from studies of the mentally ill. Monitoring of ecological change enables temporal variability to be quantified; in non-impacted, reference areas this information can act as signposts of large-scale change.

Environmental management activities can also provide the opportunity for ecologists to directly assess large-scale effects, using different management strategies in restricted areas as the basis of large experiments. With a sustained commitment by managers, resource users and scientists, this approach would allow predictions of ecological effects to be tested and much needed information on large-scale effects would be gained [96] [97].

We suggest that a more strategic long-term approach to generating information must be taken, and this needs to involve sympathetic collaboration between resource managers, planners and scientists. Consequently the differentiation of research activities into *pure* and *applied* is irrelevant to resolving these environmental questions. An approach exploiting the synthesis of a large multidisiplinary programme integrated across spatial scales will be relevant in most cases. However, it must be recognised that environmental managers often need to quickly make decisions about areas for which there is little background information or the time or money to generate it. In this case a predictive assessment of risks with a commitment to monitoring effects is probably the best that can be managed.

Fragmentary management of environmental problems, whether at regional or global scales, is no longer adequate because of the potential for cumulative and aggregative effects. An ecosystem-wide perspective is thus appropriate for the management of marine resources.

Throughout this section we have emphasised the need to produce predictions relevant to the large scale. We need to keep in mind the cost/effort balancing act which requires trade-offs between confidence and generality. This determines whether too much emphasis is placed on certainty (leading to more and more being known about less and less) or on generality (leading to less and less being known about more and more).

6. Conclusions

We hope that the current recognition of the importance of incorporating scale into studies in the ecological literature is more than a passing fad. Recognising that processes are constrained to working under certain conditions and in different ways on certain scales has important implications for developing a more integrated set of ecological theories. As many of our theories and hypotheses are based on the observation of patterns, we need to be aware of the heterogeneities in systems and how they vary with scale. Recognition of the importance of patchiness and the potential for changes of patterns with increasing spatial or temporal scale is not new and has provided important insights (e.g., [5] [9] [98]. Powell [99] considers the issue of scale to be especially important in theoretical ecology because it holds out the prospect for simplification. Our discussion, although focused on applied ecological issues, therefore relates to the most fundamental development of ecological theory and understanding.

In highlighting the importance of an explicit recognition of scale issues in marine ecology, there are many specific issues that we have not addressed (e.g., the implications of habitat fragmentation, the design and creation of marine reserves and the potential for loss of biodiversity). Scale is relevant to all of these, particularly in recognising the actual extent over which ecological processes function for different organisms in marine systems. We have been principally concerned with how issues of scale influence assessment of effects and recognising the need to integrate this factor into marine resource management. Important questions facing both ecologists and environmental managers are: identifying the limits to extrapolation from small scale studies, how processes interact across scales and how we develop new techniques for incorporating scale into ecology and environmental management.

Studies conducted to provide information for environmental management should either be designed to enable extrapolation across scales or be focused at appropriate scales. This implies that studies should be appropriately timed and sufficiently and consistently funded to encompass issues of scale. In some cases, large interdisciplinary multi-scale studies will be appropriate, while in other situations focused pertinent questions can be used to derive feasible and appropriately scaled studies.

Our conclusions beg for a commitment to broadly based studies of the environmental and ecological interactions that are degrading marine benthic systems, particularly in coastal environments.

7. Acknowledgements:

This paper has benefited from discussions with Malcolm MacGarvin, Bob Whitlatch, Tuck Hines, Paul Dayton, Pierre Legendre, Dave Schneider, Rick Pridmore, Mal Green, Dave Raffaelli, Richard Warwick, Mike Graham and Enric Sala. SFT thanks John Gray for the opportunity to discuss these ideas at the Hel workshop.

8. References

1 Livingston, R. J. (1987) Field sampling in estuaries: The relationship of scale to variability. Estuaries **10**, 194-207.

2 Livingston, R. J. (1991) Historical relationships between research and resource management in the Apalachicola River estuary. Ecol. Apps **1**, 361-382.

3 Thrush, S. F., Pridmore, R. D. and Hewitt, J. E. (1994a) Impacts on soft-sediment macrofauna: The effects of spatial variation on temporal trends. Ecol. Apps **4**, 31-41.

4 Thrush, S. F., Hewitt, J. E., Cummings, V. J., Dayton, P. K., Cryer, M., Turner, S. J., Funnell, G. A., Budd, R., Milburn, C. J. and Wilkinson, M. R. (1998) Disturbance of the marine benthic habitat by commercial fishing: impacts at the scale of the fishery. Ecol Apps. **8**, 866-879.

5 Wiens, J. A. (1989) Spatial scaling in ecology. Funct. Ecol. **3**, 385-397.

6 Giller, P. S., Hildrew, A. G. and Raffaelli, D. (1994) *Aquatic ecology: scale, pattern and process*, Blackwells Scientific, Oxford, 649 p.

7 Schneider, D. C. (1994a). *Quantitative ecology: spatial and temporal scaling*. Academic Press, San Diego, 395 p.

8 Thrush, S. F., Schneider, D. C., Legendre, P., Whitlatch, R. B., Dayton, P. K., Hewitt, J. E., Hines, A. H., Cummings, V. J., Lawrie, S. M., Grant, J., Pridmore, R. D. and Turner, S. J. (1997a) Scaling-up from experiments to complex ecological systems: Where to next? J. Exp. Mar. Biol. Ecol. **216**, 234-254.

9 Dayton, P. K. and Tegner, M. J. (1984) The importance of scale in community ecology: A kelp forest example with terrestrial analogs. in Price, P. W., C. N. Slobodchikoff and W. S. Gaud (eds), *A new ecology: Novel approaches to interactive systems*. John Wiley and Sons, New York, pp. 457-483.

10 Barry, J. P. and Dayton, P. K. (1991) Physical heterogeneity and the organisation of marine communities. in K. Kolasa, and S. T. A. Pickett (eds), *Ecological heterogeneity*. Springer-Verlag, New York, pp. 270-320.

11 Thrush, S. F. (1991) Spatial patterns in soft-bottom communities. TREE **6**, 75-79.

12 Legendre, P. (1993) Spatial autocorrelation: Trouble or new paradigm? Ecology **74**, 1659-1673.

13 Schneider, D. C. and Piatt, J. F. (1986) Scale-dependent correlation of seabirds with schooling fish in a coastal ecosystem. Mar. Ecol. Prog. Ser. **32**, 237-246.

14 Pridmore, R. D., Thrush, S. F., Hewitt, J. E. and Roper, D. S. (1990) Macrobenthic community composition of six intertidal sandflats in Manukau Harbour. N.Z. J. Mar. Freshwat. Res. **24**, 81-96.

15 Hewitt, J. E., Thrush, S. F., Cummings, V. J. and Pridmore, R. D. (1996) Matching patterns with processes: Predicting the effect of size and mobility on the spatial distributions of the bivalves Macomona liliana and Austrovenus stutchburyi. Mar. Ecol. Prog. Ser. **135**, 57-67.

16 Isaaks, E. H. and Srivastava, R. M. (1989) *Applied geostatistics*. Oxford University Press, Oxford, 561 p.

17 He, F., Legendre, P., Bellehumeur, C. and LaFrankie, J. V. (1995). Diversity pattern and spatial scale: A study of a tropical rain forest of Malaysia. Environ. Ecol. Stat. **1**, 265-286.

18 Hewitt, J. E., Legendre, P., McArdle, B. H., Thrush, S. F., Bellehumeur, C. and Lawrie, S. M. (1997) Identifying relationships between adult and juvenile bivalves at different spatial scales. J. Exp. Mar. Biol. Ecol. **216**, 77-98.

19 Legendre, P., Thrush, S. F., Cummings, V. J., Dayton, P. K., Grant, J., Hewitt, J. E., Hines, A. H., McArdle, B. H., Pridmore, R. D., Schneider, D. C., Turner, S. J., Whitlatch, R. B. and Wilkinson,

M. R. (1997) Spatial structure of bivalves in a sandflat: Scale and generating processes. J. Exp. Mar. Biol. Ecol. **216**, 99-128.

20 Schneider, D. C. (1994b). Scale-dependent patterns and species interactions in marine nekton. in P. S. Giller, A. G. Hildrew and D. Raffaelli (eds), *Aquatic ecology: scale, pattern and process*. Blackwell Scientific, Oxford, pp. 441-467.

21 Hall, S. J., Raffaelli, D. and Turrell, W. R. (1990) Predator-caging experiments in marine systems: A reexamination of their value. Am. Nat. **136**, 657-672.

22 Kneib, R. T. (1991) Indirect effects in experimental studies of marine soft-sediment communities. Am. Zool. **31**, 874-885.

23 Dayton, P. K. (1994) Community landscape: Scale and stability in hard bottom marine communities. in P. S. Giller, A. G. Hildrew and D. Raffaelli (eds), *Aquatic ecology: scale, pattern and processes*. Blackwell Scientific, Oxford, pp. 289-332.

24 Thrush, S. F., Pridmore, R. D., Hewitt, J. E. and Cummings, V. J. (1994b) The importance of predators on a sandflat: Interplay between seasonal changes in prey densities and predator effects. Mar. Ecol. Prog. Ser. **107**, 211-222.

25 Hines, A. H., Whitlatch, R. B., Thrush, S. F., Hewitt, J. E., Cummings, V. J., Dayton, P. K. and Legendre, P. (1997) Nonlinear foraging response of a large marine predator to benthic prey: Eagle ray pits and bivalves in a New Zealand sandflat. J. Exp. Mar. Biol. Ecol. **216**, 191-210.

26 Butman, C. A. (1987) Larval settlement of soft-sediment invertebrates: The spatial scales of pattern explained by active habitat selection and the emerging role of hydrological processes. Oceanogr. Mar. Biol. Annu. Rev. **25**, 113-165.

27 Schneider, D. C., Walters, R., Thrush, S. F. and Dayton, P. K. (1997) Scale-up of ecological experiments: Density variation in the mobile bivalve Macomona liliana Iredale. J. Exp. Mar. Biol. Ecol. **216**, 129-152.

28 Kolasa, J. and Rollo, C. D. (1991) Introduction: the heterogeneity of heterogeneity: A glossary. in J. Kolasa and S. T. A. Pickett (eds), *Ecological heterogeneity*. Springer Verlag, New York, pp. 1-23.

29 Levin, S. A. (1988) Pattern, scale and variability: An ecological perspective. in A. Hastings (ed.), *Community ecology, lecture notes in biomathematics*. Springer-Verlag, New York, pp. 1-12.

30 Mead, R. (1988) The design of experiments, statistical principles for practical application. Cambridge University Press, Cambridge, 620 p.

31 Hairston, S. N. G. (1989) *Ecological experiments: purpose, design and execution*. Cambridge University Press, Cambridge, 370 p.

32 Underwood, A. J. (1991) Beyond BACI: Experimental designs for detecting human environmental impacts and temporal variation in natural population. Aust. J. Mar.Freshwat. Res. **42**, 569-587.

33 Dutilleul, P. (1993) Spatial heterogeneity and the design of ecological field experiments. Ecology **74**, 1646-1658.

34 Underwood, A. J. and Peterson, C. H. (1988) Towards an ecological framework for investigating pollution. Mar. Ecol. Prog. Ser. **46**, 227-234.

35 Underwood, A. J. (1992) Beyond BACI: The detection of environmental impact on populations in the real, but variable, world. J. Exp. Mar. Biol. Ecol. **161**, 145-178.

36 Green, R. H. (1993) Application of repeated measures designs in environmental impact and monitoring studies. Aust. J. Ecol. **18**, 81-98.

37 Kareiva, P. and Andersen, M. (1988) Spatial aspects of species interactions: The wedding of models and experiments. in A. Hastings (ed.), *Community ecology, lecture notes in biomathematics*. Springer-Verlag, New York, pp. 38-54.

38 Hall, S. J., Raffaelli, D. and Thrush, S. F. (1994) Patchiness and disturbance in shallow water benthic assemblages. in P. S. Giller, A. G. Hildrew and D. Raffaelli (eds), *Aquatic ecology: Scale, pattern and processes*. Blackwell Scientific, Oxford, pp. 333-375.

39 Dayton, P. K. and Oliver, J. S. (1980) An evaluation of experimental analysis of population and community patterns in benthic marine environments. in K. R. Tenore and B. C. Coull (eds), *Marine benthic dynamics*. University of South Carolina Press, Columbia, pp. 93-120.

40 Tilman, D. (1988) Ecological experimentation: Strengths and conceptual problems. in G. E. Likens (ed.) *Long-term studies in ecology, approaches and alternatives*. Springer-Verlag, New York, pp. 136-157.

41 Brown, J. H. (1995) *Macroecology*. University of Chicago, Chicago, 269 p.

42 Thrush, S. F., Hewitt, J. E., Pridmore, R. D. and Cummings, V. J. (1996a) Adult/juvenile interactions of infaunal bivalves: contrasting outcomes in different habitats. Mar. Ecol. Prog. Ser. **132**, 83-92.

43 Smith, C. R. and Brumsickle, S. J. (1989) The effect of patch size and substrate isolation on colonization modes and rate in an intertidal sediment. Limnol. Oceanogr. **34**, 1263-1277.

44 Thrush, S. F., Whitlatch, R. B., Pridmore, R. D., Hewitt, J. E., Cummings, V. J. and Maskery, M. (1996b) Scale-dependent recolonization: The role of sediment stability in a dynamic sandflat habitat. Ecology **77**, 2472-2487.

45 Whitlatch, R. B., Hines, A. H., Thrush, S. F., Hewitt, J. E. and Cummings, V. J. (1997) Benthic faunal responses to variations in patch density and patch size of a suspension-feeding bivalve inhabiting a New Zealand intertidal sandflat. J. Exp. Mar. Biol. Ecol. **216**, 171-189.

46 Tegner, M. J., Dayton, P. K., Edwards, P. B. and Riser, K. L. (1997) Large-scale, low-frequency

oceanographic effects on kelp forest succession: A tale of two cohorts. Mar. Ecol. Prog. Ser. **146**, 117-134.

47 Emerson, C. W. and Grant, J. (1992) The control of soft-shell clam (Mya arenaria) recruitment on intertidal sandflats by bedload sediment transport. Limnol. Oceanogr. **36**, 1288-1300.

48 Commito, J. A., Thrush, S. F., Pridmore, R. D., Hewitt, J. E. and Cummings, V. J. (1995) Dispersal dynamics in a wind-driven benthic system. Limnol. Oceanogr. **40**, 1513-1518.

49 Caswell, H. and Cohen, J. E. (1991) Communities in patchy environments: A model of disturbance, competition and heterogeneity. in J. Kolasa and S.T.A. Pickett (eds) *Ecological heterogeneity*, Springer-Verlag, New York, pp. 92-122.

50 Connell, J. H. (1978) Diversity in tropical rainforests and coral reefs. Science **199**, 1302-1310.

51 Johnson, R. G. (1970) Variation in diversity within benthic marine communities. Am. Nat. **104**, 285-300.

52 De Angelis, D. L. and Waterhouse, J. C. (1987) Equilibrium and nonequilibrium concepts in ecological models. Ecol. Monogr. **57**, 1-21.

53 Pimm, S. L. (1991) *The balance of nature: ecological issues in the conservation of species and communities.* The University of Chicago Press, Chicago, 434 p.

54 de Groot, S. J. (1984) The impact of bottom trawling on benthic fauna of the North Sea. Ocean Management **9**, 177-190.

55 Dayton, P. K., Thrush, S. F., Agardy, T. M. and Hofman, R. J. (1995) Environmental effects of fishing. Aquat. Conserv. Mari. Freshwat. Ecosyst. **5**, 205-232.

56 Riesen, W. and Reise, K. (1982) Macrobenthos of the subtidal Wadden sea: Revisited after 55 years. Helgolaender Meeresunters. **35**, 409-423.

57 Holme, N. A. (1983) Fluctuations in the Benthos of the Western Channel. Oceanologica Acta, Proc. 17th European Marine Biological Symposium, 121-124.

58 Sainsbury, K. J. (1988) The ecological basis of multispecies fisheries and management of a demersal fishery in tropical Australia. in J. A. Gulland (ed.), *Fish population dynamics the implications for management*, Wiley, New York, pp. 349-82.

59 Churchill, J. H. (1989) The effect of commercial trawling on sediment resuspension and transport over the Middle Atlantic Bight Continental Shelf. Cont. Shelf Res. **9**, 841-864.

60 Mayer, L. M., Schick, D. F., Findlay, R. H. and Rice, D. L. (1991) Effects of commercial dragging on sedimentary organic matter. Mar. Env. Res. **31**, 249-261.

61 Krost, P. (1990) *The impact of otter-trawl fishery on nutrient release from the sediment and macrofauna of Kieler Bucht (Western Baltic).* Ber. Inst. fur Meereskunde Nr. 200, 167.

62 MacGarvin, M. (1994) The implications of the precautionary principle for biological monitoring. Helgolander Meeresunter **49**, 647-662.

63 Kratz, T. K., Frost, T. M. and Magnuson, J. (1987) Inferences from spatial and temporal variability in ecosystems: Long-term variation in zooplankton from lakes. Am. Nat. **129**, 830-846.

64 Carpenter, S. R. (1988) Transmission of variance through lake food webs. in S. R. Carpenter (ed), *Complex interactions in lake communities.* Springer-Verlag, New York, pp. 119-135.

65 Bellehumeur, C. and Legendre, P. (1998) Multiscale sources of variations in ecological variables: modeling spatial dispersion, elaborating sampling designs. Landscape Ecol. **13**, 15-25.

66 Allen, T. F. H. and Starr, T. B. (1982) *Hierarchy perspectives for ecological complexity.* University of Chicago Press, Chicago, 310p.

67 O'Neill, R. V. (1989) Perspectives in heirachy and scale. in J. Roughgarden, R.M. May and S.A. Levin (eds), *Perspectives in ecological theory.* Princeton University Press, Princeton, pp. 140-156.

68 Addicot, J. F., Aho, J. M., Antolin, M. F., Padilla, D. K., Richardson, J. S. and Soluk, D. A. (1987) Ecological neighborhoods: Scaling environmental patterns. Oikos **49**, 340-346.

69 Legendre, L. and Demers, S. (1984) Towards dynamic biological oceanography and limnology. Can Fish. Aquat. Sci. **41**, 2-19.

70 Hewitt, J. E., McBride, G. B., Pridmore, R. D. and Thrush, S. F. (1993) Patchy distributions: Optimizing sample size. Env. Mon. Ass. **27**, 95-105.

71 Burrough, P. A. (1983) Multiscale sources of spatial variation in soil. I. The application of fractal concepts to nested levels of soil variation. J. Soil Sci. **34**, 577-597.

72 Phillips, J. D. (1985) Measuring complexity of environmental gradients. Vegetatio 64, 95-102.

73 Milne, B. T. (1987). Lessons from applying fractal models to landscape patterns. in M. G. Turner and R. H. Gardner (eds), *The analysis and interpretation of landscape heterogeneity.* Springer-Verlag, New York, pp. 199-235.

74 Palmer, M. W. (1992) The coexistence of species in fractal landscapes. Am. Nat. **139**, 375-397.

75 Wiens, J. A., Crist, T. O., With, K. A. and Milne, B. T. (1995) Fractal patterns of insect movement in microlandscape mosaics. Ecology **76**, 663-666.

76 Burke, I. C. and Lauenroth, W. K. (1993) What do LTER results mean? Extrapolating from site to region and decade to century. Ecol. Model. **67**, 19-35.

77 Phillips, D. L., and Shure, D. J. (1990) Patch-size effects on early succession in Southern Appalachian forests. Ecology 71, 204-212.

78 Gascon, C. and Travis, J. (1992) Does the spatial scale of experimentation matter? A test with tadpoles and dragonflies. Ecology **73**, 2237-2243.

79 Brown, J. H., Davidson, D. W., Munger, J. C. and Inovye, R. S. (1986) Experimental community

ecology: The desert granivore system. in J. Diamond, T.J. Case (eds), *Community ecology*. Harper and Row, New York, pp. 41-62.

80 O'Brien, J. J. and Wroblewski, J. S. (1973) On advection in phytoplankton models. J. Theorectical Biol. **38**, 197-202.

81 Denman, K. L. and Platt, T. (1976) The variance spectrum of phytoplankton in a turbulent ocean. J. Mar. Res. **34**, 593-601.

82 Okubo, A. (1978) Horizontal dispersion and critical scales for phytoplankton patches. in J. H. Steele (ed.), *Spatial pattern in plankton communities*. Plenum Press, New York, pp. 21-42.

83 Horne, J. K. and Schneider, D. C. (1994) Analysis of scale-dependent processes with dimensionless ratios. Oikos 70, 201-211.

84 Miller, D. C., Jumars, P. A. and Nowell, A. R. M. (1984) Effects of sediment transport on deposit feeding: Scaling arguments. Limnol. Oceanogr. **29**, 1202-1217.

85 Keddy, P. A. (1991) Working with heterogeneity: An operators's guide to environmental gradients. in J. Kolasa and S. T. A. Pickett (eds), *Ecological heterogeneity*. Springer-Verlag, New York, pp. 181-201.

86 Ellis, J. I. and Schneider, D. C. (1997) Evalution of a gradient sampling design for environmental impact assessment. Env. Mon. Ass. **48**, 157-172.

87 Wiens, J. A., Stenseth, N. C., Van Horne, B. and Ims, R. A. (1993) Ecological mechanisms and landscape ecology. Oikos **66**, 369-380.

88 Thrush, S. F., Cummings, V. J., Dayton, P. K., Ford, R., Grant, J., Hewitt, J. E., Hines, A. H., Lawrie, S. M., Legendre, P., McArdle, B. H., Pridmore, R. D., Schneider, D. C., Turner, S. J., Whitlatch, R. B. and Wilkinson, M. R. (1997b). Matching the outcome of small-scale density manipulation experiments with larger scale patterns: An example of bivalve adult/juvenile interactions. J. Exp. Mar. Biol. Ecol. **216**, 153-169.

89 Rastetter, E. B., King, A. W., Cosby, B. J., Hornberger, G. M., O'Neill, R. V. and Hobbie, J. E. (1992) Aggregating fine-scale ecological knowledge to model coarser-scale attributes of ecosystems. Ecol. Apps **2**, 55-70.

90 Root, T. L. and Schneider, S. H. (1995) Ecology and climate: Research strategies and implications. Science **269**, 334-340.

91 Fairweather, P. G. (1993) Links between ecology and ecophilosophy, ethics and the requirements of environmental management. Aust. J. Ecol. **18**, 3-19.

92 Horne, J. K. and Schneider, D. C. (1997) Spatial variance of mobile aquatic organisms: capelin and cod in Newfoundland coastal waters. Phil. Trans.Roy. Soc. B **352**, 633-642.

93 Costanza, R., Wainger, L., Folke, C. and Maler, K.-G. (1993) Modeling complex ecological economic systems. BioSci. **43**, 545-555.

94 Holling, C. S. (1996) Surprise for science, resilience for ecosystems, and incentives for people. Ecol. Apps **6**, 733-735.

95 Loehle, C. (1991) Managing and monitoring ecosystems in the face of heterogeneity. in J. Kolasa and S. T. A. Pickett (eds), *Ecological Heterogeneity*, Springer-Verlag, New York, pp. 144-159.

96 Hilborn, R. and Walters, C. J. (1981) Pitfalls of environmental baseline and process studies. Env. Impact Ass. Rev. **2**, 265-278.

97 Walters, C. J. and Holling, C. S. (1990) Large-scale management experiments and learning by doing. Ecology **71**, 2060-2068.

98 Watt, A. S. (1947) Pattern and process in the plant community. J. Ecol. **35**, 1-22.

99 Powell, T. M. (1989) Physical and biological scales of variability in lakes, estuaries, and the coastal ocean. in J. Roughgarden R. M. May and S. A. Levin (eds), *Perspectives in ecological theory*. Princeton University Press, Princeton, pp. 157-176.

UNDERSTANDING THE SEA FLOOR LANDSCAPE IN RELATION TO IMPACT ASSESSMENT AND ENVIRONMENTAL MANAGEMENT IN COASTAL MARINE SEDIMENTS

ROMAN N. ZAJAC
Department of Biology and Environmental Science, University of New Haven 300 Orange Ave., West Haven, CT 06516, U.S.A.
zajacrn@charger.newhaven.edu

Abstract

It is becoming increasingly clear that in order to adequately assess both the natural ecological dynamics of soft-sediment communities and their responses to human impacts it is necessary to have detailed knowledge of the spatial structure and dynamics of the sea floor. Benthic landscape structure can be mapped and quantified at different spatial and temporal scales using technologies such as side-scan sonar, in conjunction with conventional bottom sampling methods. Concurrently, it is critical to develop a theoretical and empirical framework to focus research on relationships between sea floor structure and the benthic communities. The elements of a benthic landscape (or benthoscape) ecology are presented, and key areas of research are suggested to include: 1) relationships between physical and biotic patch structure, 2) the interaction of seascape and benthic landscape mosaics, 3) the degree of variation in benthic ecological dynamics at different spatial and temporal scales, 4) how populations interact across the benthic landscape and 5) how benthic communities respond to different scales of disturbance. The increase in the number of studies which focus on issues of spatial and temporal scale in soft-sediment environments over the past decade indicates that a benthoscape ecology is slowly emerging and that it may provide critical information for the management and restoration of coastal environments.

1. Introduction

Coastal habitats are some of the world's most beleaguered environments. They form the interface between the terrestrial and oceanic realms of earth and, as such are exceptionally dynamic, contain a diversity of organisms, and have borne much of man's long history of development and associated environmental impacts. Despite centuries of harvesting and development, coastal habitats still nurture humankind in multi-faceted ways. However, in many areas these environments are now experiencing ecosystem failures due to accumulated physical and chemical impacts (primarily from the terrestrial realm) and/or over harvesting of resources. Many of these impacts have been local, occurring within a particular embayment, harbor or sound, or along a local portion of the shoreline. But increasingly the affected areas are becoming larger in size and/or the impacts are becoming more severe. For example, hypoxic and anoxic conditions have increased in areal extent and intensity in a number of coastal systems, including bays and sounds along the eastern coast of North America [1], intertidal areas of the Wadden Sea [2], and the Baltic Sea [3] [4]. In some environments, overall ecosystem disturbance and contamination have increased due to changes in hydrography and increased inputs of toxics in waste effluents as described, for example, by Nichols et al. [5] for San Francisco Bay and by Pavoni et al. [6] for the Lagoon of Venice. In offshore

J.S. Gray et al. (eds.), Biogeochemical Cycling and Sediment Ecology, 211–227.

regions of the coastal zone, more of the sea floor is being trawled [7] [8] due to a combination of higher consumer demand, overfishing of existing fishing grounds, and the development of new trawling equipment and practices. In short, the spatial mosaic of human disturbance to the coastal sea floor has increased in scale.

Due to the increase in the spatial extent of various disturbances to the sea floor, it is imperative that we likewise increase the scale at which we study such disturbances in order to assess their impacts and develop and implement mitigation and management efforts to reduce and/or stop negative impacts. A number of large-scale approaches have been proposed and initiated. These include, for example, Large Marine Ecosystems (LME) [9], the National Estuary Program [10] and Environmental Monitoring and Assessment Program [11] in the United States and the Trilateral Commission on the Wadden Sea [12]. The success of multi-scale initiatives to understand and manage coastal systems, especially with respect to impacts to the sea floor, will turn on the extent to which we understand the physio-chemical and ecological structure and dynamics of the sea floor, and what role the benthos plays in the overall functioning of the coastal zone. Although benthic studies are done at multiple scales and provide much valuable information, progress in achieving the level of understanding that is needed to attain management goals is hampered by the lack of explicit knowledge of the spatial heterogeneity of the sea floor. In this paper, I discuss some of the difficulties of impact assessment in coastal environments, and argue for the concerted development of a benthic landscape (or benthoscape) ecology in order to focus research on these systems by directly addressing spatial structure and heterogeneity. A framework for a benthoscape ecology is presented, incorporating information from studies that have begun to utilize this type of approach, as well as suggestions of research needs.

2. Detecting Impacts in Coastal Sediments in Relation to Spatial Heterogeneity

Over the past decade there has been an increasing effort to develop assessment programs that accurately identify and quantify impacts to coastal systems [e.g. 13-17]. The level of accuracy is dependent on the extent to which natural spatial and temporal variation in the system is accounted for relative to changes due to disturbance. Many assessments are based on temporal changes in the system following disturbance, but as Thrush et al. [18] point out, heterogeneity in the spatial distribution of benthic organisms can confound temporal trends. Before-After-Control-Impact (BACI) and related designs [13] [16] [17] [19] explicitly attempt deal with spatial variation by establishing control and impact sites to partition natural and disturbance-based variation. Recognizing that spatial variation in coastal systems can be high, several ways to account for spatial variation and increase the power of detecting changes due to disturbance have been proposed such as using nested, hierarchical designs to assess within-site variation [13], and multiple control sites in BACI designs [19]. Underwood [19] has also proposed regular monitoring of a series of sites in order to establish Before data sets that would estimate the magnitude of spatial variation in a given region. He states that these sites should be in randomly chosen replicate habitats, and need not be identical, as this is impractical and unnecessary [19].

Be it for a single Control site, multiple Control sites, or a series of Before monitoring sites, the selection of such sites can be very difficult. The criteria for site selection include "appropriate features of physical characteristics, mix of species, abundance of the target species," and represent the range of habitats of the one that might be disturbed (the Impact location)" [19]. It is easy to see that a key problem in coastal assessment is identifying replicate habitats and the selection of appropriate sites within and among these habitats. It is likely that no two potential habitats would ever be identical, and in terms of large-scale monitoring projects [e.g. 11], extrapolating the degree of spatial variation from coastal area to another may be unjustified.

Just how similar should the sites be? To answer this question it may be necessary, to compare sites on a more quantitative basis as opposed to assessing general features, in order to select control / monitoring sites that would adequately account for spatial variation. For example, sediment grain-size may be an important factor when selecting sites in soft-sediment environments. If an impact is expected to occur in a mud habitat, then it may be appropriate to select Control sites that were in mud habitats, assuming that this would decrease the variability in community characteristics among sites . Site selection might then proceed by examining general sediment distribution maps and related data bases. However, it is likely that these will be inadequate. Researchers developing the Environmental Monitoring and Assessment Program found that sediment type in over 50% of selected monitoring sites along the eastern coast of the United States was misclassified in an extensive sediment data base [11]. Many existing databases that provide information on sea floor characteristics are probably inadequate because of the reliance on point samples to extrapolate to larger areas. What are needed are more accurate (quantitative) data sets on continuous portions of the sea floor in areas where impacts are anticipated and those that are candidates for Control sites. As discussed below, this objective may very well be practical at this time due technological advances and the emergence of a benthic landscape ecology.

3. Marine Benthic Landscapes

Landscape ecology focuses on understanding the structure and dynamics of kilometers-wide areas which are comprised of mixtures of interacting ecosystems [20]. Because of the focus on pattern and process, landscape ecology provides a particularly useful framework for investigating the responses of ecological systems to human impacts at multiple spatial and temporal scales relative to natural environmental heterogeneity. Landscape ecology has developed within the terrestrial realm of ecology, and at this time has a fairly well established theoretical and empirical framework (see for example texts by Forman and Godron [20] and by Forman [21]).

Ecologists working in the marine / coastal realm have long studied these systems at different scales in order to understand the relationships between habitat structure and dynamics. More recently, marine ecologists have begun to incorporate landscape-level approaches into their work [9, 21-27]. However, as of yet there is no cohesive framework for the study of marine benthic landscapes that can a) help focus research by recognizing the basic structural and dynamic components of marine landscapes, and b) allow for direct and quantitative comparisons among coastal environments in order to determine similarities and differences in how ecological dynamics are shaped by benthic landscape structure. A better understanding of the structure and dynamics of benthic environments across multiple spatial and temporal scales may also help reduce difficulties with impact assessment, fuse water column and benthic research, and allow for comparisons to terrestrial landscapes in order to recognize spatial and temporal dynamics that are. or are not, common to these two realms of the biosphere.

In order to develop a benthic landscape ecology we need to consider: a) the physical and ecological structure of the sea floor, b) the flow of materials and organisms in relation to sea floor structure, and c) and how structure and dynamics respond to disturbances and longer-term changes. The latter is germane to the theme of this NATO workshop. It is obvious that a better understanding of SEA FLOOR structure and dynamics is needed to more effectively address coastal impacts and develop management and conservation plans. My main argument here is that by developing a theory of benthic landscape ecology and increasing our empirical knowledge of pattern and process in coastal benthic environments we will attain the needed level of understanding to meet these goals more rapidly.

4. A Benthic Landscape Prospectus

In this section of the paper, I present an overview of the possible components of benthic landscape ecology and identify what aspects of these I feel are in need of further research. I include examples of previous and current work in a variety of environments that collectively are beginning to form a landscape framework for marine and coastal benthic ecology.

4.1. BENTHIC LANDSCAPE STRUCTURE

To understand benthoscape structure, it is imperative to determine the spatial relationships among the distinctive ecosystems, or landscape elements that comprise a particular coastal region. This includes quantifying the distribution of species, materials and energy in relation to sizes, shapes, numbers, kinds and configurations of the landscape elements. Several structural components need to be considered, including:

a) Patches - the various landscape elements or large-scale habitats making up the benthoscape,

b) Corridors- specific types of patches which are linear, and because of their structure have critical effects on landscape function [21], and

c) Matrix - the patch type that has the most influence on landscape function.

In addition to these structural elements, the boundaries among them may be ecologically important [e.g. 21]. Boundaries can also be referred to as edges transition zones, or ecotones, depending on their characteristics.

Perhaps one of the largest difficulties that has previously hampered our ability to adequately assess benthic dynamics and changes in response to disturbance over multiple scales has been our dependence on point samples to determine sea floor structure. Researchers have had on the most part to extrapolate continuous sea floor characteristics from point samples taken at various, but usually coarse spatial resolutions [e.g. 28-31]. Organismal abundances and their relationship to sea floor characteristics determined from point samples alone can be inaccurate, depending on the level of heterogeneity in the area and the number of samples taken [32] [33]. Fortunately, advances in oceanographic technology are now allowing benthic ecologists to overcome this difficulty, and our knowledge of sea floor structure is increasing rapidly with the burgeoning use of various sea floor imaging devices. These include high-resolution swath bathymetry [34], side scan sonar [35], RoxAnn technology [36] and satellite imaging, aerial photography and/or video in intertidal [37] and shallow subtidal systems where water clarity is high [26].

Using these methods, detailed information on large and contiguous portions of the sea floor can be obtained. When used in conjunction with standard bottom grab samples, not only can sea floor structure be mapped, but also patch characteristics at different scales can be determined. For example, side scan sonar images, combined to form a mosaic, provide information both on the large-scale (10s of km^2) and meso-scale (hundreds of m^2 to km^2) structure of the sea floor (Figures 1 & 2). Smaller-sized sea floor elements can be resolved if high-resolution side scan sonar is used. Once patch structure is defined, additional landscape metrics can be generated such as patch size, shape factor and patch isolation [21, and references therein], and boundaries among the patches characterized (e.g. width, convolution, heterogeneity). The side scan records can also be further analyzed using image analysis to obtain information on within- and among-patch heterogeneity (Figure 3)

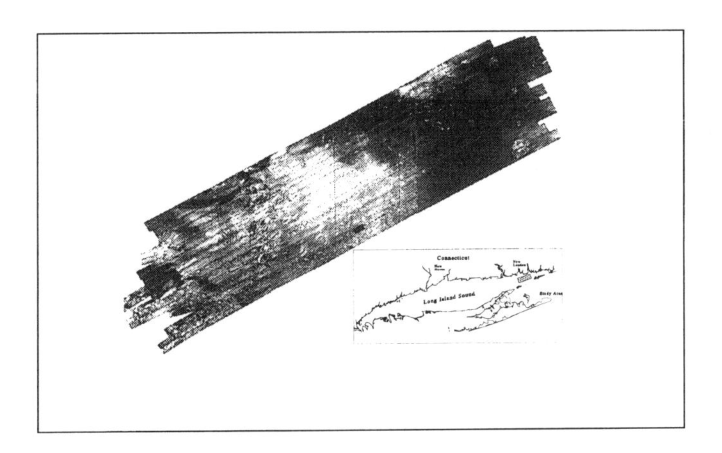

Figure 1. Side scan mosaic of a 2.5 x 8 km area in eastern Long Island Sound (insert). Depth increases from 10 to 15 m in the northeast section to 20 -30 m in the southwest section.

and to identify specific sediment types [27] [33]. Small-scale patch characteristics (on the order of < m^2 to tens of m^2) can be further quantified by analyzing video and photo records [33] [38], and conventional bottom grab samples. These technologies are also being effectively used to determine the extent of disturbance to the sea floor from activities such as trawling [39] and sediment disposal [40].

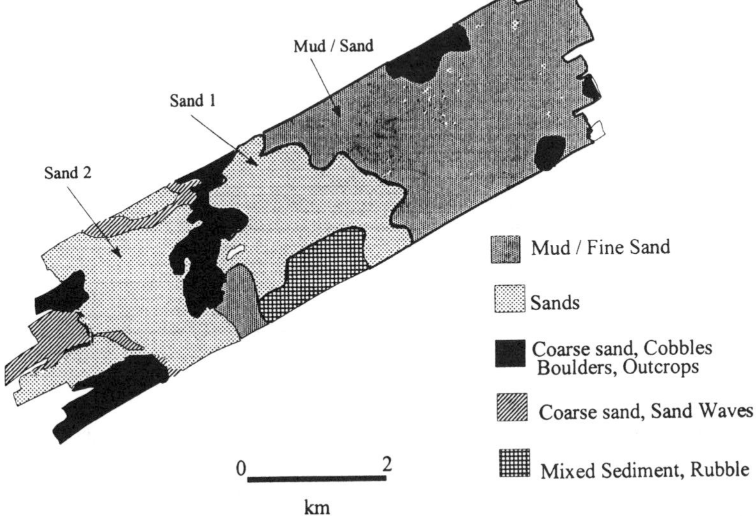

Figure 2. Interpretation of side scan mosaic shown in Figure 1. Mud / Sand, Sand 1 and Sand 2 refer to the largest benthic landscape elements in the study area, as discussed in the text and subsequent figures.

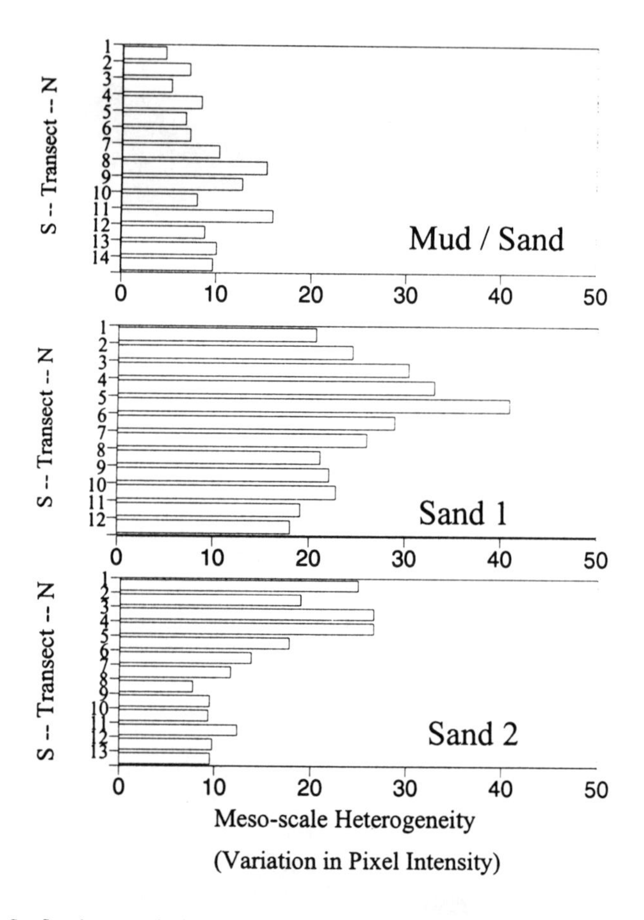

Figure 3. Sea floor heterogeneity in the three largest benthoscape elements shown in Figure 2 as measured by pixel variation. 'Transects' were placed along acoustically clean sections of a digital image of each section of the mosaic, running parallel to the track lines. The transect swaths along which pixel intensity (256 level gray scale) was measured were approximately 55 m wide by 500 to 2000 m long. Each pixel was 6.61 m on a side. Variation in pixel intensity is the standard deviation of gray scale levels among all pixels in the transect. Note that the Sand 1 and Sand 2 elements have much high variation at this scale of measurement than the Mud / Sand element, but that within-element variation in the sand elements is also high. The transect numbers are in a north (N) to south (S) direction.

4.1.1. Distinctive Aspects of Benthic Landscape Structure

Most soft-sediment landscapes are defined by their geomorphologic structure. That is, patches are most easily recognized based on sediment characteristics and geomorphologic topographic features. This is in contrast to most terrestrial landscapes where patch structure is primarily defined by vegetation and/or human structures [20] [21]. Where sufficient conditions exist, vegetated areas such as seagrass meadows [26] and kelp beds [25] can be a major component of the landscape (i.e. the matrix) and benthoscape structure and dynamics may be more similar to terrestrial landscapes. In tropical areas, coral reefs also comprise biogenic patches that can define and control benthoscape function. In some benthoscapes, mussel beds and/or oyster reefs may be prominent patch types [41]. But for the most part, benthic landscape elements are low relief sediment patches defined by sediment type and small- to meso-scale (< m to tens of meters), non-living, topographic features such as sandwaves, shell-hash deposits, rock outcrops and ledges. Within these physically defined landscape elements will be patches of different assemblages of infauna, but the latter are difficult to map accurately without extensive conventional sampling. Therefore, in most cases benthic landscape structure will be physically defined rather than based on biological characteristics.

Corridor-type patches (e.g. hedgerows and power line cuts) which are prominent elements in many terrestrial landscapes [21], may not be as widespread in coastal soft-sediment benthoscapes. Analogous patch types would comprise, for example, intertidal drainage channels and in sublittoral areas, human disturbance patches such as navigation channels, anchor drag lines, sea floor pipelines and trawl marks. However, other patch types distinct to marine benthic landscapes, with significant effects on benthic dynamics, probably exist but await to be recognized after more coastal benthic environments are studied within a landscape framework.

The structure, variety and characteristics of benthic transition zones are also not well known. This is because boundaries among patches in soft-sediment environments are almost always extrapolated from point samples and their true location and features are not known. The extensiveness of transition zones a particular area will vary depending on overall sea floor heterogeneity and the widths of the transition zones. For example, the width of transition zones in an area in Long Island Sound (Figure 1) was estimated to range from approximately 50 to 200 m. Using this range, it was estimated that approximately 25% to 47% of the total area of the sea floor shown in Figure 1 consists of transition zones [33]. It is likely that in many heterogeneous nearshore environments, the amount of transition zone area will be large, and as such, transitions may play key roles in benthic dynamics.

4.1.2. Questions Regarding Benthic Landscape Structure

There is much that we do not know yet about benthic landscape structure. However, there are several key questions that should be addressed inn order to improve our understanding of coastal systems at the landscape level. These are:

* How stable is the landscape, and in particular transition zones? Side scan imaging provides a large-scale depiction of the sea floor, but it can be time-consuming and costly. Most surveys are only done once and at varying time of the year. Because of this we know very little about the extent to which benthoscape structure changes in response to natural processes and over what temporal scales.

* What is the structure of the inshore landscape? Aerial photographs and satellites can be used to map intertidal areas, whereas acoustic devices can be used in deeper waters (usually > 2.5 m). However, it is difficult to apply either of these technologies in shallow waters. This is an important portion of the coastal zone, providing critical habitat and being subjected to a variety of human disturbances. Without good knowledge of this zone an important piece of the puzzle will be missing. Technology development will be a key element in our efforts to map and understand this portion of the coastal zone.

To what extent can we recognize distinct biotic patches within sedimentary landscape elements, what is their scale of coincidence and what is the level of biotic patch heterogeneity?

Although the physical patch structure of the benthos can be imaged relatively easily, the biotic patches within these physical patches are difficult to map at multiple scales. In some cases biogenic structures such as mussel beds and amphipods mats will be evident, but in most cases fauna will be not be evident in side scan images, and, for infauna also not in video and photographs, other than any physical traces such as burrows or tubes. Combining technologies such as side scan, sediment profiling and photography / video can begin to provide some of this information [33] [40]. However, it is likely that only by applying spatial grid sampling programs in well-defined sediment patches will this question be adequately addressed. I feel that this is an important question because the relationships between physical and biotic patch structure will define the dynamics of the system and determine how the system may respond to different types and sizes of disturbance.

4.2. BENTHIC LANDSCAPE DYNAMICS

Detailed information on benthic landscape structure should lead to a better understanding of the dynamics of soft-sediment populations and communities by providing an accurate spatial framework research on dynamics. There are four key areas which should be addressed in order to enhance our knowledge of soft-sediment dynamics at the landscape level: 1) responses of populations and communities to benthic landscape structure, 2) spatial and temporal scales of variability, 3) links to the water column seascape and 4) metapopulation dynamics in relation to sea floor structure.

4.2.1. Responses of infauna to benthic landscape structure

Once patches, corridors and transition zones are identified and characterized, benthic communities within these landscape elements should be sampled using designs that explicitly address the spatial structure that is present. The extent to which abundances and other ecological characteristics correlate to different types of structural elements will provide insights into the dynamics of populations and communities and the role sea floor structure plays at different scales.

For example, local (< 1 m^2) sediment characteristics and depth explained little (usually < 30%) of the variation in infaunal population abundances in eastern Long Island Sound, irrespective of benthic landscape structure [33]. However, when the structure of the benthic landscape (Figure 1) was taken into account the spatial patterns and potential dynamics of the populations became more apparent. The spionid polychaete *Prionospio steenstrupi* was the most abundant infaunal species at the time this benthic landscape was sampled (June 1992), and the population was comprised of many newly settled juveniles [33]. Local sediment factors and depth accounted for only 20% of the variation in abundance across the site, but significant differences were detected among the benthic landscape elements. Analyses based on a nested sampling design in just the three largest landscape elements (Figure 1) indicated no significant differences in abundance among these patches, but that there were significant meso-scale differences within the patches (Figure 4). Furthermore, when the samples were grouped based on whether they were located in transition zones or in the interior of patches, it was found that *Prionospio steenstrupi* had significantly higher abundances in transition zones than in the interior of large patches (Figure 5). This suggests that recruitment into the population was higher in the transition zones.

These results suggest that soft-sediment communities and populations respond to the structure of the benthic landscape and not just to sediment characteristics and other factors such as depth. Responses to landscape structure will likely be complex, as might be expected. Some species will be found throughout the landscape, but with clear affinities to certain patch types and/or habitat characteristics within patches; other species will be found only in certain patch types. Transition zones also appear to be areas of interesting, and perhaps significant, benthic dynamics.

4.2.2. Spatial and Temporal Scales of Variation

Determining the scales at which significant variation in benthic community characteristics and dynamics exist is critical in developing accurate assessment protocols. There has been much recent work addressing spatial variation in benthic communities. Distinct, large-scale trends (on the order of tens to hundreds of kilometers) in infaunal community structure in coastal areas are well-known [28-31]. These trends are generally associated with changes in depth, sediment characteristics and large-scale hydrodynamic features. The degree of meso- and small-scale variation in these studies is difficult to assess given the scale of sampling, which is usually coarse (> 1 km). However, recent studies are increasingly noting the importance of meso-scale structure patterns in benthic communities.

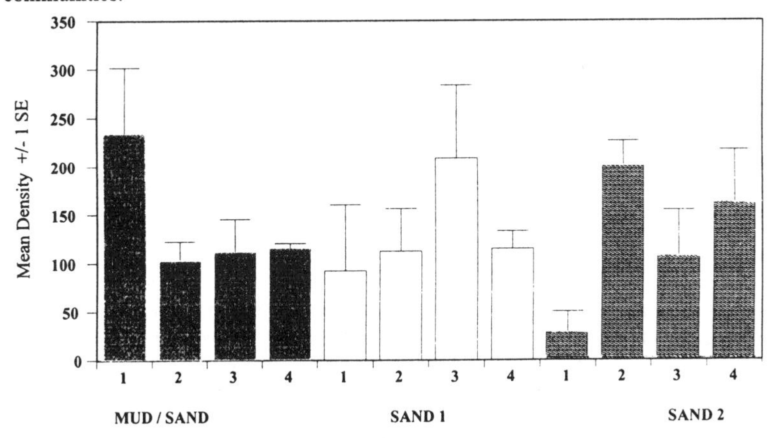

Figure 4. Differences in the abundance (mean density per 6 cm diameter core) of the spionid polychaete Prionospio steenstrupi within (represented by four subdivisions, 1-4) and among (Mud/Sand, Sand 1 and Sand 2) the largest benthic landscape elements of the mosaic area shown in Figures 1 and 2.

Rogal et al. [42] found significant heterogeneity in infaunal community structure at scales of 100 m, in an apparently homogenous habitat at 15 m depth in the western Baltic Sea, and concluded that "even within a small area inconspicuous sediment changes can influence faunal composition significantly." Polychaetes and bivalves exhibited patches of elevated abundances at scales of 1 to 100 m in an intertidal sand flat in New Zealand, with varying types of spatial pattern [43]. Hodda [44] investigated variation in populations of nematodes over three spatial scales, among estuaries, among sites within estuaries and within the sites themselves, and found that 52% of the variation in their numbers could be accounted for by meso-scale (among sites within estuaries) differences in habitat characteristics. These included sediment grain size and organic content and availability of certain types of food associated with surface topography and pools of water. A similar study conducted by Morrisey et al. [45] in southeastern Australia compared differences in the abundance of macroinfauna at 5 spatial scales ranging from 1 m between replicate samples to 3.5 km representing separate locations of sampling sites within the study area.

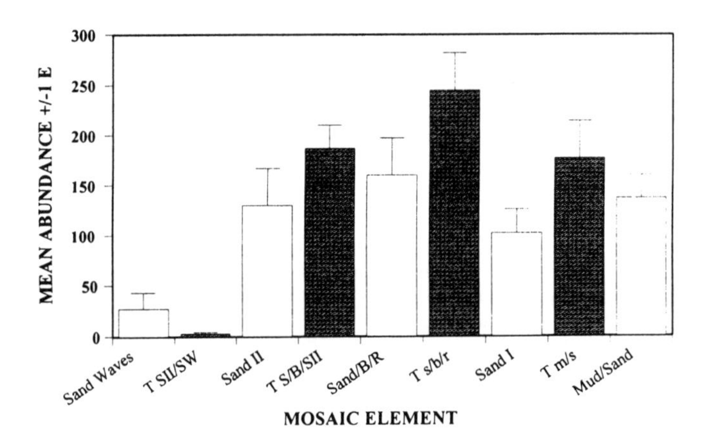

Figure 5. Differences in the abundance (per 6 cm diameter core) of the spionid polychaete Prionospio steenstrupi among transition zones (hatched bars indicated by a T) and the interior portions (open bars) of landscape elements in the side scan study area shown in Figures 1 and 2.

They found significant spatial variation at spatial scales from 10 m to 3.5 km, with taxonomic groups exhibiting significant variation at several spatial scales but component families in some instances displayed different patterns. McArdle and Blackwell [46] found that physical characteristics in a sandy lagoon ranging from km to tens of meters significantly affected the density of the bivalve *Chione stutchburyi*. Detailed studies revealed that locally elevated abundances of the bivalve were in response to factors operating on a scale of 600 m^2 and that temporal changes were also important relative to this spatial scale [46]. The results of research in eastern Long Island Sound [33] are in general agreement with the studies noted above, showing that patterns of infaunal spatial variation in relation to habitat characteristics are quite varied across several scales, but that significant variation does occur at meso-scales.

Although these studies have provided significant insights as to the patterns of spatial variation in soft-sediment communities, it still not clear how differences in the spatial scale of community structure and population attributes relate to the specific structure of the benthic environment. This type of information is needed in order to develop and test hypotheses concerning the factors that shape abundances and distributions, and variation in these.

4.2.3 Links to the Water Column Seascape

The dynamics of benthic landscapes are a function of within- and among-patch population dynamics as mediated by the hydrographic seascape. The structure and dynamics of the water column are important determinants of several aspects of benthic dynamics such as production and distribution of food resources and dispersal of larvae and adults. The water column also shapes various types of disturbances that impact bottom communities. The water column seascape itself is a mosaic of different physical and biological patches spanning multiple scales [23] [47]. Understanding the relationships between water column and benthic processes has long been a focus of coastal research, and significant insights have been made over the past decade [23] [25]. This work continues, but suprisingly few studies have attempted to explicitly compare the patch structure and dynamics of the water column with that of the benthic landscape. This is not an easy task and can be quite costly. However, key insights obtain by attempting to do so. For example, Warwick and Uncles [48] used information from side scan records, bottom grabs and hydrographic data to show that there is a strong relationship between the distribution of community types and tidal current patterns in the Bristol Channel.

4.2.4. Metapopulation Dynamics

The spatial scales over which populations of a particular soft-sediment species interact in coastal environments is not well known. Most studies are conducted in only one habitat within a larger system, and it is not clear to what extent the dynamics of a particular population can be applied to other populations. Several aspects of this problem need to be considered. Because of the potential for larval transport, it is often assumed that coastal populations are fairly well mixed over large spatial scales. To what extent is this the case is not well known. Larval mixing and the degree to which populations interact via larval transport can differ depending on the type of coastal environment [49]. Even if the populations are relatively well connected via larval dispersal, their local dynamics may differ considerably. Such differences need to be recognized in order to determine at what scales and how populations interact across the benthic landscape. For example, Zajac and Whitlatch [50] [51] found that populations of the polychaete *Nephtys incisa* exhibited significant variation in individual growth, size-structure and fecundity among five sites 200 to 3 km apart, but that despite these differences the populations formed a relatively coherent demographic unit having similar population growth rates. However, disturbance resulted in altering a number of population characteristics in populations in the disturbance area and up to 200 m away, but not at greater distances from the disturbance site [51].

Further research should be directed towards understanding the population structure and dynamics of benthic fauna relative to sea floor structure and the interactions among populations occupying different landscape elements (patches). Whether the populations form metapopulations and can be studied within a metapopulation framework [52] is not clear. However, such efforts could focus on a few key, or indicator, species with different life modes and life histories.

4.3 DISTURBANCE AND CHANGE IN BENTHIC LANDSCAPES

Most detailed studies of the responses of benthic communities to disturbance have been conducted at spatial scales of $<25\ m^2$. Current disturbance-succession models are primarily based on small-scale experiments and/or do not explicitly address the spatial characteristics of the sea floor. Our knowledge of recovery processes when a large portion of a coastal environment is disturbed is relatively rudimentary. A cursory review of studies which have looked at recovery from large-scale disturbances show that response time is quite variable (Figure 6). A similar degree of variation appears to exists for recovery times following smaller-scale disturbances as well [53]. This variation may be due, in part, to differences in the structure of the benthic landscapes in which the disturbances took place, the characteristics of the disturbance (including structural aspects) and hydrodynamic conditions.

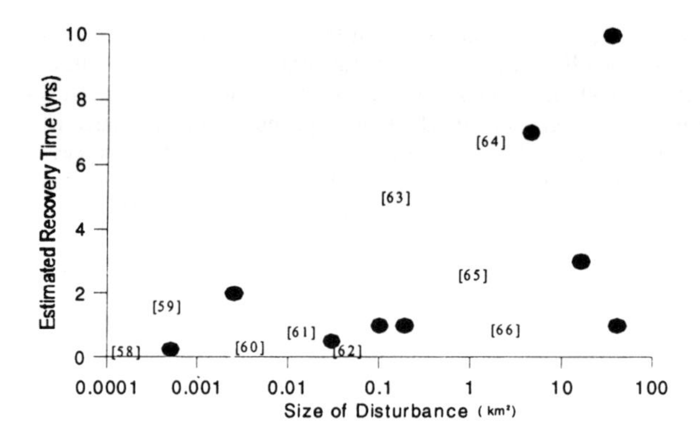

Figure 6. Relationship between estimated recovery time of infaunal succession relative to different sizes of disturbance. Numbers next to each data point are references from which data were obtained.

Sea floor imaging can help considerably to map certain kinds of disturbances such as trawl marks and disposal areas, and allow researchers to not only quantify the area and intensity of disturbance, but also the geometry and spatial characteristics (e.g. the direction of trawl marks relative to current patterns). In turn, this information can help develop more powerful sampling and analysis designs to assess disturbance impacts and recovery relative to the spatial structure and heterogeneity of the sea floor.

Current models of succession in soft-sediment do not explicitly take the spatial scale of disturbance into account, nor benthoscape structure. Predicted patterns of initial recolonization by opportunists followed by steady increase of later stage (equilibrium) species [54, 55] may not hold depending on the size and type of disturbance relative to benthoscape structure and dynamics. For example, succession models generally assume that disturbance extirpates the extant benthic community, allowing for recolonization by opportunists. However, in work being carried out in the western portion of Long Island Sound, it was found that although infaunal densities declined significantly across a broad area (approximately 20 km^2) during a hypoxic period, the remnant communities were highly variable across several spatial scales (Figure 7). Thus, this large-scale disturbance increases spatial heterogeneity in community structure and subsequent succession may be quite different across the affected portion of the benthoscape due to the different starting points (Figure 8).

5. Conclusions

The importance of knowing the detailed spatial structure and dynamics of the sea floor is becoming clear in our efforts to adequately assess both natural ecological dynamics of soft-sediment communities and how these communities respond to human impacts. Using technologies to map and characterize the sea floor, in conjunction with conventional bottom sampling methods, we can begin to develop a general view of benthic landscape structure and dynamics at different spatial and temporal scales. As noted by Butman et. al. [56], in relation to sea floor geological characteristics, "Remote sensing mapping techniques are essential to adequately determine the complex spatial variability of the bottom morphology and sediment texture: accurate maps cannot be prepared from the analysis of sediment grab samples alone." To this end it is imperative that systematic benthic mapping and characterization programs be established and supported for coastal systems in different regions (e.g. European and North American coastal waters) and in different environments (e.g. temperate vs. tropical). Several such program are already in place, such as the Sea Floor Mapping Project being conducted by the United States Geological Survey [57].

But mapping the sea floor is just the first step. What is also critical is the development of a theoretical and empirical framework to analyze and conduct more focused research on relationships between sea floor structure and the biotic communities that inhabit the sea floor. The increase in the number of studies which focus on issues of spatial and temporal scale over the past decade indicate that a benthic landscape ecology is slowly emerging. There are several key areas that need more work so as to develop this avenue of benthic research. These include the relationship between physical and biotic patch structure, the interaction of seascape and benthic landscape mosaics, identification of how much variation in benthic ecological dynamics exists at different spatial and temporal scales, how populations interact across the benthic landscape and how these communities respond to different scales of disturbance. This will be no small task and the effort itself will probably have a long temporal signature. However, the gains in understanding soft-sediment systems I feel will be substantive, and will allow for much more accurate assessments of human impacts to the coastal zone. In addition, knowledge of the benthic landscape and its dynamics can help in the management of coastal resources and aid in the selection of conservation areas.

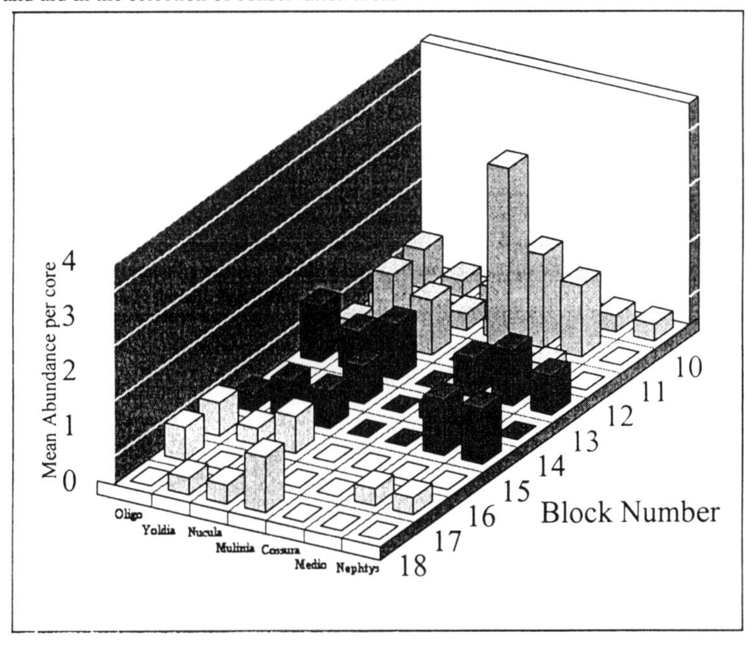

Figure 7. Remnant communities in a deep water section in western Long Island Sound sampling area within a particular landscape element. Blocks with the same shading were located in the same element and were approximately 0.5 to 1 km apart. There were three landscape elements. The species areas follows: Oligo (an unidentified oligochaete), Yoldia (*Yoldia limatula*), Nucula (*Nucula annulata*), Mulinia (*Mulinia lateralis*), Cossura (*Cossura longocirrata*), Medio (*Mediomastus ambiseta*), Nephtys (*Nephtys incisa*). Note the variability among and within block groups.

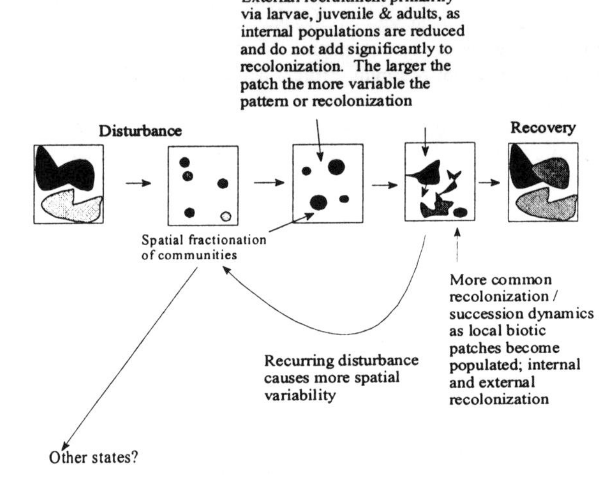
External recruitment primarily via larvae, juvenile & adults, as internal populations are reduced and do not add significantly to recolonization. The larger the patch the more variable the pattern or recolonization
Disturbance
Recovery
Spatial fractionation of communities
More common recolonization / succession dynamics as local biotic patches become populated; internal and external recolonization
Recurring disturbance causes more spatial variability
Other states?

Figure 8. Model of potential responses to large-scale disturbances in soft-sediment benthic landscapes. Because of increased spatial variation in remnant community structure, initial recolonization patterns may also be variable and not follow the patterns predicted by models developed from small-scale experimental work. There may be some temporal lag as to when more typical successional patterns are evident as biotic patches reach critical abundance levels so as to make significant contributions to recolonization of the overall disturbed portion of the benthic landscape.

6. Acknowledgements

The benthic landscape work done in Long Island Sound used in this paper to illustrate several points has been done in conjunction with Ralph Lewis (Connecticut Dept. of Environmental Protection), Larry Poppe and Dave Twichell (United States Geological Survey, Woods Hole) and Joe Vozarik (Northeast Utilities Environmental Laboratory, Waterford, Connecticut), and was funded in part by grants from the State of Connecticut Long Island Sound Research Fund. My thoughts on benthic landscape ecology have profited from discussions with Susan Bell, Brad Robbins, Ralph Lewis, Larry Poppe, Peter Auster, Bob Whitlatch and Simon Thrush. Finally, many thanks, as always, to Fran, Julia and Katya.

7. References

1. Parker, C.A. and O'Reilly, J.E. (1991) Oxygen depletion in Long Island Sound: A historical perspective, *Estuaries* **14**, 248-264.
2. Common Wadden Sea Secretariat (1996) Black Spots in the Wadden Sea, *Wadden Sea Newsletters No. 2.*
3. Prena, J. (1994) Oxygen depletion in Wismar Bay (western Baltic Sea) 1988, *Arc. Fish. Mar. Res.* **42,** 77-87.
4. Gamenick, I., Jahn, A., Vopel, K. and Giere, O. (1996) Hypoxia and sulphide as structuring factors in a macrozoobenthic community on the Baltic shore: colonization studies and tolerance experiments, *Marine Ecology Progress Series* **144**, 73-85.
5. Nichols, F.H., Cloern, J.E., Luoma, S.N. and Peterson D.H. (1986) The modification of an estuary, *Science* **231**, 567-573.

6. Pavoni, B., Marcomini, A., Sfriso, A., Donazzolo, R. and Orio, A. (1992) Changes in an estuarine ecosystem: The lagoon of Venice as a case study, in D.A. Dunnette and R.J. O'Brien (eds.), *The Science of Global Change*, American Chemical Society, pp. 287-305.
7. Raloff, J. (1996) Fishing for answers: Deep trawls leave destruction in their wake - but for how long? *Science News* **150**, 268-271.
8. Dayton, P.K., Thrush, S.F., Agardy, M.T., and Hofman, R.J. (1995) Environmental effects of marine fishing, *Aquatic Conservation* **5**, 205-232.
9. Sherman, K. (1991) The large marine ecosystem concept: research and management strategy for living marine resources, *Ecological Applications* **1**, 349-360.
10. Imperial, M.T., Robadue, D. and Hennessey, T.M. (1992) An evolutionary perspective on the development and assessment of the National Estuary Program, *Coastal Management*, **20**, 311- 341.
11. Weisberg, S.B., Frithsen, J.B., Holland, A.F., Paul, J.F., Scott, K.J., Summers, J.K., Wilson, H.T., Valente, R. Heimbuch, D.G., Gerritsen, J., Schimmel, S.C. and Latimer, R.W. (1992) EMAP-Esturaies Virginian Privince 1990 Demonstration Project Report, EPA 600/R-92/100. U.S. Environmental Protection Agency, Environmental Research Laboratory, Narragansett, RI.
12. The Trilateral Cooperation on the Protection of the Wadden Sea. Internet address: http://194.45.109.50/cwss/index.html
13. Green, R.H. (1979) *Sampling Design and Statistical Methods for Environmental Biologists*, Wiley and Sons, New York.
14. Underwood, A.J. and Peterson, C.H. (1988) Towards an ecological framework for investigating pollution, *Marine Ecology Progress Series* **46**, 227-234.
15. National Research Council (1990) *Managing Troubled Waters: The Role of Marine Environmental Monitoring*, National Academy Press, Washington, D.C.
16. Underwood, A.J. (1992) Beyond BACI: the detection of environmental impacts on populations in the real, but variable, world, *J. Experimental Marine Biology and Ecology* **161**, 145-178.
17. Schmitt, R.J. and Osenberg, C.W. (1996) *Detecting Ecological Impacts - Concepts and Applications in Coastal Habitats*, Academic Press, San Diego.
18. Thrush, S.F., Pridmore, R.D. and Hewitt, J.E. (1994) Impacts on soft-sediment macrofauna: effects of spatial variation on temporal trends, *Ecological Applications* **4**, 31-40
19. Underwood, A.J. (1994) On beyond BACI: sampling designs that might reliably detect environmental disturbances, *Ecological Applications* **4**, 3-15.
20. Forman, R.T.T. and M. Godron (1986) *Landscape Ecology*, J. Wiley & Sons, New York.
21. Forman, R.T.T. (1995) Land Mosaics: The Ecology of Landscapes and Regions. Cambridge University Press. Cambridge
22. Ray, G.C. (1991) Coastal-zone biodiversity patterns. BioScience 41: 490 - 498
23. Barry, J. P., and Dayton, P.K. (1991) Physical heterogeneity and the organization of marine communities, in Kolasa, J. and S.T.A. Pickett (eds.), *Ecological Heterogeneity* Springer-Verlag, New York, pp. 270-320.
24. 24. Angel, M.V. (1994) Spatial distribution of marine organisms: patterns and processes, in P.J. Edwards, R.M. May and N.R. Webb (eds.), *Large-Scale Ecology and Conservation Biology*, Blackwell Scientific Publications, Oxford, pp. 59-109.
25. Dayton, P. K. (1994) Community landscape: scale and stability in hard bottom marine communities, in P.S. Giller, A.G. Hildrew and D.G. Rafaelli (eds.), *Aquatic Ecology: Scale, Pattern, and Process*, Blackwell Scientific Publications, Oxford, pp. 289-332.
26. Robbins, B.D. and Bell, S.S. (1994) Seagrass landscapes: a terrestrial approach to the marine subtital environment. *Trends in Ecology and Evolution* **9**, 301-303.
27. Auster, P.J., Michalopoulos, C., Robertson, R., Valentine, P.C., Joy, K. and Cross, V. (in press) Use of acoustic methods for the classification and monitoring of seafloor habitat complexity: description of approches, in *Linking Protected Areas with Working Landscapes*, Science and Management of Protected Areas Association.
28. Reid, R.N., Frame, A.B. and A.F. Draxler (1979) Environmental baselines in Long Island Sound, 1972-1973, *NOAA Technical Report* NMFS SSRF-738.
29. Duineveld, G.C., Kunitzer, A.A., Niermann, U., De Wilde, P.A.W.J. and Gray, J. S. (1991) The macrobenthos of the North Sea, *Netherlands Journal of Sea Research* **28**, 53-65.
30. Gambi, M.C. and Giangrande, A. (1986) Distribution of soft-bottom polychaetes in two coastal areas of the Tyrrhenian Sea (Italy): structural analysis. *Estuarine, Coastal and Shelf Science* **23**: 847-862
31. Josefson, A.B. (1987) Large-scale patterns of dynamics in subtidal macrozoobenthic assemblages in the Skagerrak: effects of a production-related factor, *Marine Ecology Progress Series* **38**, 13-23.
32. Bonsdorff, E. (1988) Zoobenthos and problems with monitoring; and example from the Aland Sea, Kieler Meeresforsch., 6, 85-98.
33. Zajac R.N. (1996) Ecologic Mapping and Management-Based Analyses of Benthic Habitats and Communties in Long Island Sound, Final Report, Office of Long Island Sound Programs, State of Connecticut Dept. of Environmental Protection, Hartford
34. Spitzak, S.E., Caress, D.W. and S.P. Miller. (1997) Advances in multibeam Survey. *Sea Technology*, **38** (6), 45-49.

35. Belderson, R.H., Kenyon, N.H., Stride, A.H. and Stubbs, A.R. (1972) *Sonographs of the Sea Floor*, Elsevier, London.
36. RoxAnn
37. Bartholdy, J. and Folving, S. (1986) Sediment classification and surface type mapping in the Danish Wadden Sea by remote sensing, *Netherlands J. Sea Research* **20**, 337-345.
38. Auster, P.J., Malatatesta, R.J., LaRosa, S.C., Cooper, R.A., and Stewart. L.L. (1991) Microhabitat utilization by the megafaunal assemblage at a low relief outer continental shelf site - Middle Atlantic Bight, USA, *J. Northwest Atlantic Fisheries* **11**, 59-69.
39. Krost, P., Bernhard, M., Werner, F. and Hukriede, W. (1990) Otter tracks in Kiel Bay (Western Baltic) mapped by side-scan sonar, *Meeresforschung* **32**, 344-353.
40. Menzie, C.A., Ryther, J., Boyer, L.F., Germano, J.D. and Rhoads, D.C. (1982) Remote methods of mapping seafloor topography, sediment type, bedforms and benthic biology, *Oceans* **197**, 1046 - 1051.
41. Dealteris, J.T. (1988) The application of hydroacoustics to the mapping of subtidal oyster reefs. *J. Shellfish Research* **7**, 41-45.
42. Rogal, U., Anger, K., Schriever, G. and Valentin, C. (1978) In-situ investigations on small-scale local and short-term changes of sublittoral macrobenthos in Lubeck Bay (western Baltic Sea). *Helgolander wiss. Meeresunters* **31**, 303-313.
43. Thrush, S.F., Hewitt, J.E. and Pridmore R.D. (1989) Patterns in the spatial arrangements of polychaetes and bivalves in intertidal sandflats, *Marine Biology* **102**, 529-535.
44. Hodda, M. (1990) Variation in estuarine littoral nematode populations over three spatial scales, *Estuarine, Coastal and Shelf Science* **30**, 325-340.
45. Morrisey, D.J., Howitt, L., Underwood, A.J. and Stark, J.S (1992) Spatial variation in soft-sediment benthos, *Marine Ecology Progress Series* **81,** 197-204.
46. McArdle, B.H. and Blackewell, R.G. (1989) Measeurement of density variability in the bivalve *Chione stutchburyi* using spatial autocorrelation. *Marine Ecology Progress Series* **32**, 245-252.
47. Steele, J.H. (1989) The ocean 'landscape', *Landscape Ecology* **3**, 185-192.
48. Warwick, R.M. and Uncles, R.J. (1980) Distribution of benthic macrofauna associations in the Bristol Channel in relation to tidal stress, *Marine Ecology Progress Series* **3**, 97-103.
49. Keough, M.J. and Black, K.P. (1996) Predicting the scale of marine impacts; Understanding planktonic links between populations, in R.J. Schmitt and C.W. Osenberg (eds.), *Detecting Ecological Impacts - Concepts and Applications in Coastal Habitats,* Academic Press, San Diego, pp. 199-234.
50. Zajac, R.N. and Whitlatch, R.B. (1988) The population ecology of *Nephtys incisa* in central Long Island Sound prior to and following disturbance, *Estuaries* **11**, 117-133.
51. Zajac, R.N. and Whitlatch, R.B. (1989) Natural and disturbance induced demographic variation in an infaunal polychaete, *Nepthys incisa*, *Marine Ecology Progress Series* **57**, 89-102.
52. Hanski, I.A. and Gilpin, M.E. (1997) *Metapopualtion Biology: Ecology, Genetics and Evolution*, Academic Press, San Diego
53. Whitlatch, R.B. Personal communication
54. Pearson, T.H. and Rosenberg, R. (1978) Macrobenthic succession in relation to organic enrichment and pollution of the marine environment, *Oceanography Marine Biology Annual Review* **16**, 229-311.
55. Rhoads, D.C., McCall, P.L. and Yingst, J.Y. (1978) Production and disturbance on the estuarine seafloor, *American Scientist* **66**, 577-586.
56. Butman, B., Bothner, M.H., Hathaway, J.C., Jenter, H.L., Knebel, H.J., Manheim, F.T. and Signell, R.P. (1992) Contaminant transport and accumulation in Massachusetts Bay and Boston Harbor: A summary of U. S. Geological survey studies. U. S. Geological Survey., Open- File Report 92-202 Woods Hole, Mass.
57. UGGS Sea Floor Mapping Internet Site: http://kai.er.usgs.gov/surveys
58. Bonsdorff, E. (1983) Recovery potential of macrozoobenthos from dredging in shallow brackish waters, *Oceanologica Acta*, Proceedings 17th European Marine Biology Symposium, pp. 27-32.
59. Mattsson, J. and Linden, O. (1983) Benthic macrofauna succession under mussels, *Mytilus edulis* L. (Bivalvia), cultured on hanging long lines, *Sarsia* **68**, 97-102.
60. Conner, W.G. and Simon, J.L. (1979) The effects of oyster shell dredging on an estuarine benthic community, *Estuarine and Coastal Marine Science* **9**, 749-758.
61. Jones, A.R. (1986) The effects of dredging and spoil disposal on macrobenthos, Hawkesbury Estuary, N.S.W., *Marine Pollution Bulletin* **17**, 17-20.
62. Kaplan, E.H., Welker, J.R., Kraus, M.G. and McCourt, S. (1975) Some factors affecting the colonization of a dredged channel, *Marine Biology* **32**, 193-204.
63. Rosenberg, R. (1976) Benthic faunal dynamics during succession following pollution abatement in a Swedish estuary, *Oikos* **27**, 414-427.
64. Elmgren, R., Hansson, S., Larsson, U., Sundelin, B. and Boehm, P.D. (1983) The 'Tsesis' oil spill: acute and long-term impact on the benthos, *Marine Biology* **73**, 51-65.
65. Dean, D. and Haskin, H.H. (1964) Benthic repopulation of the Raritan River estuary following pollution abatement, *Limnology and Oceanography* **9**, 551-564.

66. Parulekar, A.H., Dhargalkar, V.K. and Singbal, S.Y.S. (1980) Benthic studies in the Goa estuaries: Part III - Annual cycle of macrofaunal distribution, production & trophic relations, *Indian J. Marine Sciences* 9, 189-200.

CONCLUSIONS AND RECOMMENDATIONS

JOHN S. GRAY[1], W. AMBROSE JR[2], ANNA SZANIAWSKA[3]
[1] *Biologisk Institutt, Universitetet i Oslo, P.b. 1064 Blindern, 0316, Oslo, Norway*
[2] *Deparment of Biology, Bates College, Lewistpn, Maine, USA*
[3] *Institute of Oceanography, Gdansk University, Al. Marszalka J. Pilsudskiego 46, 81-378 Gdynia, Poland*

1. Background

The ARW recognised that physical boundaries in the ocean support unusually high biological productivity (e.g. frontal systems, gyres, upwelling areas). Such areas have been the focus of much research in recent years [1]. The interface between water and sediment, however, is the largest of these boundary areas in the ocean but has not received the research attention appropriate to its importance. Marine sediments play key roles in the marine ecosystem since they not only provide direct benefits for man, (e.g. food), but more importantly provide "ecosystem goods and services" ([2] Ehrlich & Mooney,1983) which include carbon, nutrient and contaminant cycling, and coastal protection. Coastal and shelf sediments are also the site for carbon and contaminant burial which is of major importance in the debate on global climate change since the carbon will be removed from the carbon cycle for periods of up to thousands of years. Additionally, coastal and near-shore sediments are sites for accumulation of contaminants. The economic value of services for coastal areas has recently been estimated by Constanza et al [3] to be in the order of \$12,568 10^9 y^{-1}. For the world these services are far in excess of global GNP.

The ARW observed that the frequency and magnitude of disturbance events in the oceans have increased in recent years. These may be natural disturbances such as the effects of the El Nino-Southern Oscillation on global scales and effects of hurricanes and storms over regional areas. Man-made disturbances in the coastal zone are common, such as changes in coastal habitats and consequent loss of biological diversity, anoxic conditions (e.g. in the Baltic Sea, the coasts of the North Sea, the Baltic Sea, estuaries of North America, fjords of Norway and many other areas globally), and effects of the fishing industry on the fauna and biogeochemical cycling. All of these may affect larger scale global processes.

The ARW felt that it was important to try and define what it understood as disturbance effects on marine sediments. One definition discussed at the workshop was:

> "Disturbance is the impact of an unpredictable event which affects individuals, populations, assemblages, or habitats and/or functioning of a system."

In a recent book discussing disturbance effects on biodiversity Huston p 215 [4] defines disturbance as: "any process or condition external to the natural physiology of living organisms that results in the sudden mortality of biomass in a community on a time scale significantly shorter (e.g. several orders of magnitude faster) than that of accumulation of the biomass."

Having defined disturbance, the working groups then reviewed the major types of disturbance.

J.S. Gray et al. (eds.), Biogeochemical Cycling and Sediment Ecology, 229–232.

2. Scientific conclusions.

The ARW concluded that:

Disturbance events affecting the benthic boundary layer probably have led to changes in species compositions, species losses and consequences for biogeochemical cycling and the effects on large scale carbon, nutrient and contaminant fluxes.

Whilst it was acknowledged that long-term studies of natural disturbance events are needed key research questions were then developed focusing on what the ARW judged to be the major anthropogenic disturbances in the marine environment.

There are other initiatives in the coastal area such as the International Geosphere-Biosphere's Programme on Land Ocean Interactions in the Coastal Zone (LOICZ) and the EU's European Land Ocean Interaction Studies, (ELOISE) these are primarily concerned with river-sea and ocean-sea interactions and not least on the socio-economic aspects. **The ARW agreed to develop a programme of research on restoration of disturbed marine habitats.** This programme is much more specific and has a narrower focus than LOICZ and ELOISE.

The main foci agreed were:

2.1 DEMERSAL FISHING AND FISHING GEAR IMPACTS.

The group recognised that the fishing industry imposed many different types of disturbance on marine sediments, but felt that trawling was perhaps the most severe and widespread.

Three types of area were identified for experimental studies, non-fished areas, set aside areas where controlled fishing could be done and gradients (or mosaics) of fishing intensity. Manipulative experiments and quantitative observational data are needed examining the spatial structure and bentho-pelagic coupling in benthic systems in relation to varying rates of disturbance. Rather than traditional species lists size-structure, age structure and productivity measurements should be explored as measures of restoration state. Effects on the genetic structure of affected populations is also of major interest. Modelling is an integral part of the project and in particular metapopulation analysis seems a promising tool. The initial site identified by the ARW at the Gulf of Gdansk offers excellent opportunities since the benthic system has a simplified structure. Experiments also need to be done in species rich areas so that comparisons may be made in biogeochemical cycling processes and rates and restoration capabilities.

2.2 *RECOVERY AND RESTORATION OF BENTHIC SYSTEMS AFTER HYPOXIA AND ANOXIC EVENTS*

It is important to distinguish between hypoxia and anoxia since their effects on benthic systems are different. Anoxia may be permanent or intermittent and result from organic enrichment or other physico-chemical processes such as physical transport of oxygen. Thus it is important to distinguish between the effects of these forcing functions. Microorganisms are dominant in the initial phase of recovery and key biogeochemical processes driven by them are measured. Sulphide pools as well as pools of other reduced substances are important in this phase. Once macrofauna are established the recovery process proceeds rapidly.

The basic hypothesis that will be tested is that in systems without macrofauna organic matter degradation and biogeochemical cycling will occur at reduced rates. Combined field and mesocosm studies are recommended and by addition of macrofauna the restoration process can be studied. The Gulf of Gdansk offers varying degrees of hypoxia and anoxia (both spatially and temporally) and makes an useful field study site As with benthic fishing impacts, there is a need for comparative data from this area to sites containing deep-burrowing bioturbating species such as decapod crustacea. Modelling is an intrinsic part of the project and population and diagenetic modelling efforts will proceed in parallel to the experiments.

2.3 *THE RESILIENCE OF BENTHIC SYSTEMS TO DREDGING AND DREDGED MATERIAL*

The reasons for considering this as a priority research area is that it is a universal problem, many harbours are extremely contaminated, yet the spatial scale of the contamination is confined which allows experimentation at manageable scales (from laboratory microcosm to field). Other disturbance events (such as storms) may have large impacts on these contaminated sediments and thus it is an urgent problem to determine how various disturbances interact and the environmental implications posed by contaminated sediments. The project is inevitably interdisciplinary since it requires hydrographers, water- and sediment chemists, microbiologists and benthic ecologists. The ARW believe that since the restoration process is likely to be highly site specific it is unlikely that generalisations will result from one single study.

The experimental approach is similar to that for anoxia using manipulations by adding bioturbating organisms and testing the efficiency of different species of bioturbators. Key processes such as the rate of carbon burial, studies of chemoclines as an integrative measure of recovery and the importance of bentho-pelagic coupling will be measured. Experiments will start with simple systems and move to more complex. The Gulf of Gdansk has the advantage of a simplified benthic system that is well-understood and where contaminant gradients are largely known.

2.4 KEY AREAS

From the analysis above it is clear that an ideal area in which to start a study aimed at developing strategies for the restoration of coastal sediments is the Gulf of Gdansk. It is a shallow sedimentary environment with clear problems of seasonal anoxia in the Bay of Puck and with contaminated sediments at the mouth of the River Vistula. The area is trawled frequently. The biota has been well studied over many years and studies of the effects of anoxia has been a major research task at the University of Gdansk. The fauna has few species and thus is an ideal place for initial experiments. In addition good laboratory facilities are available at Hel.

The ARW wants to achieve practical results and were agreed that the best way to proceed was to do collaborative research and to this end the **ARW elected the following to serve on a steering committee on *Restoration of Disturbed Marine Sediments***: Dr W. Ambrose, (USA), Professor G. Graf, (Germany), Professor J.Gray (Norway), Professor A. Lisitzin, (Russia), Dr D. Raffaelli, (UK), Professor R. Rosenberg, (Sweden), Professor A. Szaniawska, (Poland), Dr K. Skora (Poland).

We believe that there is the political will and that there are international funding possibilities to conduct such a study. **We recommend that NATO fund a Collaborative Research Grant to the Steering Committee on *Restoration of Disturbed Marine Sediments.***

A very much more ambitious and long term goal will be the restoration of the Caspian Sea (The Caspian is in an arid area and has had an increasing sea level (1 m y^{-1}). Known disturbances are dredging, dumping of contaminants, local anoxia on the shelf and oil pollution and eutrophication. What we envisage here is first a collation of the relevant Russian literature and translation of key documents into English, followed by an ARW on the Caspian Sea and its problems. The steering committee on *Restoration of Disturbed Marine Sediments* would be charged with developing a proposal for an ARW. We plan expert visits and hope to establish collaborative links (see below).

2.5 NEW TECHNOLOGIES

Having defined the key research questions, the technological needs, in particular which new technologies could be used to improve understanding, and the need for parallel development of new types of models of key processes such as diagenesis.

The ARW recognised the potential and need for new technological developments such as: fluorosensors, benthic landers, sampling tools with high depth resolution (due to the fact that we want to study interfaces where the gradients in microbiology and chemistry etc. are extremely sharp), heat flux measurements at small scales, side-scan sonars and acoustic methods for landscape-scale applications, and the as yet almost untapped potential for the application of molecular techniques to sediment biota. There has been a strong tendency for sediment ecologists to rely on sampling methods developed in the early 1900's!

2.6 NEW INITIATIVES RESULTING FROM THE WORKSHOP

The most positive initiative was the recognition that with a wide range of disciplines and expertise the ARW wished to continue the work and ideas generated by formation of a network of scientists. In particular the ARW provided a forum for microbiologists, geochemists, modellers and ecologists to develop joint research initiatives. We wish to continue this fruitful development and **recommend that NATO provide funds for the establishment of a network *on Marine Sediment Ecology and Biogeochemical Cycling.***

The project will initially focus on an interdisciplinary research project aimed at developing strategies for the restoration of the Gulf of Gdansk. Should this project prove successful then the long-term goal would be to apply the methods and strategies to restoration of other areas such as the Gulf of Riga, the Caspian Sea and other areas with similar problems.

3. Recommendations

We recommend that NATO provide:

Funds for the establishment of a network in Marine Sediment Ecology and Biogeochemical Cycling.
A Collaborative Research Grant to the Steering Group on *Restoration of Disturbed Marine Sediments* with the primary goal to initiate the restoration of the Gulf of Gdansk.
Linkage Grants for consolidating the research links established at this ARW.

4. References

1. Mann, K.H. & Lazier, J.R.N. 1991. *Dynamics of marine ecosystems: Biological-physical interactions in the ocean.* Blackwell. 466 pp.
2. Ehrlich, P. & Mooney, H. 1983. Extinction, substitution and ecosystem services. BioSci. 33: 248-254.
3. Constanza, R. et al. 1997. The value of the world' ecosystem services and natural capital. Nature 387: 253-260.
4. Huston, M.A. 1994. *Biological Diversity: The coexistence of species on changing landscapes*. Cambridge. 681 pp.

Index

A

B

C

D

E

F

G

H

I

K

L

M

N

O

P

R

S

T

U

W

Y

Z